Computational Methods in Multiphase Flow VI

WIT*PRESS*

WIT Press publishes leading books in Science and Technology.
Visit our website for new and current list of titles.
www.witpress.com

WIT*eLibrary*

Home of the Transactions of the Wessex Institute.
Papers presented at Multiphase Flow VI are archived in the WIT eLibrary in volume 70 of WIT Transactions on Engineering Sciences (ISSN 1743-3533).
The WIT eLibrary provides the international scientific community with immediate and permanent access to individual papers presented at WIT conferences.
http://library.witpress.com

Sixth International Conference on Computational and Experimental Methods in Multiphase and Complex Flow

Multiphase Flow VI

Conference Chairmen

A.A. Mammoli
University of New Mexico, USA

C.A. Brebbia
Wessex Institute of Technology, UK

International Scientific Advisory Committee

Organised by
Wessex Institute of Technology, UK
University of New Mexico, USA

Sponsored by
WIT Transactions on Engineering Sciences

WIT Transactions

Transactions Editor

Carlos Brebbia
Wessex Institute of Technology
Ashurst Lodge, Ashurst
Southampton SO40 7AA, UK
Email: carlos@wessex.ac.uk

Editorial Board

Computational Methods in Multiphase Flow VI

EDITORS

A.A. Mammoli
University of New Mexico, USA

C.A. Brebbia
Wessex Institute of Technology, UK

WIT*PRESS* Southampton, Boston

Editors:

A.A. Mammoli
University of New Mexico, USA

C.A. Brebbia
Wessex Institute of Technology, UK

Published by

WIT Press
Ashurst Lodge, Ashurst, Southampton, SO40 7AA, UK
Tel: 44 (0) 238 029 3223; Fax: 44 (0) 238 029 2853
E-Mail: witpress@witpress.com
http://www.witpress.com

For USA, Canada and Mexico

Computational Mechanics Inc
25 Bridge Street, Billerica, MA 01821, USA
Tel: 978 667 5841; Fax: 978 667 7582
E-Mail: infousa@witpress.com
http://www.witpress.com

British Library Cataloguing-in-Publication Data

A Catalogue record for this book is available
from the British Library

ISBN: 978-1-84564-518-2
ISSN: 1746-4471 (print)
ISSN: 1743-3533 (on-line)

The texts of the papers in this volume were set individually by the authors or under their supervision. Only minor corrections to the text may have been carried out by the publisher.

Printed in Great Britain by Martins the Printers.

Preface

The study of multiphase flows is ever increasing in scope, primarily because of the increasing importance that these flows have in manufacturing, minerals extraction, environmental remediation, and medicine to name a few, and because of the technological advances that are constantly occurring. Despite experimental, theoretical and computational efforts that have spanned decades, we are still far from a full understanding of the complex behaviors inherent in multiphase flows. There are several causes for the difficulty in formulating accurate models even for the simplest of systems. The interaction of phenomena occurring at several different length scales, which in general cannot be decoupled from one another, is the primary difficulty. At the smallest (nano) scales, processes occur which are neither continuum nor molecular, and therefore cannot be treated either at the molecular level (the problem is too big) or at the continuum level (the model may miss important non-continuum effects). Examples of such situations are interface phenomena at engineered interfaces which involve both a vapor and a liquid phase, or surface-to-surface interactions via an intervening fluid. One length scale up (micro), structure formation can cause difficulties in formulating boundary conditions and constitutive equations. Most investigators have approached the problem by applying direct numerical simulation at some level, and using models or approximations to treat smaller-scale phenomena, with the ultimate aim of finding out enough about the flow behavior to be able to construct a constitutive model which captures the essential physics.

Presently, a "theory of everything" is not considered a viable option, because of the many possible combinations of flow parameters. Therefore, it is still necessary to categorize multiphase flow problems into various categories (e.g. bubbly flows, fluidized beds, suspensions). Within each of these categories, developing accurate descriptions of the flows is possible.

The overall focus of this series of conferences is the combination of experimental and computational techniques to gain a better understanding of individual classes of multiphase flow, The goal of the conference is that of facilitating the exchange of ideas and experiences directly and interactively, thereby promoting the development of knowledge in this increasingly important topic. The contents of this book reflect the quality of the submissions and the diligence of the reviewers, whom we wish to thank.

The Editors
Kos, Greece 2011

Contents

Section 1: Multiphase flow simulation

Section 2: Turbulent flow

Section 3: Bubble and drop dynamics

Section 4: Heat transfer

Section 5: Incline flows

Section 6: Energy applications

Section 7: Flow in porous media

Section 1
Multiphase flow simulation

Hydrodynamics behaviour of slot-rectangular spouted beds: assessment of slot width effect

E. Esmaili[1], N. Mahinpey[1] & C. J. Lim[2]
[1]*Dept. of Chemical and Petroleum Engineering, Schulich School of Engineering, University of Calgary, Canada*
[2]*Dept. of Chemical and Bio-Resources Engineering, University of British Columbia, Vancouver, Canada*

Abstract

Spouted beds, originally invented in Canada (1955), have been widely used in drying, granulation, catalytic polymerization, residue treatment and coating of several materials. Understanding of gas and particle hydrodynamics behavior in a slot-rectangular spouted bed is important for evaluation of particle circulation rate and gas-solid contact efficiency and also for overcoming scale-up difficulties. Although have been extensively applied, the mechanisms of solid movement in spouted beds are still not completely understood. In this research, computational fluid dynamics (FLUENT 6.3) has been used to perform a complete study on the hydrodynamic behavior of spouted bed using an Eulerian-Eulerial two-phase model. The Eulerian model assumes that both phases can be considered as fluid and also take the interpenetrating effect of each phase into consideration by using inter-phase drag models. The computational work has been significantly reduced using a 2D axisymmetric mesh for the bed, operating at different conditions (i.e. different superficial gas velocities, static bed heights, size of particle). Furthermore, the effect of slot width on the hydrodynamic behavior of the bed has been predicted. The results predicted by numerical simulation have been validated with the earlier experiments conducted. The comparisons have been made in terms of fountain height for various operating conditions. The results show a good agreement between the experimental data and the numerical simulation.
Keywords: spouted beds, slot-rectangular, computational fluid dynamics, FLUENT.

WIT Transactions on Engineering Sciences, Vol 70, © 2011 WIT Press
www.witpress.com, ISSN 1743-3533 (on-line)
doi:10.2495/MPF110011

1 Introduction

Since its discovery by Mathur and Gishler [2], cylindrical-conical spouted beds have been used in a number of applications such as gas-solid catalytic reactions, coal gasification, drying of pastes and grains, particle mixing, tablet coating and granulation [3]. However, the spouted bed technique has seldom been applied in large scale industrial processes due to certain limitations, in particular scale-up difficulties [4] such as instability to achieve good quality spouting in large scale vessels, and difficulties in predicting the performance of spouted beds larger than 0.3 m in diameter [1].

Conventional spouted beds consist of a cylindrical column with a conical base where the bed of particles is spouted by the injection of air through a circular orifice at the bottom. The flow structure of spouted beds is quite different from fluidized beds. At stable spouting, a spouted bed consists of three regions, a spout in the center, a fountain above the bed surface and an annulus between the spout and the wall, Figure 1. The behavior of spout and the fountain is similar to fluidized beds with particles dynamically suspended, while the annulus region is more like a packed bed or moving bed. It is generally believed that in the annulus, particles move vertically downward and the radially inward, following approximately parabolic paths [3]. The pressure drop of a conical spouted bed under stable spouting condition is about 40% of a fluidized bed with the same static bed height [5, 6]. This suggests that particles are not in a fully suspended state at stable spouting.

As an alternative to units of circular cross-section, Mujumdar and colleagues [4, 7, 8] suggested a spouted bed of rectangular cross-section to eliminate the

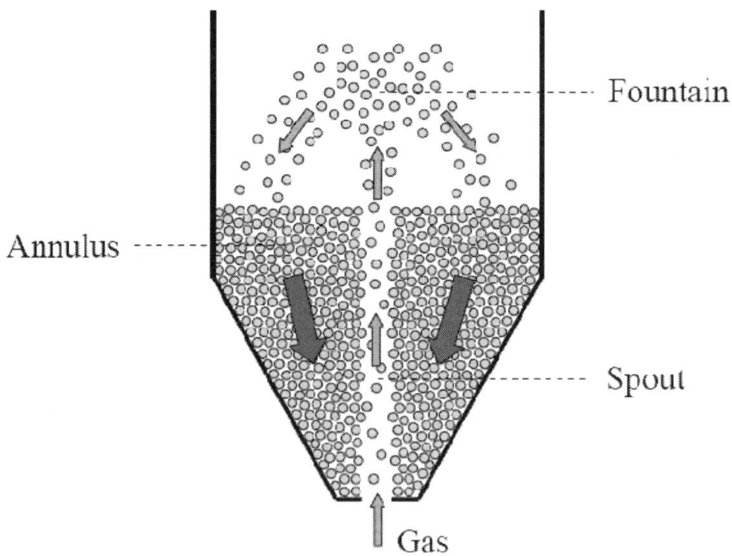

Figure 1: Schematic of a spouted bed operating in stable condition.

scale-up disadvantages of conventional spouted beds. In this configuration the column has vertical plane walls and a diverging base on two opposite sides, while the other two opposing faces are both completely vertical. The spouting air enters through a horizontal slot at the bottom of the two diverging plane surfaces. Scaling up of the slot-rectangular columns can be achieved by simply moving the facing vertical walls further apart, or combining small units by sharing either the front and back faces or the sidewalls. A schematic diagram of this geometry is shown in Figure 2.

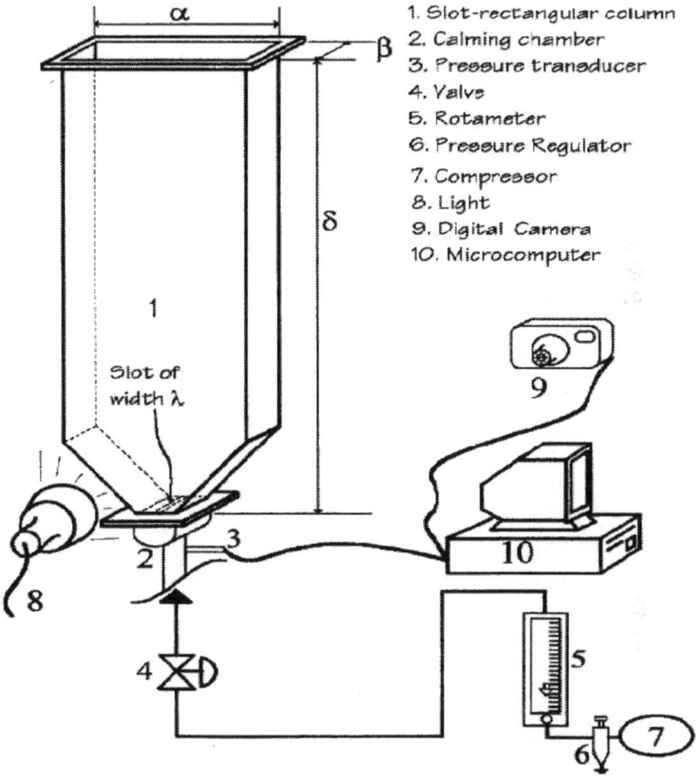

Figure 2: Schematic of the experimental apparatus.

The numerical simulation technique and computational fluid dynamics (CFD) studies have become popular in the field of gas-solid two-phase flow. Many researchers have applied the numerical simulation as a useful tool to obtain detailed information about flow phenomena without disturbing them [9–14]. Shirvanian *et al.* [15] developed a three dimensional Eulerian simulation to predict the manner in which particle size, density, loading, and inlet jet velocity affect solids circulation in a rectangular spouted vessel. Their model is shown to be able to successfully predict the experimentally observed phenomenon of particle ''choking'' where the particle recirculation rate remains constant with increasing particle loading once a ''critical loading'' is achieved. Their simulations also showed ''mounding'' of the particles on the outside of the draft duct walls near the inlet at high solids loadings, which would tend to reduce the effectiveness of metal recovery. This condition can be minimized or prevented by reducing the solids loading and operating at lower fluid flow rates to increase the particle circulation rate.

Bettega *et al.* [16] performed a numerical study of the scale-up of spouted beds. Their results verified that the scale-up relationships of He *et al.* [17] were well represented by CFD simulations, indicating the capability of this tool to evaluate similitude method. They concluded that, when the full set of scaling relationships are not satisfied, important disturbances appears in the hydrodynamic of scaled equipments and make the spouted bed to extinguish. These behaviours are the result of an overestimation or underestimation of velocity in the entrance, respectively, when the whole set of dimensionless parameters is not matched.

Sobieski [18] developed a computer model for an actual spouted bed grain dryer to study the response of the Eulerian multiphase model to changes in values of its parameters. The fountain height was assumed to be the basic value characterizing the bed during operation. His results verified that the Eulerian multiphase model can be used for the simulation of the hydrodynamics of spouted-bed grain dryers. He also concluded that higher quantitative match of experimental results with computations (particularly as concerns the internal bed dynamics) can be obtained by appropriate choice of the turbulence model.

The present work is based on the earlier experiments conducted by Dogan *et al.* [1]. A numerical simulation of slot rectangular spouted bed has been presented to study the effect of slot width on the hydrodynamic behavior of the bed. FLUENT 6.3.26 has been used to perform the numerical simulations based on an Eulerian-Eulerian two phase model. Adjusted Di Felice drag model [19] has been used to take the interpenetrating effect of each phase into consideration. The results have been presented in terms of the fountain height for various operating conditions (i.e. different superficial gas velocities, static bed heights, and size of particle).

2 Experimental setup

The experiments were carried out by Dogan *et al.* [1] in a Plexiglas rectangular column of cross-sectional dimensions $\alpha = 150\text{mm}$ and $\beta = 29\text{mm}$ with an overall height $\delta = 700\text{mm}$ (see Figure 2). The included angle of the diverging lower section, θ, was 30°. Air entry slots of length equal to β and widths, $\lambda = 2, 6, 10$ and 20 (mm) were used in the experiments. A fine wire mesh was inserted across these slots to prevent particles from falling into the air pipe. A calming chamber was installed below the air entry slots to regulate and smooth the air flow. The fountain height, H_F, was measured with a ruler taped to the column wall. A summary of the experimental conditions studied in this work is presented in Table 1. The particles used in their experiments were closely sized glass beads of three different diameters and low density polyethylene. The particle characteristics and key dimensions are listed in Table 2.

Table 1: Experimental conditions.

Column width, α (m)	0.15
Column thickness, β (m)	0.029
Slot width, λ (m)	0.002-0.02
Bed height, H_o (m)	0.03-0.3
Air flow rate, (m^3/s)	0-0.015

Table 2: Particle properties.

Particles	d_P (mm)	$\rho_S\ (kg/m^3)$	ε_b (-)	$U_{mf}\ (m/s)$	$U_T\ (m/s)$
Glass beads	0.86	2520	0.385	0.44	7.8
Glass beads	1.44	2520	0.385	0.62	10.2
Glass beads	2.28	2520	0.390	1.48	12.8
Polyethylene	3.77	940	0.392	0.80	8.3

3 Governing equations

A ''two-fluid'' (fluid and solid phases), Eulerian–Eulerian model was developed to simulate the hydrodynamics of a rectangular spouted bed [19]. The governing equations of the model have been summarized in Table 3.

Table 3: Governing equations.

Continuity:

$$\frac{\partial}{\partial t}\left(\varepsilon_q\rho_q\right)+\nabla.\left(\varepsilon_q\rho_q\vec{V}_q\right)=0,\qquad q=s\ or\ g \tag{1}$$

Momentum equations:

$$\frac{\partial}{\partial t}\left(\varepsilon_g\rho_g\vec{V}_g\right)+\nabla.\left(\varepsilon_g\rho_g\vec{V}_g\vec{V}_g\right)=\nabla.\bar{\bar{\tau}}_g-\varepsilon_g\nabla P+\varepsilon_g\rho_g g+K_{sg}\left(\vec{V}_s-\vec{V}_g\right) \tag{2}$$

$$\frac{\partial}{\partial t}\left(\varepsilon_s\rho_s\vec{V}_s\right)+\nabla.\left(\varepsilon_s\rho_s\vec{V}_s\vec{V}_s\right)=\nabla.\bar{\bar{\tau}}_s-\nabla P_s-\varepsilon_s\nabla P+\varepsilon_s\rho_s g+K_{sg}\left(\vec{V}_s-\vec{V}_g\right) \tag{3}$$

Stress tensors

$$\bar{\bar{\tau}}_{g,eff}=\varepsilon_g\left(\bar{\bar{\tau}}_g+\bar{\bar{\tau}}_t\right) \tag{4}$$

$$\bar{\bar{\tau}}_g=\mu_g\left(\nabla\vec{V}_g+\left(\nabla\vec{V}_g\right)^T\right)-\tfrac{2}{3}\mu_g\bar{\bar{I}}.\nabla.\vec{V}_g \tag{5}$$

$$\bar{\bar{\tau}}_t=\mu_t\left(\nabla\vec{V}_g+\left(\nabla\vec{V}_g\right)^T\right)-\tfrac{2}{3}\mu_t\bar{\bar{I}}.\nabla.\vec{V}_g-\tfrac{2}{3}\varepsilon_g\rho_g k\bar{\bar{I}} \tag{6}$$

$$\bar{\bar{\tau}}_s=\varepsilon_s\mu_s\left(\nabla\vec{V}_s+\left(\nabla\vec{V}_s\right)^T\right)+\varepsilon_s\left(\lambda_s+\tfrac{2}{3}\mu_s\right)\nabla.\vec{V}_s\bar{\bar{I}}, \tag{7}$$

$$\mu_s=\frac{4}{5}{\varepsilon_s}^2\rho_s d_s g_0(1+e)\sqrt{\frac{\Theta}{\pi}}+\frac{\frac{5\sqrt{\pi}}{48}\rho_s d_s\sqrt{\Theta}}{(1+e)g_0}\left[1+\tfrac{4}{5}g_0\varepsilon_s(1+e)\right]^2 \tag{8}$$

$$\lambda_s=\frac{4}{3}{\varepsilon_s}^2\rho_s d_s g_0(1+e)\sqrt{\frac{\Theta}{\pi}} \tag{9}$$

Table 3: Continued.

Radial distribution function	
$g_0 = \frac{3}{5}\left[1 - \left(\frac{\varepsilon_s}{\varepsilon_{s,max}}\right)^{\frac{1}{3}}\right]^{-1}$	(10)
Granular temperature	
$\frac{3}{2}\left[\frac{\partial}{\partial t}(\varepsilon_s\rho_s\Theta_s) + \nabla.\left(\varepsilon_s\rho_s\vec{V}_s\Theta_s\right)\right] = \bar{\bar{\tau}}_s : \nabla\vec{V}_s - \nabla . q_s - \gamma_{\Theta_s} - 3K_{sg}\Theta_s,$	(11)
$q_s = k_{\Theta_s}\nabla\Theta_s,$	(12)
$k_{\Theta_s} = \frac{150 d_s\rho_s\sqrt{\Theta_s\pi}}{384(1+e)g_0}\left[1 + \frac{6}{5}(1+e)\varepsilon_g g_0\right]^2 + 2d_s\rho_s\varepsilon_s{}^2(1+e)g_0\sqrt{\frac{\Theta_s}{\pi}},$	(13)
$\gamma_{\Theta_s} = \frac{12(1-e^2)g_0}{d_s\sqrt{\pi}}\rho_s\varepsilon_s^2\sqrt{\Theta_s^3},$	
Inter-phase momentum exchange	
$K_{sg} = \frac{3}{4}C_D\frac{\varepsilon_s\rho_g}{d_s}\left\lvert\vec{V}_s - \vec{V}_g\right\rvert f(\varepsilon_s)$	(14)
$f(\varepsilon_s) = (1-\varepsilon_s)^{-x}$	(15)
$x = P - Q.exp[-(1.5-\beta)^2/2]$	(16)
$\beta = \log(Re_s)$	(17)
P and Q will be adjusted to satisfy:	
$\varepsilon_s(\rho_s - \rho_g)g = \frac{K_{sg}}{\varepsilon_g}\left\lvert\vec{V}_s - \vec{V}_g\right\rvert,$	(18)
$\varepsilon_{s,mf}(\rho_s - \rho_g)g = \frac{K_{sg}}{\varepsilon_{g,mf}}U_{mf}^{experiment},$	(19)

4 Numerical simulation

Governing equations of mass and momentum conservation as well as the granular temperature equation are solved using finite volume method employing the Phase-Coupled Semi Implicit Method for Pressure Linked Equations (PC-SIMPLE) algorithm, which is an extension of the SIMPLE algorithm to multiphase flow. A two-fluid Eulerian-Eulerian model, which considers the conservation of mass and momentum for each phase, has been applied. The adjusted Di Felice drag model [19] has been implemented to include the momentum transfer between the phases. The 2D axisymmetric computational domain has been meshed with unstructured triangular cells in the cone base and structured square cells in the rest of the domain. GAMBIT 2.13 has been used for meshing the geometry, Figure 3. The mesh was generated using node spaces of larger than the particle diameter. Volume fraction, density, and pressure are stored at the main grid points that are placed in the center of each control volume. A staggered grid arrangement is used, and the velocity components are solved at the control volume surfaces. A pressure correction equation is built based on total volume continuity. Pressure and velocities are then corrected so as to satisfy the continuity constraint. Second-order upwind discretization scheme was used for discretizing the partial differential equations. An adaptive time-stepping algorithm with 100 iterations per each time step and a minimum value of 2×10^{-4} (s) for the lower domain of time steps was used to ensure a stable convergence. The adaptive determination of the time step size is based on the estimation of the truncation error associated with the time integration scheme (i.e., first-order implicit or second-order implicit). If the truncation error is smaller than a specified tolerance, the size of the time step is increased; if the truncation error is greater, the time step size is decreased. The convergence criteria for other residual components associated with the relative error between two successive iterations has been specified in the order of 10^{-3}. The discretized equations along with the initial and boundary conditions are solved using

Table 4: Computational model parameters.

Parameter	Value/description	
Bed dimension	$0.15 \times 0.029 \times 0.7\ (mm), \theta = 30°$	
	Slot width	$0.002 - 0.02\ (m)$
	Static bed height	$0.06 - 0.21\ (m)$
Solver	Pressure based, unsteady, 2nd order Imp., axisymmetric	
Multiphase flow model	Eulerian, 2-phase (air, particles)	
Grid interval spacing		$0.0025(m)$
Time steps	0.0002 (s)	
Discretization	Momentum	2nd order upwind
	Volume fraction	QUICK
	Turbulent	2nd order upwind
Materials	Particle density	$2520(kg/m^3)$
	Gas density	$1.225(kg/m^3)$
	Gas viscosity	$1.7894 \times 10^{-5}(kg/m.s)$
	Mean particle diameter	$0.86 - 3.77\ (mm)$
Phases	Gas-solid drag model	Adjusted Di Felice
	Res. coeff.	0.95
	Initial solid packing	$0.385 - 0.392$
	Max. packing limit	0.63
Boundary conditions	Inlet Velocity	$0.5 - 10\ (m/s)$
	Outlet	Pressure-outlet
	Axis	Axisymmetric
	Wall	Non-slip for the fluid
Under-relaxation factors	Pressure	0.3
	Momentum	0.4
	Volume fraction	0.3
	Granular temperature	0.2
Convergence criterion	0.001	

FLUENT 6.3 on Colfax CXT5000 PSC GPU Processing with 960 cores and 32GB RAM. The results have been saved after 5 (sec) flow time simulation to make sure that the fountain reaches a statistically steady state height. Computational model parameters and boundary conditions are summarized in Table 4.

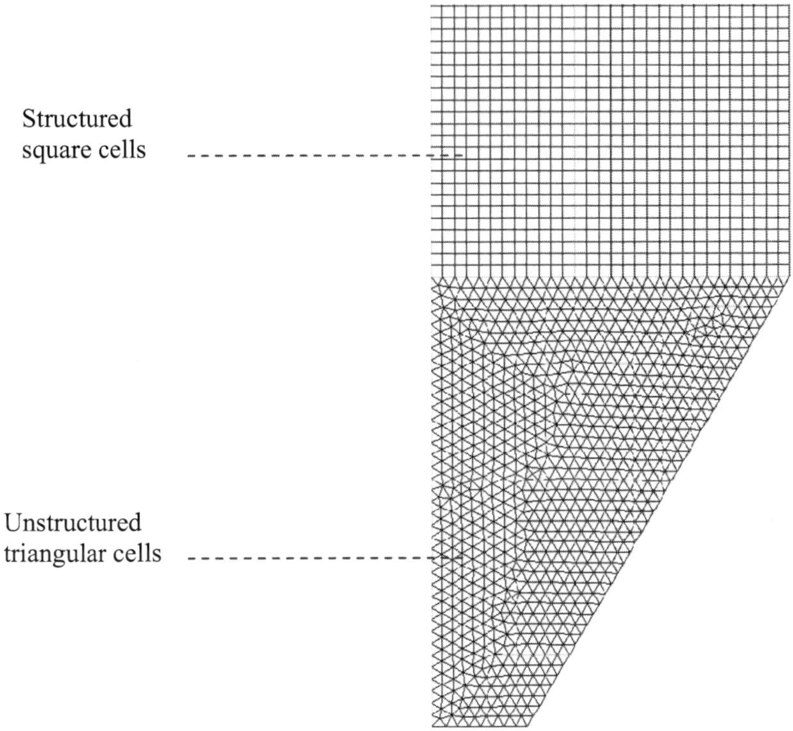

Figure 3: Computational mesh.

5 Results and discussion

Simulation results have been compared to the experimental data [1] in order to validate the model. Figure 4 compares the experimental and simulation results for the fountain height as a function of superficial air velocity for a specific operating condition, $(\mathrm{H} = 0.11\ (\mathrm{m}), \mathrm{dp} = 0.86(\mathrm{mm}), \lambda = 0.006(\mathrm{m}))$. As for cylindrical columns [20], the fountain height increases with increasing superficial gas velocity. The figure shows that the slope of the best fit line to data has been well predicted by the model, (Error < 2%). However, the numerical simulation overestimates the fountain height at a specific superficial air velocity, (Error < 10%). Assumptions and simplifications made in the numerical simulation is one of the main sources of this error. 2D simulation neglects the

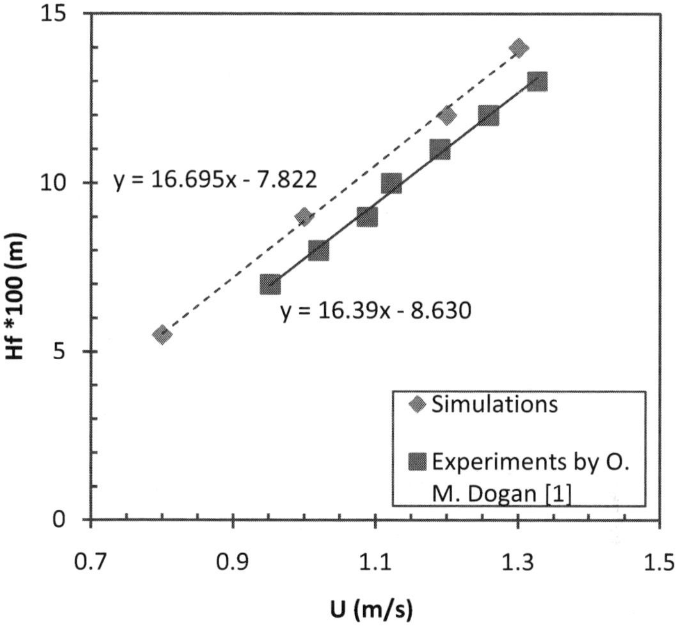

Figure 4: Fountain height as a function of superficial air velocity $(H = 0.11\ (m), dp = 0.86(mm), \lambda = 0.006(m))$.

Upgrading the computational domain for 3D simulations of the bed, variable particle size and density can decrease the computational error to a minimum.

Figure 5 shows the model predictions of fountain height as a function of superficial air velocity for three different slot width, and $(H = 0.11\ (m), dp = 1.44(mm))$. The effect of slot width on the minimum spouting velocity is also shown in this figure. It can be seen that U_{ms}, increased with increasing λ. An increase in U_{ms}, with inlet orifice size was previously reported for conical cylindrical columns [3], While, for conical beds, some authors found a decrease in U_{ms} with increasing orifice diameter [3]. The figure also shows that fountain height increases with decreasing slot width which is in a good agreement with experimental data reported by Dogan *et al.* [1]. This implies that with smaller slot width, higher superficial air velocity will be required to obtain a desired fountain height. Comparing the slope of each line shows that the rate of fountain height increase with superficial air velocity also decreases with increasing slot width.

Figure 6 shows the contour plots of solid volume fraction at different superficial air velocities after 5 (sec) real-time simulations. It is characteristic for the spouted beds that the fountain forms only after exceeding a certain value of velocity of the gas supplying the device. The transition from the established spouted condition to the fixed bed condition, described by Mathur and Epstein

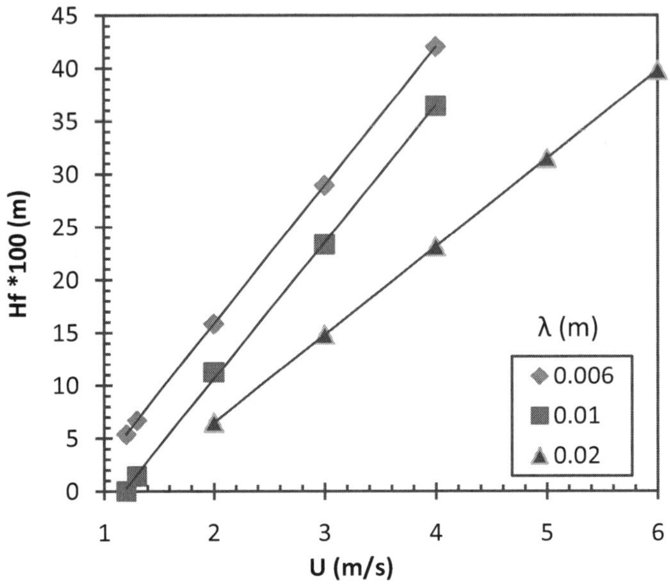

Figure 5: Fountain height as a function of superficial air velocity and slot thickness $(H = 0.11\ (m), dp = 1.44(mm))$.

[3], is presented in this figure. The plots also show highest concentration of solid particles in the annulus region, while the fountain and the spout regions have lower concentrations, respectively. These computations allowed determination of the fountain height and the volume distribution of particles in the model. The studies presented here on the influence of supply gas velocity on fountain dynamics are consistent at the qualitative level with the experimental results described in literature [1].

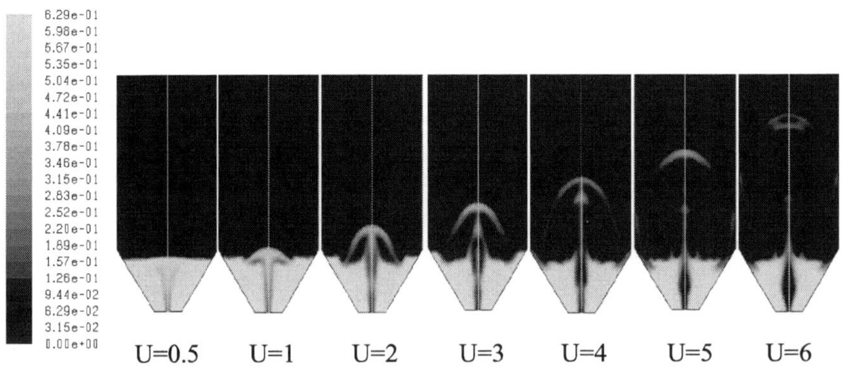

Figure 6: Contours of solid volume fraction for different superficial gas velocity $(H = 0.11\ (m), dp = 1.44\ (mm), \lambda = 0.010(m))$, (Velocities in (m/s)).

6 Conclusion

Numerical simulation of a slot rectangular spouted bed have been performed in a 2D axisymmetric solution domain using the Eulerian-Eulerian approach in order to investigate the effect of slot width on the hydrodynamic behaviour of the bed. FLUENT 6.3 was used to perform the calculations. The results show that the fountain height decreases by increasing the slot width at a specific superficial air velocity. The results show a linear relationship between the fountain height and superficial gas velocity. The results predicted by numerical simulation have been validated with the earlier experiments conducted by Dogan *et al.* [1]. The comparisons have been made in terms of fountain height for various operating conditions, (i.e. different superficial gas velocities, static bed heights, size of particle). The results show a good agreement between the experimental data and the numerical simulation. However, further modeling efforts are required to study the effect of bed thickness using a three dimensional model, and variable size and density distribution of particles in the bed.

Nomenclature

C_d	Single particle drag function, dimensionless
d_s	Solid diameter, m
e	Restitution coefficient, dimensionless
g	Acceleration due to gravity, m/s^2
g_0	Radial distribution coefficient, dimensionless
K_{sg}	Gas/solid momentum exchange coefficient, dimensionless
P	Pressure, Pa
P_s	Solid pressure, Pa
$\vec{V}$	Velocity, m/s
Re	Reynolds number, dimensionless
k_{Θ_s}	Conductivity of granular temperature, $kg/(m.s)$
γ_{Θ_s}	Dissipation of granular temperature, $kg/(m.s^3)$
	Greek symbols
ρ	Density, kg/m^3
Θ_s	Granular temperature, m^2/s^2
$\bar{\bar{\tau}}$	Stress tensor, Pa
μ	Shear viscosity, kg/sm
λ	Bulk viscosity, kg/sm
ε	Volume fraction, dimensionless
	Subscripts
g	Gas
s	Solid

References

[1] Dogan, O. M., Freitas, L. A. P., Lim, C. J., Grace, J. R., Luo, B., "*Hydrodynamics and Stability of Slot-Rectangular Spouted Beds. Part I: Thin Bed*", Chem. Eng. Comm., (2000), Vol. 181, pp. 225-242.

[2] Mathur, K. B. and P. E. Gishler. "*A technique for contacting gases with coarse solid particles*", AIChE J. (1955) 1 (2), 157-164.

[3] Mathur, K. B. and N. Epstein, "*Spouted Beds*", Academic Press, New York, (1974).

[4] Anderson, K., Raghavan, G.S.V, Mujumdar, A.S. "*Drying '84*", Hemisphere Pub., New York, (1984), pp. 205-209.

[5] Wang, Z. G., Bi, H. T., Lim, C. J., "*Numerical simulations of hydrodynamic behaviors in conical spouted beds*", Particuology (2006). 4(3-4), 194-203.

[6] Mukhlenov, I. P. and Gorshtein, A. E., "*Hydrodynamics of reactors with a spouting bed of granular material*", Vses. Konf. Khim. Reactrom Novosibirsk (1965), (3) 553-562.

[7] Kalwar, M. I., Raghavan, G. S. V. and Mujumdar, A. S., Can. J. Chml. Eng. (1992), 70, RR7.

[8] Passos, M. L., Mujumdar, A. S. and Raghavan, G. S. V., Powder Technol. (1993), 74, 97.

[9] Du, W., Bao, X. J., Xu, J., Wei, W. S., "*Computational fluid dynamics (CFD) modeling of spouted bed: Assessment of drag coefficient correlations*". Chem. Eng. Sci. (2006), 61(5), 1401-1420.

[10] Duarte, C. R., Murata, V. V., Barrozo, M. A. S., "*A study of the fluid dynamics of the spouted bed using CFD*", Brazilian J. of Chem. Eng. (2005), 22(2), 263-270.

[11] He, Y. R., Zhao, G. B., Bouillard, J., Lu, H. L., "*Numerical simulations of the effect of conical dimension on the hydrodynamic behavior in spouted beds*", Can. J. Chem. Eng. (2004), 82(1), 20-29.

[12] Huilin, L., Yongli, S., Yang, L., Yurong, H., Bouillard, J., "*Numerical simulations of hydrodynamic behaviour in spouted beds*", Chem. Eng. Research and Design (2001), 79(A5), 593-599.

[13] Kawaguchi, T., Sakamoto, M., Tanaka, T., Tsuji, Y., "*Quasi-three-dimensional numerical simulation of spouted beds in cylinder*", Powder Technol. (2000), 109 (1-3), 3-12.

[14] Lu, H. L., He, Y. R., Liu, W. T., Ding, J. M., Gidaspow, D., Bouillard, J., "*Computer simulations of gas-solid flow in spouted beds using kinetic-frictional stress model of granular flow*", Chem. Eng. Sci. (2004), 59(4), 865-878.

[15] Shirvanian, P.A., Calo, J.M., Hradil, G., "*Numerical simulation of fluid-particle hydrodynamics in a rectangular spouted vessel*", International J. of Multiphase Flow (2006), 32, 739-753.

[16] Bettega, R., Correa, R. G., and Freire, J. T., "*Scale-Up Study of Spouted Beds Using Computational Fluid Dynamics*", The Can. J. of Chem. Eng. (2009), 87, 193-203.

[17] He, Y. L., Lim, C. J., and Grace, J. R., "*Scale-up Studies of Spouted Beds*" Chem. Eng. Sci. (1997), 52, 329–339.

[18] Sobieski, W., "*Numerical Analysis of Sensitivity of Eulerian Multiphase Model for a Spouted-Bed Grain Dryer*", Drying Technol. (2008), 26:12, 1438-1456.

[19] Esmaili, E., "*Adjustment of Drag Coefficient Correlations in Three Dimensional CFD Simulation of Gas-Solid Bubbling Fluidized Bed*", MS.c Thesis, Dept. of Chem. and Petroleum Eng., University of Calgary, December 2009.

[20] Grace, J. R. and Mathur, K. B. (1978), *Height and structure of the fountain region above spouted beds.* The Canadian Journal of Chemical Engineering, 56: 533–537. doi: 10.1002/cjce.5450560501.

Analogues of Rayleigh-Taylor and Richtmyer-Meshkov instabilities in flows with nonuniform particle and droplet seeding

P. Vorobieff[1], M. Anderson[1], J. Conroy[1], R. White[1], C. R. Truman[1] & S. Kumar[2]
[1]Department of Mechanical Engineering, The University of New Mexico, Albuquerque, USA
[2]Department of Engineering, The University of Texas, Brownsville, USA

Abstract

The well-known Rayleigh-Taylor (RT) and Richtmyer-Meshkov (RM) instabilities characterize the behavior of flows where two gases (or fluids) of different densities mix due to gravity (RT) or due to impulsive acceleration (RM). Recently, analogous instabilities have been observed in two-phase flows where the seeding density of the second phase, e.g., particles or droplets in gas, and the resulting average density, is initially non-uniform. The forcing causes the second phase to move with respect to the embedding medium. With sufficient seeding concentration, this leads to entrainment of the embedding phase. The resulting movement is similar to the movement that would evolve in a mixing flow with no second phase seeding, but with non-uniform density (not unlike a mixture of lighter and heavier gases), where RT and RM instabilities develop in the case of gravity-induced and impulsive acceleration, respectively. The hydrocode SHAMRC has been used in the past to study the formation and growth of the RM instability. Here we attempt to use it to model the first order formation and growth phenomena of the new class of instability in two-phase flows first, by approximating the second phase as a continuous fluid with an averaged density, and second, by taking the relative motion of particles (droplets) into account explicitly. The initial conditions are varied to provide a wide range of instability growth rates. Comparison of the numerical results with experiment shows good agreement.
Keywords: compressible flow, two-phase flow, instability, shock wave.

WIT Transactions on Engineering Sciences, Vol 70, © 2011 WIT Press
www.witpress.com, ISSN 1743-3533 (on-line)
doi:10.2495/MPF110021

1 Introduction

There exist two well-known hydrodynamic instabilities that develop in a fluid (gas) of initially non-uniform density.

The first of these two is the Rayleigh-Taylor instability (RTI), first generally described by J.W. Strutt (Lord Rayleigh) [1] for a continuous density interface, appreciably later described analytically for a sharp interface by Taylor [2], and shortly thereafter observed experimentally by Lewis [3]. Most commonly, this instability develops on a density interface between heavy and light fluid in a gravity field, when the heavy fluid is above the light fluid. The unstable (heavy above light) stratification leads to growth of perturbations on the density interface and formation of vortices, followed by development of secondary instabilities, transition to turbulence, and mixing of the heavy and light fluids. It is this interfacial instability that makes it possible, for example, to pour beer from a bottle (otherwise, the liquid would be held in by atmospheric pressure). RTI is also responsible for the formation of many natural features on Earth, such as salt domes [4] and cold plumes at subduction zones in the mantle [5]. The same instability plays an important role in stellar evolution, including such violent phenomena as Type Ia supernovæ [6].

The second instability, known as Richtmyer-Meshkov instability (RMI), also characterizes a density interface, but subjected to an impulsive acceleration, for example, due to the passage of a shock wave. It was first described analytically by Richtmyer [7]. Several years later, an experimental study of the same phenomenon by Meshkov [8] followed. RMI differs from RTI in several features. First, it develops both when the impulsive acceleration of the interface is directed from the "light" side toward the "heavy" side and in the opposite direction (in the latter case, initial perturbations will begin growing in amplitude after undergoing a phase inversion). Second, while RTI evolution is characterized by a continued supply of energy to the deforming interface (e.g., via gravity field), energy input for RMI occurs over a finite (and usually short – e.g., passage of a shock wave) time interval. Subsequently, RMI and RTI are characterized by different temporal relationships that describe perturbation amplitude growth. Like RTI, RMI is responsible for formation of some astrophysical features, e.g., in supernova remnants [9]. RMI is also important for several engineering applications, such as scramjet design [10] and inertial confinement fusion (ICF), where RMI can affect the fusion target stability [11].

Problems involving RTI and RMI are often chosen as benchmarks for validation of numerical codes [12–14] – both because they present a computational challenge and because comparison with many well-characterized experiments (or, in some cases, data describing natural phenomena) is available.

All the studies described above pertain to RTI and RMI developing on an interface (sharp or diffuse) between two liquids or gases. However, there is also a question of considerable practical (as well as theoretical) interest: what happens if, instead of a density interface between two gases, sustained or impulsive acceleration is applied to an *average density* interface in a multiphase medium, e.g., a gas non-uniformly seeded with small liquid droplets or solid particles? Modeling such a problem directly could present an even greater challenge than numerical simulation of RMI and RTI, because of the necessity to account for the interaction between the embedding (gas) and embedded (droplets or particle) phases. The simple assumption that the embedded phase can be modeled by "smearing" the mass that it adds across the appropriate volume and treating the resulting medium as a single-phase might work for some cases (especially for gravity-driven flows), but it is explicitly invalid, for example, after the passage of a shock front. The latter would accelerate the embedding phase, while particles or droplets of the embedded phase will lag behind, interacting with the gas in a possibly non-trivial way. Along with Stokes drag, compressibility effects have also to be taken into account [15, 16]. Accounting for the interaction between each particle (droplet) and the gas could soon become computationally prohibitive.

Our recent experiments [17, 18] confirm formation of large-scale vortices in a gaseous medium non-uniformly seeded with micron- or submicron-sized droplets or particles after shock acceleration. The flow structure superficially resembles one that would emerge after shock acceleration of continuous medium with *average* initial density matching that of our two-phase, non-uniformly seeded medium. However, several important differences also exist between vortex formation due to classical RMI and the roll-up we observe.

In the following sections, we will describe two experiments – multiphase "analogues" of RTI and RMI, present some data from these experiments, and discuss results from numerical modeling of the second experiment.

2 Experiment

Two different experimental arrangements were built. The setup for the modeling of two-phase RTI (Fig. 1a) was extremely simple. Experiments were conducted in a tall (about 1 m), fully enclosed rectangular acrylic container. In the middle of the container, horizontal grooves and a slot in the walls were machined for a removable steel bottom plate. In the top cover of the container, an opening was left for a pipe attached to a commercial theatrical fog machine, producing a stream of submicron-sized droplets of propylene glycol carried by air. Once the top section of the container was filled with droplets and the temperature of the air-droplet mix was found to be equal to the ambient (to avoid any effects due to air buoyancy), the removable bottom plate was gently extracted, allowing the seeded and unseeded air to mix. The setup was illuminated from the side with an LED panel, the flow evolution recorded with a 720×480 pixel resolution digital camera at 30 frames per second.

The arrangement for the shock-driven flow studies was by necessity much more complex (Fig. 2). At the heart of it lay a shock tube with a 75 mm square interior cross-section. The shock tube consisted of a 1.2 m long high-pressure driver section, a 2.9 m long driven section, a 0.8 m long test section (made from transparent polycarbonate), and a 0.8 m long runoff section. The initial conditions studied here were produced as follows. Above the test section, a 75 l enclosed

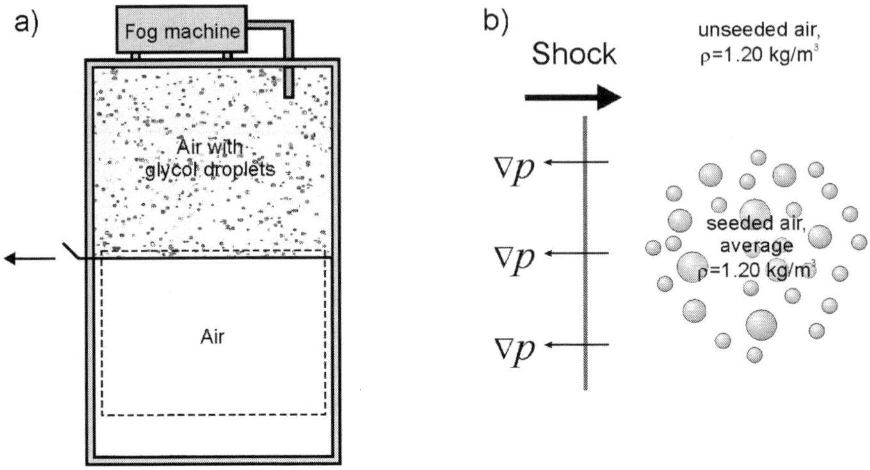

Figure 1: a) Experimental arrangement for the studies of the multiphase analogue of Rayleigh-Taylor instability. Dashed rectangle represents the field of view of the flow visualization camera. b) Schematic of initial conditions for shock-accelerated experiment.

settling tank was installed. On top of the tank, a commercial fog machine (of the same type as for the previous experiment) was mounted. The mixture of air and propylene glycol fog was allowed to reach thermal equilibrium with the surroundings before the experiments (again, to exclude air buoyancy effects). Subsequently, it was directed vertically downward across the test section through a concentric nozzle. The inner nozzle (diameter 6.35 mm) contained the glycol-seeded flow, while the outer concentric nozzle (outer diameter 12.7 mm) carried unseeded air in the same direction. As the result, inside the test section, a laminar cylindrical jet of droplet-seeded air formed, exiting through a 12.7 mm hole in the bottom (Fig. 1a).

This jet formed the non-uniform droplet seeding for the initial conditions of the experiment. The shock accelerating these initial conditions was produced by pressurizing the driver section with helium and then mechanically puncturing the membrane separating the driver section from the driven section. Two pressure transducers mounted downstream in the driven section recorded pressure traces to confirm the shock speed and trigger a high-speed camera used for flow visualization (DRS Imaging IMACON-200) and the four lasers (New Wave Research Gemini) used as light sources for flow visualization. The light from each laser passed through a cylindrical lens, a spherical lens, and a flat mirror, to illuminate the same horizontal cross-section of the flow. The camera, installed on the opposite side of the test section from the lasers, was aimed at a 45° mirror above the test section, recording a sequence of four frames corresponding to the four laser pulses per experiment.

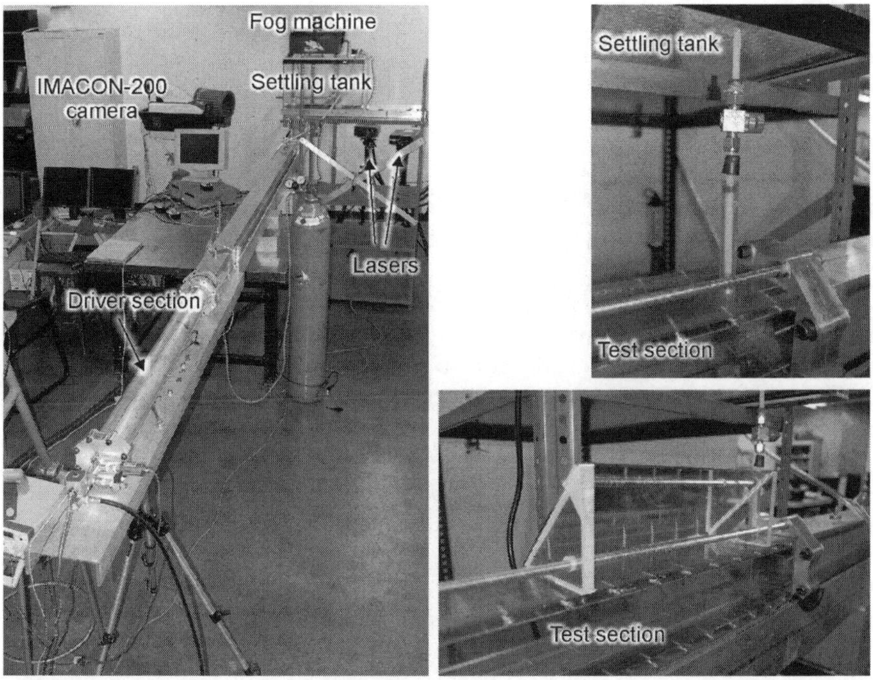

Figure 2: View of the shock tube from the upstream direction (with respect to the propagation of the shock) and close-ups of the concentric-flow nozzle and the test section with the mirror.

For the analogues of RTI and RMI described in the following section, the *average* initial density of the glycol-droplet mixture was measured to be 1.26 kg/m^3, compared to 1.20 kg/m^3 for the unseeded air.

3 Numerical setup

To expand upon knowledge gained from the RMI experiments, the Second-order Hydrodynamic Automatic Mesh Refinement Code (SHAMRC [19]) was used to model the shocks and initial conditions found in the RMI experiments. SHAMRC is a two- and three-dimensional finite-difference hydrodynamic computer code. It is a descendant of SHARC (Second-order Hydrodynamic Advanced Research Code). It is used to solve a variety of airblast related problems which include high explosive (HE) detonations, nuclear explosive (NE) detonations, structure loading, thermal effects on airblast, cloud rise, conventional munitions blast and fragmentation, shock tube phenomenology, dust and debris dispersion and atmospheric shock propagation. The code has the capability to run with a single Eulerian grid or with the Automatic Mesh Refinement (AMR) option that divides the calculational domain into smaller Eulerian grids at several levels of refinement to provide high-resolution results.

A two-dimensional Cartesian mesh was constructed out of square elements approximately 0.005 cm on each side. The size of the elements was chosen so that there would be exactly 128 zones across the diameter of the cylindrical initial conditions. The mesh extents were set to the width of the shock tube test section in the y direction and a distance of 51 cm in the x direction. The x-extent was chosen so that approximately 2 ms of instability growth could be observed for shock waves with a Mach number 1.66. The shock waves were generated by a feed-in boundary condition located at the left boundary. This boundary condition provides a set of constant material properties and satisfactorily recreates the shock wave which accelerates the initial conditions.

The experimental initial conditions consisted of discrete glycol droplets suspended in ambient air. Two approaches were taken in modeling these droplets. The first was to create a pseudo-glycol fluid that has the same average mass as the droplet/air mixture. This material was modeled as a perfect gas designed to match pressure and temperature equilibrium with the surrounding ambient atmosphere. The second approach was to model droplets as discrete particles. The SHAMRC particle model treats each computational particle as a cloud of particles with the same radius. Depending of the mass of the computational particles, it may represent millions of physical particles. These particles are accelerated by the surrounding gas via drag and are allowed to heat an cool through conduction. As each computational particle has a distinct radius, a particle distribution can be generated by randomly varying the particle radius when the particles are generated. The exact distribution of the glycol droplets is an important factor, as drag is a strong function of the particle radius. Differences in particle radii will cause differences in acceleration rates and in turn cause larger particles to lag behind smaller ones. The initial droplet distribution was only approximately known (ranging from submicron-sized to several microns in diameter), so for the initial simulation a distribution was chosen with particle diameters on the order of 0.7 μm.

4 Results of experiment

Figure 3 shows a sequence of images from the video describing the two-phase RTI analogue. The most prominent features on the disturbed interface are "mushroom caps" very similar to these observed in classical RTI. From these images, the growth rate of the perturbation amplitude is well described by a quadratic function. RTI amplitude growth as the function of time t is usually characterized [20, 21] as

$$h(t) = \alpha A g t^2,$$

where A is the Atwood number

$$A = \frac{\rho_1 - \rho_2}{\rho_1 + \rho_2}, \tag{1}$$

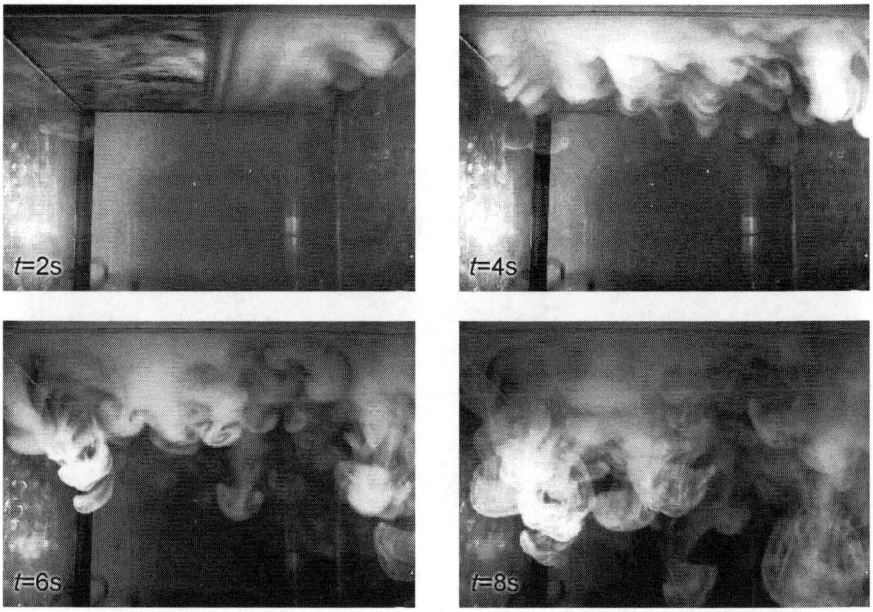

Figure 3: Flow visualization of the two-phase analogue of Rayleigh-Taylor instability. Time $t = 0$ corresponds to the moment when the sliding bottom plate is set in motion. Image extent is about 50 cm in the horizontal direction, intervals between images 2 s.

with ρ_1 and ρ_2 being the densities of the heavier and lighter fluid (gas) correspondingly, and g is the acceleration due to gravity. In our case, ρ_2 would be the density of unseeded air, while ρ_1 would represent the *average* density of air seeded with droplets, leading to an effective Atwood number $A \sim 0.03$. This is a rather low value (for example, RTI with sulfur hexafluoride SF_6 above air would be characterized by $A \sim 0.67$).

Our measurements of amplitude $h(t)$ combined with this value for A yield an estimate $\alpha = 0.05 \pm 0.01$, which agrees well with results for "classic" RTI [22]. The corresponding characteristic velocity can be estimated as $v = 3.8 \pm 0.5$ cm/s. Based on this velocity, characteristic droplet size d =1 μm, and kinematic viscosity of air $\nu = 1.56 \times 10^{-5}$ m^2/s, the Reynolds number $Re_d = dv/\nu$ for a drop is about 0.002. At the same time, the Reynolds number of the macroscopic flow based on the representative size of the vortex structure $l \sim 2$ cm is $Re_l \sim 50$. This disparity strongly suggests that the relative motion of the droplets with respect to the embedding air is negligible, and, at least from the point of view of the initial instability growth, the seeded air can indeed be regarded as continuous medium with a density ρ_1 slightly greater than that of the unseeded air ρ_2.

Thus the notion of the existence of the two-phase analogue of RTI is unambiguously confirmed. However, modeling this laboratory flow would be

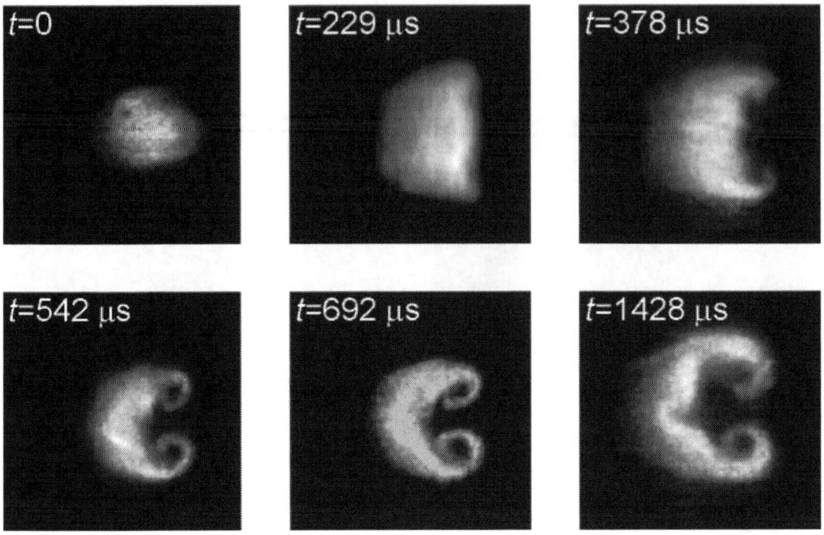

Figure 4: Experimental images acquired after Mach 1.66 shock acceleration (left to right) of a laminar, initially cylindrical air jet seeded with glycol droplets. Horizontal image extent is 20 mm, time $t = 0$ corresponds to moment immediately before shock wave arrival. Times after shock acceleration are labeled in the images. Images show a laser-illuminated horizontal section of the flow (normal to the axis of the jet). Note that the field of view follows the droplets, which are being advected by the shock-accelerated air with a piston velocity of about 300 m/s.

trivial, since continuum approach can be used. The analogue of RMI is another matter. Fig. 4 shows an image sequence acquired in several experimental runs after the shock passes through the initial conditions (droplet-seeded cylindrical jet) at Mach number $M \sim 1.66$. The bulk of the unseeded air behind the shock front in these images moves with a piston velocity of about 300 m/s. The droplets clearly do not follow the flow perfectly – they lag behind the shock-accelerated air, as the images in Fig. 4 clearly show. The "tail" of lagging droplets of larger sizes is particularly apparent in the first image following the shock passage (top row, middle, $t = 229$ μs). As these droplets lag behind the air, they exchange momentum with it, thereby slowing it down. This results in the air-droplet mixture reaching some equilibrium velocity that is lower than the piston velocity of air not seeded with droplets. That,in turn, leads to shear and roll-up of counter-rotating vortices.

The morphology shown evolving in Fig. 4 is both similar (counter-rotating vortex pairs) and different (droplet tail) from that evolving in the case of "classical" RMI.

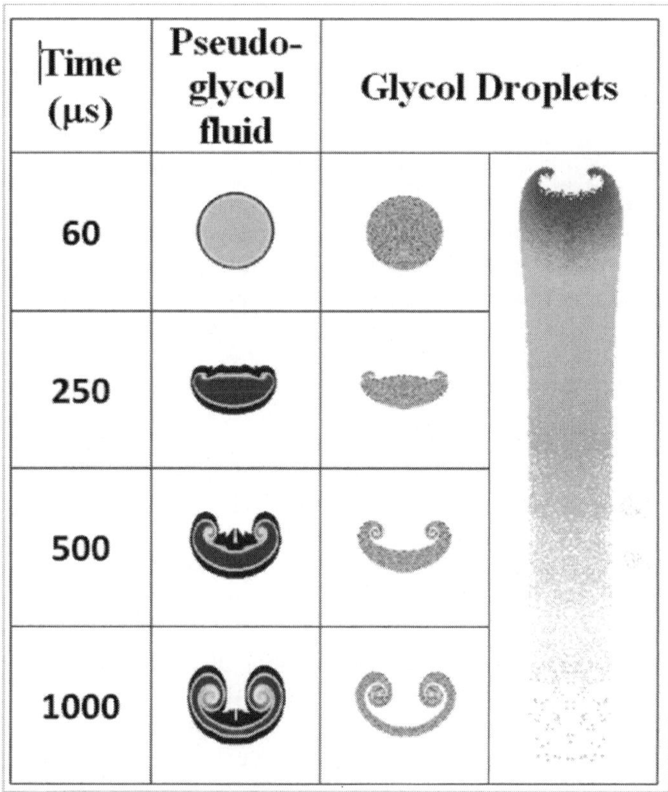

Figure 5: Numerical images of the two modeling approaches at selected times. Column 1 shows times, column 2 – results of continuum modeling, column 3 – result of modeling with computational particles of characteristic size 0.7 μm, column 4 – a single snapshot of the modeling result with computational particles of characteristic size 7 μm corresponding to time 1000 μs. The shock acceleration direction is from bottom to top, horizontal cell extent is 12.7 mm.

5 Modeling results

Columns two and three of Figure 5 show a series of time images from the two modeling approaches. The second column ("Pseudo-glycol fluid") represents the continuum approach, while the third and the fourth show results from modeling with particles. The difference between columns 3 and 4 is in characteristic particle size, which is increased by a factor of 10 for the last column. It is clear that both methods produce instabilities that are similar in morphology to what is observed in experiments. One aspect that is missing from the numerical results is the tail of lagging particles. This is not surprising for the pseudo-glycol fluid approach, but the tail should be visible in the discrete particles modeling approach. One possibility is that the particle sizes chosen for the simulation are too small. The right column of Figure 5 shows results from a run with a different particle size distribution at 1 ms. The distribution has particle sizes an order of magnitude larger than the distribution in column three. In this new calculation, a large tail of trailing particles is formed. The tail is now much larger than that found in experiments, but this calculation demonstrates the importance of obtaining the correct particle size distribution. The roll-up of the vortices is also different in this calculation. More work is required to determine the actual particle distribution that makes up the initial conditions in the experiments and to match it with numerics.

6 Discussion

For "classic" RTI vs. its two-phase analogue, only in the former case is the physical gas-gas density gradient present, but this appears to be a distinction without a difference, as the embedded droplets faithfully follow the embedding air, and on the relatively short (seconds) time scale characterizing the mixing of seeded air with the unseeded air in our experiments, the flow behaves macroscopically like RTI, manifesting mixing zone growth consistent with earlier RTI experiments. The two-phase analogue of RMI is much more interesting, because after impulsive acceleration, droplets interact with the embedding phase in a complex way which presents a challenge to model accurately.

Does what we observe represent a new hydrodynamic instability? All the physical mechanisms involved are well-known, yet their combined manifestation, especially in the case of two-phase shock-accelerated flow, leads to unexpected results, whose importance may be considerable in astrophysical problems (e.g., shock propagation through dusty plasma), as well as in many engineering applications.

7 Conclusion

We have presented an experimental and numerical study of multiphase analogues of hydrodynamic instabilities emerging on a density gradient subject to sustained (Rayleigh-Taylor) or impulsive (Richtmyer-Meshkov) acceleration. Our work shows that vortices similarly form in a gaseous medium initially seeded with microscopic droplets so that its average density is non-uniform, and the gradient of this average density plays the role similar to the gaseous (or fluid) density gradient in classical RMI and RTI.

Future work will involve experiments with a better characterization of the initial conditions in terms of droplet sizes and distribution, as well as numerical simulations using this information. The goal of numerical exercises here is twofold. First, the code must be validated by quantitative comparison with a well-characterized experiment. Second, after such validation provides confidence that the numerics faithfully reproduce the relevant physics, numerical modeling can be applied to problems of practical interest, for example, involving propagation of a shock wave through dusty gas or plasma.

Acknowledgments

This research is funded by NNSA (US National Nuclear Security Agency) through DOE (US Department of Energy) Grant DE-PS52-08NA28920 and by DTRA (US Defense Threat Reduction Agency) awards HDTRA1-07-1-0036 and HDTRA1-08-1-0053.

References

[1] J.W. Strutt. Investigation of the character of the equilibrium of an incompressible heavy fluid of variable density. *Proceedings of the London Mathematical Society*, 14:170–177, 1883.

[2] G.I. Taylor. The instability of liquid surfaces when accelerated in a direction perpendicular to their planes. I. *Proceedings of the Royal Society of London. Series A, Mathematical and Physical Sciences*, 201(1065):192–196, 1950.

[3] D.J. Lewis. The instability of liquid surfaces when accelerated in a direction perpendicular to their planes. II. *Proceedings of the Royal Society of London. Series A, Mathematical and Physical Sciences*, 202(1068):81–96, 1950.

[4] M.P.A. Jackson and C.J. Talbot. External shapes, strain rates, and dynamics of salt structures. *Geological Society of America Bulletin*, 97:305–323, 1986.

[5] T.V. Gerya and D.A. Yuen. Rayleigh-Taylor instabilities from hydration and melting propel cold plumes at subduction zones. *Earth and Planetary Science Letters*, 212:47–62, 2003.

[6] W.H. Cabot and A.J. Cook. Reynolds number effects on RayleighTaylor instability with possible implications for type Ia supernovae. *Nature Physics*, 2:562–568, 2006.

[7] R.D. Richtmyer. Taylor instability in shock acceleration of compressible fluids. *Communications on Pure and Applied Mathematics*, 13:297–319, 1960.

[8] E.E. Meshkov. Instability of the interface of two gases accelerated by a shock wave. *Izvestiya Akademii Nauk SSSR, Mekhanika Zhidkosti i Gaza*, 4:151–159, 1969.

[9] A. Burrows, J. Hayes, and B.A. Fryxell. On the nature of core-collapse supernova explosions. *Astrophysics Journal*, 450:430–450, 1995.

[10] J. Yang, T. Kubota, and E.E. Zukoski. Applications of shock-induced mixing to supersonic combustion. *AIAA Journal*, 31:854–862, 1993.

[11] V.N. Goncharov. Theory of the ablative Richtmyer-Meshkov instability. *Physical Review Letters*, 82:2091–2094, 1999.

[12] R.M. Baltrusaitis, M.L. Gittings, R.P. Weaver, R.F. Benjamin, and J.M. Budzinski. Simulation of shock-generated instabilities. *Physics of Fluids*, 8:2471–2483, 1996.

[13] A.W. Cook and Y. Zhou. Energy transfer in Rayleigh-Taylor instability. *Physical Review E*, 66:026312–1 026312–12, 2002.
[14] A. Palekar, P. Vorobieff, and C.R. Truman. Two-dimensional simulation of Richtmyer-Meshkov instability. *Progress in Computational Fluid Dynamics*, 7(8):427 – 438, 2007.
[15] M.K. Parmar, A. Haselbacher, and S. Balachandar. Prediction and modeling of shock-particle interaction. AIAA Paper 2009-1124, 2009.
[16] M.K. Parmar, A. Haselbacher, and S. Balachandar. Improved drag correlation for spheres and application to shock-tube experiments. *AIAA Journal*, 48:1273–1276, 2010.
[17] M. Anderson, P. Vorobieff, S. Kumar, J. Conroy, R. White, C. Needham, and C.R. Truman. Numerical simulation of a shock-accelerated multiphase fluid interface. To be published in Proceedings of 28th International Symposium on Shock Waves, 2011.
[18] P. Vorobieff, M. Anderson, J. Conroy, R. White, C.R. Truman, and S. Kumar. Vortex formation in shock-accelerated gas induced by particle seeding. Submitted to Physical Review Letters, 2011.
[19] J.E. Crepeau. SHAMRC, second-order hydrodynamic automatic mesh refinement code, vol. 1: Methodology. Applied Research Associates, Albuquerque, NM, USA, 1999.
[20] G. Birkhoff. Taylor instability and laminar mixing. University of California Report No. LA-1862, 1955.
[21] D.L. Youngs. Numerical simulation of turbulent mixing by Rayleigh-Taylor instability. *Physica D: Nonlinear Phenomena*, 12:32–44, 1984.
[22] P. Ramaprabhu, G. Dimonte, and M.J. Andrews. A numerical study of the influence of initial perturbations on the turbulent RayleighTaylor instability. *Journal of Fluid Mechanics*, 536:285–319, 2005.

Comparison of numerical schemes for computational simulation of liquid wavy film flow on vertical wall

J. K. Min[1] & I. S. Park[2]
[1]*Rolls-Royce University Technology Center, Pusan National University, Korea*
[2]*School of Mechanical Engineering, Kyungpook National University, Korea*

Abstract

Film flows are classified into no-wavy laminar, wavy laminar and turbulence along the Reynolds number or the flow stability. Since the wavy motions of the film flows are so intricate and nonlinear, the studies have largely been dependent upon the experimental way. The numerical approaches have been limited on the no-wavy flow regime. In this study, the film's full wavy motion is numerically investigated. The present numerical approach is based on the VOF interface capturing methods. The implicit and explicit schemes for the interface time marching are compared with each other. The HRIC (High Resolution Interface Capturing) and the geometry reconstruction method are tested as the treatment of convective terms at the liquid-air interface. The dependency of the grid density and the time step size are investigated in the aspect of the probability density function of the film thickness. All results are compared with Takamasa's experimental results.
Keywords: wavy film flow, free surface, VOF.

1 Introduction

The film flows are generally applied to the phase-change heat transfer (condensation or evaporation) because of their advantages for heat transfer like the high latent heat, the thin film thickness and the large heat transfer area.

WIT Transactions on Engineering Sciences, Vol 70, © 2011 WIT Press
www.witpress.com, ISSN 1743-3533 (on-line)
doi:10.2495/MPF110031

Many previous studies [1–5] proved that the film thickness is the most essential parameter governing heat transfer. Thus a lot of researchers have attempted to apply more reasonable film thicknesses for their applications. Nusselt suggested the exact analytic solution for the film flows in a laminar governing equation set. It is still considered as a very powerful and useful result for simple application problems. Some researchers [6–8] have solved only the convective and diffusive heat transfer equation with the film thickness and velocity distribution from the Nusselt solution as the flow information. They couldn't be applied to the accurately predicting the local heat and mass transfer in the developing region. Thereafter a lot of researchers [3–5] attempted to numerically solve the parabolized Navier-Stokes equation as the governing equation with the marching technique. This has been evaluated as the more accurate numerical approach by considering the convective heat transfer effect by the normal velocity component which has been neglected in the thin film flow so far. Because the parabolized governing equations were solved, this method was applicable to the laminar as well as the turbulence flow regimes and found the average film thickness reasonably. But it could not solve the time changing phenomena such as the film's wavy motion which happened due to the flow instability in higher flow rate conditions.

There have been attempts to solve the elliptic type full Navier-Stokes equations [9–12]. Since this method could additionally consider the diffusion in the primary flow direction, it was advantageous for the falling film on the horizontal circular tube. The numerical free surface capturing or tracking methods such as MAC (Marker and Cell) [13], VOF (Volume Of Fluid) [14], moving grid technique [15] and so on, were applied to investigate the film's free surface location precisely.

The analysis for the three dimensional film flows with secondary flows in the cross section like the grooved, fluted or spiral tube was conducted [11, 12]. The approaches using the elliptic governing equations which were investigated with a flourish in the late 1990s could make no more evolutions because of their weakness of high computational cost comparing with the advantages from their accurate calculation.

It is known that the film flow is getting unstable as the flow rate increases. Thus the laminar wavy motion starts to appear over Reynolds number 30 and the flow transits to the turbulence between Reynolds 1,600 and 3,200. Most applications belonged in the laminar wavy regime to apply the heat-transfer-boosting effects by the interfacial wave and the thin film thickness. Since waves in the turbulence regime are highly frequent and short waves, the heat and mass transfer enhancement by the wavy motion can be somewhat simulated by adjusting the turbulence intensity. But the waves of long wavelength and high amplitude in the low Reynolds number need the wavy motion itself to be accurately solved. Because the wavy motion of film flow means the temporary and spatially fluctuating film thickness, its height (amplitude), frequency, and velocity are as important as the average film thickness.

Complex interactions between the liquid turbulence structure in the film and gas-liquid interfaces govern the physical properties of such flows. The resulting interfacial waves show fascinating nonlinear phenomena, including solitary waves and complex disordered patterns. To elucidate these phenomena in detail, spatial-temporal knowledge of the interfacial waves is essential.

Due to the limitation of the numerical approaches, the experimental techniques measuring film flow such as electrical-resistance or electrical-capacitance methods have widely developed [16–22]. The electrical methods can measure the film's spatially averaged thickness, but not the film's wavy behaviors (wave amplitude, frequency, and so on). A laser displacement sensor [21] and a supersonic echo method [22] have a higher spatial resolution and thus it becomes possible to measure the film's wavy characteristics. But for the cases of large curvature gas-liquid interface, the measuring is impossible because the reflected laser beam and the sonic wave cannot reach the detector. Takamasa and Hazuku [23] successfully measured the wavy characteristics of the films flowing down a vertical plate wall and a vertical tube wall with a sensitivity of 2 μm and 1kHz. They used the laser focus displacement method with the high spatial and temporal resolution. A lot of studies have been conducted with the above experimental ways to reveal not only the film's wavy motion itself (such as wave celerity, amplitude and frequency) but also the relationship with the heat transfer.

In this study, various numerical schemes are tested for the laminar wavy film flow region. The present numerical approach is based on the VOF interface capturing methods. The implicit and explicit schemes for the interface time matching are compared with each other. The HRIC (High Resolution Interface Capturing) and the geometry reconstruction method are tested as the treatment of convective terms at the liquid-air interface. The dependency of the grid density and the time step size are investigated in the aspect of the probability density function of the film thickness. All results are compared with Takamasa's experimental results [23]. In addition the results obtained from turbulence flow simulation are compared together.

2 Numerical analysis

We consider a vertical wall on which the wavy laminar film flows as shown in Fig. 1. For constant fluid properties, the continuity and momentum equations for two-dimensional incompressible unsteady laminar film flows are written as:

$$\frac{\partial}{\partial x_i}(\rho u_i) = 0 \tag{1}$$

$$\frac{\partial(\rho u_i)}{\partial t} + \frac{\partial}{\partial x_j}(\rho u_i u_j) = -\frac{\partial p}{\partial x_j} + \frac{\partial}{\partial x_j}\left[\mu\left(\frac{\partial u_i}{\partial x_j} + \frac{\partial u_j}{\partial x_i}\right)\right] + \rho g_i \tag{2}$$

where, u_i is the velocity vector, p pressure, g_i gravity vector, ρ density and μ viscosity of the fluid. To describe the motion of free surface, the VOF (Volume Of Fluid) approach adopts the volume fraction function α which has a value between zero to unity according to the fraction of liquid inside each control volume. The corresponding equation for the volume fraction function is:

$$\frac{\partial \alpha}{\partial t} + \frac{\partial}{\partial x_j}(\alpha u_j) = 0 \cdot \tag{3}$$

Using this volume fraction α, the effective material properties in Eqs. (1) and (2) can bc evaluated as follows:

$$\rho = \alpha \rho_l + (1-\alpha)\rho_g \tag{4}$$

$$\mu = \alpha \mu_l + (1-\alpha)\mu_g \tag{5}$$

where the subscripts l and g represents the liquid and gaseous phases, respectively.

The boundary conditions for this problem are also shown in Fig. 1. On the liquid inlet boundary, the uniform velocity u_0 and fixed film thickness δ_0 are specified according to the corresponding Reynolds numbers. No-slip condition is applied on the vertical wall and gas inlet. Pressure outlet conditions are used for the other boundaries. On the free surface, the surface tension force is taken into consideration to satisfy the following force balance equation:

$$p_l - p_g = \sigma \frac{1}{R} \tag{6}$$

where σ represents the surface tension coefficient and R the radius of curvature of the free surface. In the numerical procedure, the CSF (Continuum Surface Force) model by Brackbill *et al.* [24] has been used. In this method, the surface curvature κ can be evaluated by the divergence of surface normal which is defined as the normalized gradient of volume fraction α. To impose the pressure jump across the surface, the volume force defined by the following equation should be applied for the control volume at the free surface.

$$F_{vol} = \sigma \frac{\rho \kappa \nabla \alpha}{\frac{1}{2}(\rho_l + \rho_g)} \tag{7}$$

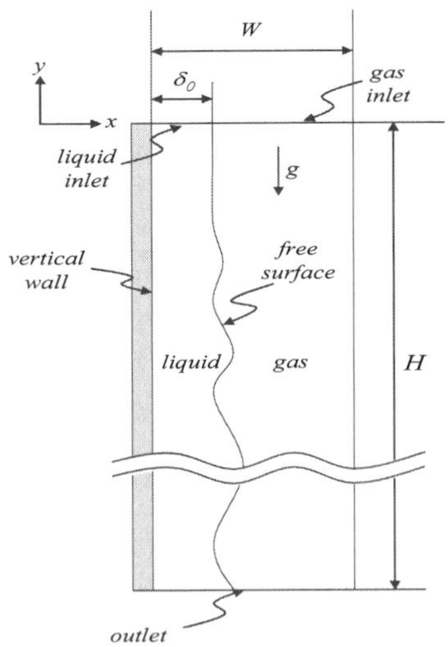

Figure 1: Schematic of wavy film flow.

For the numerical computations, the commercial CFD code FLUENT 6.3 has been used. Second order upwind scheme for the spatial discretization, implicit time marching and explicit time discretization for the VOF equation were adopted. In order to capture the location of interface precisely, the volume fraction α should vary sharply near the free surface of the liquid. VOF approach has the possibility of smearing the fraction function due to the false diffusion of numerical scheme. This problem is largely related to the numerical evaluation of flux values at the control surface of partially filled cell near the interface. Numerous computational techniques to improve the accuracy of surface tracking by VOF have been studied. Hirt and Nichols [14] suggested the donor and acceptor method in their initial VOF technique. According to this method, a cell near the free surface can be identified as a donor and the neighbouring cell as the acceptor depending on the flow direction at the control surface. The amount of the fluid from each phase across the control surface is calculated using the volume fraction value at the donor cell and the acceptor cell is assumed to receive the same amount of fluid from the donor cell. This method also determines the orientation of free surface using the gradient value of volume fraction α at the cell near the free surface, which can be either horizontal or vertical. Using the determined orientation and the velocity of the interface, the flux values across the control surface can be evaluated by modified upwinding technique in order to inhibit the false diffusion. Youngs [25] has suggested the PLIC (piecewise linear interface calculation) technique. In this method, the slope of the interface at the cell can be evaluated using the volume fraction values of surrounding neighboring cells. With this linear slope and the motion of the interface, the fluid flux across the control surface can be calculated as well.

Water is used for the liquid film flow and air for the gaseous phase in the present study. The width of liquid film δ_0 at the inlet is set to 0.4–0.6mm based on average film thickness, the height of the domain $H = 500$mm and the width $W = 4.6$mm. Corresponding dimensionless parameters are defined as follows:

$$Re = \frac{\rho u_0 \delta_0}{\mu} \tag{8}$$

$$We = \frac{\rho \delta_0 u_0^2}{\sigma} \tag{9}$$

where Re is the Reynolds number and We the Weber number. The Reynolds number has the range of 200 to 1000 and the inlet velocity for each Re can be calculated using the definition, eq. (8).

A 40×3000 grid was generated for the computational domain, which was clustered to the wall and uniform in the flowing direction. Several test showed that the Courant number $Co = \Delta t / (\Delta x_{cell} / v_{fluid})$ should be less than 0.5 in order to get a stable solution. As a result, the time step was selected to have $\Delta t = 10^{-4}$ sec, which showed $Co = 0.43$ for $Re = 1000$ case. For the computation, machines having Intel Xeon processors with eight nodes were used and it took 10–24 hours of CPU time to get a solution for 15000 time steps.

3 Results and discussions

For the discretization scheme at the cell near the free surface, the HRIC (High Resolution Interface Capturing) scheme is known as an improved scheme providing better accuracy. This is a blending scheme of upwind and downwind differencing. In this study, however, the use of HRIC scheme produced less realistic results. Table 1 shows the cases for testing the numerical schemes for the time marching and the interfacial cell treatment. Fig. 2 shows the distributions of the probability density function for the film thickness for each case. Comparing with Takamasa's experimental result, the explicit time marching and geometry reconstruction method is best fit with.

Fig. 3 shows the time variations of film thickness calculated using explicit HRIC scheme, implicit HRIC scheme and geometric reconstruction scheme, respectively. Geometric reconstruction scheme is a generalized Youngs' method and used with second-order discretization scheme in this calculation. It is hard to identify a solitary wave when explicit and implicit HRIC schemes were used, whereas a distinct solitary wave is observed with geometric reconstruction scheme, which is closer to previous experimental findings [23].

For the comparison of the effects by the grid and time step sizes, two grid systems (grid spacing= 0.5 and 0.2 mm) and three time step sizes $(2\times10^{-4}, 1\times10^{-4}, and\, 0.5\times10^{-4}\,\text{sec.})$ are tested. Fig. 4 shows the distributions of the probability density functions for each case. For the cases of grid size, 0.5 mm, much wider ranges of film thickness are predicted than the cases of grid size, 0.2 mm. As the time step sizes become smaller, the distribution becomes sharper. But the gap with the experimental result is narrowed no more. The average film thickness is about 0.5 mm as shown in the probability density function. This means that finer grid spacing than the average film thickness is required for the accurate calculation. The results of grid size 0.2 mm are more realistic comparing with Takamasa's experimental data and invariant with the time step size.

The film Reynolds number handled in this study is ranged between 200 and 1,000. This is known as the laminar wavy film flow region. However, the wavy motion is a transient phenomenon to the turbulence, and the wave itself is a large fluctuation and originated by the disturbance from outside. Such generated waves are amplified and sometimes damped out according to the mass flow rate. In this study, to investigate the film flow's turbulence effect, we adopt the two

Table 1: Test cases for inspecting numerical scheme for time differencing and interfacial cell treatment.

Case	Time Differencing	VOF	Viscosity
1	Explicit	Geometry Reconstruction	Laminar
2	Explicit	HRIC	Laminar
3	Implicit	HRIC	Laminar

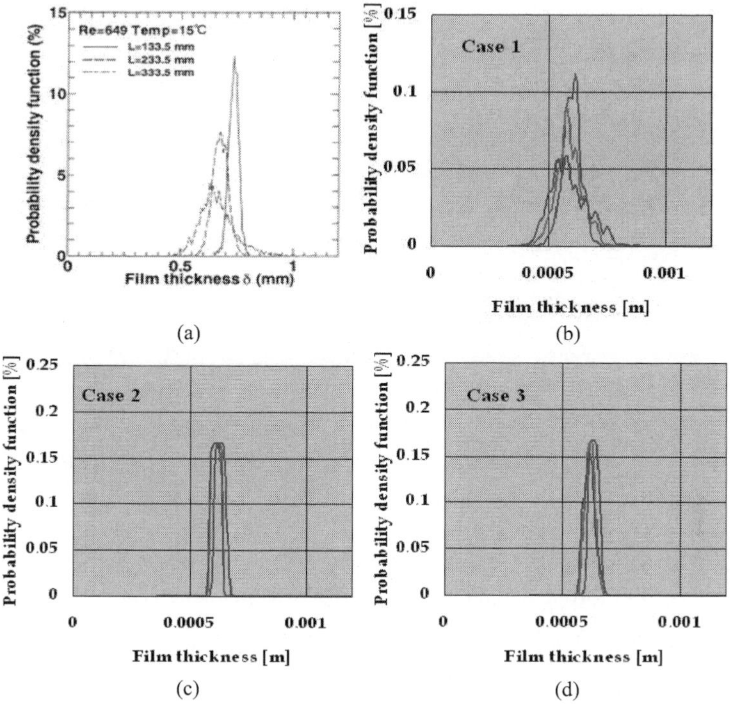

Figure 2: Probability density functions for film thickness at Re=649, (a) Takamasa's experiment, (b) Explicit geometry reconstruction, (c) Explicit HRIC, and (d) Implicit HRIC.

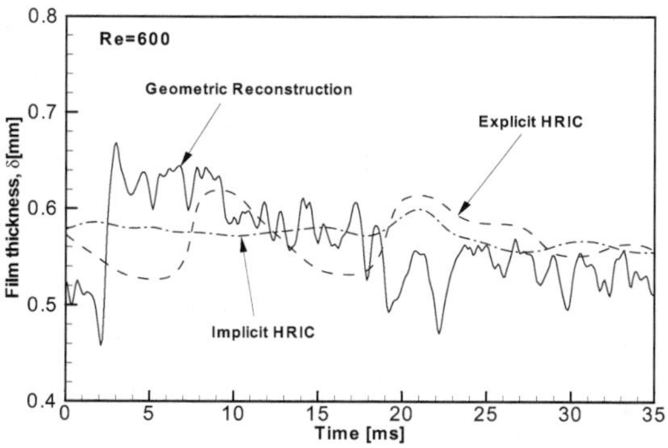

Figure 3: Comparison of interface cell treatment-geometry reconstruction scheme and HRIC scheme.

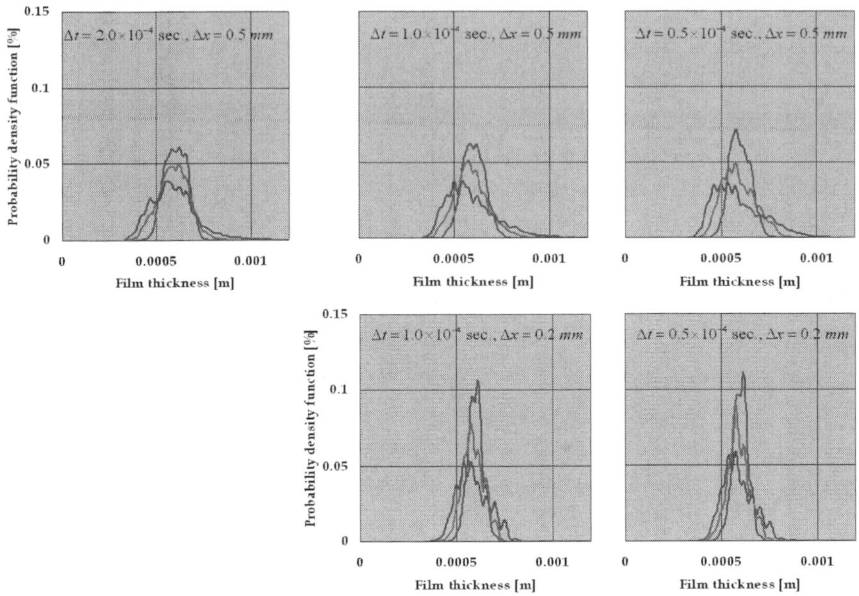

Figure 4: Probability density functions for various time step and grid sizes at Re=649.

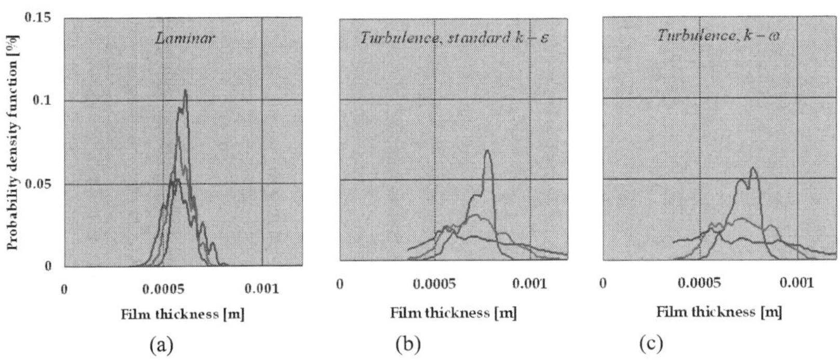

(a) (b) (c)

Figure 5: Comparison with turbulence simulations at Re=649, (a) Laminar, (b) Standard $k-\varepsilon$, and (c) $k-\omega$ turbulence model.

generalized 2-equation turbulence models (standard $k-\varepsilon$ model and $k-\omega$ model) and compared the results with the laminar simulation. Fig. 5 compares the results one another. The Reynolds number of test case is 649. The turbulence simulations generate more severe wave fluctuations because of the turbulence fluctuation. Both turbulence analyses show a quite different wavy motions comparing with the real case. Thus in the present Reynolds number range (200–1,000), when the flow is simulated by the laminar assumption, more reasonable results can be obtained.

Fig. 6 shows the wave's time-propagation processes for Reynolds number 200. To accurately observe and compare the wave shapes, the figures are made with x-y aspect ratio 5:1. The solitary waves with high wave amplitude appear and the foregoing small waves are merged at several points by the big wave approaching from upstream. The small amplitude ripples are observed among the solitary waves.

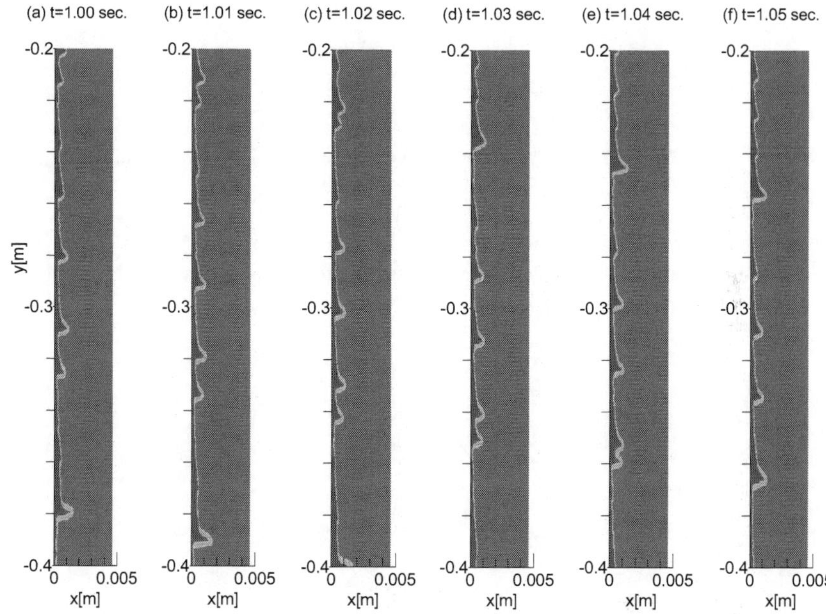

Figure 6: Propagation of wave along time at Re=200.

Fig. 7 shows the relationship between average film thickness and the Reynolds number. The average film thickness is defined as the arithmetic mean value of the instantaneous film thicknesses and calculated as follows.

$$\bar{\delta} = \frac{1}{n}\sum_{i=1}^{n}\delta_i \tag{10}$$

The Nusselt's theoretical result for laminar flow condition is

$$\bar{\delta} = \left(\frac{3\nu^2 Re}{g}\right)^{1/3} \tag{11}$$

The universal velocity profile is Karman's velocity profile (Eq. (12)) calculated for turbulent flow condition.

$$\left(3.0 + 2.5\ln\bar{\delta}^+\right)\bar{\delta}^+ = Re + 64 \tag{12}$$

where

$$\bar{\delta}^+ = u^* \frac{\bar{\delta}}{\nu} \tag{13}$$

$$u^* = \sqrt{g\bar{\delta}} \tag{14}$$

The results by Brauer (Eq. (15)), Takahama (Eq. (16)), and Aragaki (Eq. (17) shown in the figure are represented as follows:

$$\bar{\delta} = 0.302\left(\frac{3\nu^2}{g}\right)^{1/3} Re^{0.526} \tag{15}$$

$$\bar{\delta} = 0.473\left(\frac{\nu^2}{g}\right)^{1/3} Re^{0.526} \tag{16}$$

$$\bar{\delta} = \left(\frac{\nu^2}{g}\right)^{1/3} \left[8.92 Re^{5/2} + 4.04 \times 10^{-5} Re^{9/2}\right]^{2/15} \tag{17}$$

Takamasa's result is the experimental one obtained by their laser focus displacement method. The present results agree well with Nusselt and Takamasa's results especially in the far downstream region.

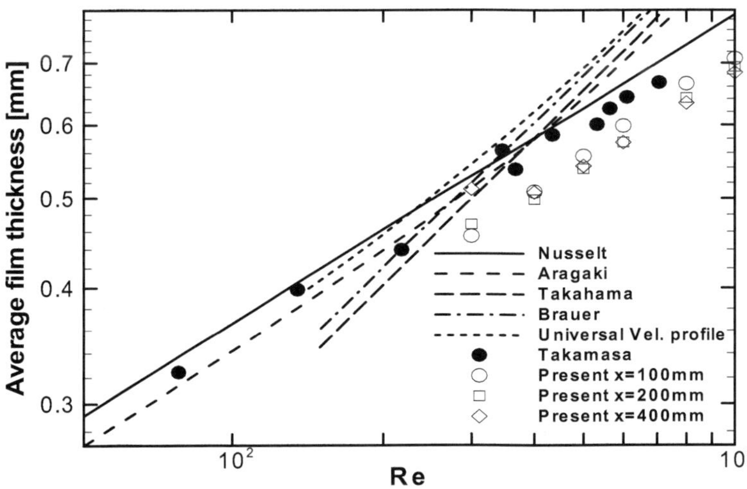

Figure 7: Average film thickness for various Reynolds numbers and various positions from liquid inlet.

4 Conclusions

For a film flow heat transfer which is widely applied due to its benefits for heat transfer. The film flow becomes unstable with the wave propagation over the Reynolds number 30. In this study, the wavy motions of the film flows were numerically investigated, which have largely depended on the experimental ways. The interfaces between liquid and gas were sharply captured by the VOF and PLIC method. The surface tension acting on the interfaces was correctly calculated from a well defined interface shape by the VOF and PLIC. Such correctly estimated surface tension forces were able to propagate waves on the interface correctly. The present numerical results showed a good agreement with the experimental one in the aspects of the average film thickness. It is clearly demonstrated that the current numerical Navier-Stokes procedure and precise free-surface capturing method with consideration of surface-tension effect are essential in order to predict the wavy film motion accurately.

Acknowledgement

This research was supported by Basic Science Research Program through the National Research Foundation of Korea (NRF) funded by the Ministry of Education, Science and Technology (the Grant 2010-0024619).

References

[1] Moon-Hyun Chun & Kyun-Tae Kim, Assessment of the new and existing correlations for laminar and turbulent film condensations on a vertical surface, *Int. Commun. Heat Mass Transfer*, 17, pp. 431-441, 1990.
[2] A. Faghri & R, A. Seban, Heat transfer in wavy liquid films, *Int. J. Heat Mass Transfer*, 28 (2), pp. 506-508, 1985.
[3] M. Feddaoui, A. Mir & E. Belahmidi, Cocurrent turbulent mixed convection heat and mass transfer in falling film of water inside a vertical heated tube, *Int. J. Heat Mass Transfer*, 46, pp. 3497-3509, 2003.
[4] M. K. Groff, S. J. Ormiston & H. M. Soliman, Numerical solution of film condensation from turbulent flow of vapor-gas mixtures in vertical tubes, *Int. J. Heat Mass Transfer*, 50, pp. 3899-3912, 2007.
[5] I. S. Park & M. Y. Kim, Numerical investigation of the heat and mass transfer in a vertical tube evaporator with the three-zone analysis, *Int. J. Heat Mass Transfer*, 52, pp. 2599-2606, 2009.
[6] N.I. Grigoreva & V.E. Nakoryakov, Exact solution combined heat and mass transfer problem during film absorption, *Journal of Engineering Physics*, 33(5), pp. 1349-1353, 1977.
[7] G. Grossman, Simultaneous heat and mass transfer in film absorption under laminar flow, *Int. J. Heat Mass Transfer*, 25 (3), pp. 357-371, 1982.
[8] G. Grossman & M.T. Heath, Simultaneous heat and mass transfer in film absorption in turbulent liquid films, *Int. J. Heat Mass Transfer*, 27(12), pp. 2365-2376, 1984.
[9] E. Stuhlträger, Y. Naridomi, A. Miyara & H. Uehara, Flow dynamics and heat transfer of a condensate film on a vertical wall-i. Numerical Analysis and Flow Dynamics, *Int. J. Heat Mass Transfer*, 36, pp. 1677-1686, 1993.
[10] J.K. Min & D.H. Choi, Analysis of the absorption process on a horizontal tube using Navier-Stokes equations with surface-tension effects, *Int. J. Heat Mass Transfer*, 42 (24), pp. 4567-4578, 1999.

[11] I.S. Park & D.H. Choi, Heat- and mass-transfer analysis for the condensing film flow along a vertical grooved tube, *Int. J. Heat Mass Transfer*, 44(22), pp. 4277-4285, 2001.
[12] I.S. Park, Numerical analysis for flow, heat and mass transfer in film flow along a vertical fluted tube, *Int. J. Heat Mass Transfer*, 53(1-3), pp. 309-319, 2010.
[13] F.H. Harlow & J.E. Welch, Numerical calculation of time-dependent viscous incompressible flow of fluid with free surface, *Physics of Fluids*, 8 pp. 2182-2189, 1965.
[14] C.W. Hirt & B.D. Nichols, Volume of fluid (vof) method for the dynamics of free boundaries, *Journal of Computational Physics*, 39, pp. 201-225, 1981.
[15] S. Muzaferija & M. Peric, Computation of free-surface flows using the finite-volume method and moving grids, *Numerical Heat Transfer, Part, B*, 32, pp. 369-384, 1997.
[16] S. Poltalski & A.J. Clegg, An experimental study of wave inception on falling liquid film, *Chemical Engineering Science*, 27, pp. 1257-1265, 1972.
[17] K. Mori, T. Matsumoto & H. Uematsu, Time-spatial interfacial structures and flow characteristics in falling liquid film, *Proceedings of Third International Conference on Multiphase Flow*, CD-ROM, #514, 1998.
[18] A. Elsäßer, W. Samenfink, J. Ebner, K. Dullenkopf & S. Wittig, Effect of variable liquid properties on the flow structure within, *Proceedings of Ninth International Symposium on Application of Laser Techniques to Fluid Mechanics*, 1(3), pp. 1-10, 1998.
[19] A.E. Dukler & O.P. Bergelin, Characteristics of flow in falling liquid film, *Proceeding of Chemical Engineering Progress Symposium Series*, 48(11), pp. 557-563, 1952.
[20] C.F. Hewitt & N.S. Hall-Taylor, *Annular Two-phase Flow*, Pergamon Press, Oxford, 1970.
[21] M. Nasr-Esfahany & M. Kawaji, Turbulence structure under a typical shear induced wave at a liquid/gas interface, *Proceedings of AIChE Symposium Series*, 310(92), pp. 203-210, 1996.
[22] A. Serizawa, T. Kamei, I. Kataoka, Z. Kawara, T. Ebisu & K. Torikoshi, Measurement of dynamic behavior of a liquid film flow with liquid droplets in a horizontal channel, *Proceedings of Second International Conference on Multiphase Flow*, 2, pp. 27-34, 1995.
[23] T. Takamasa & T. Hazuku, Measuring interfacial waves on film flowing down a vertical plate wall in the entry region using laser focus displacement meters, *Int. J. Heat Mass Transfer*, 43, pp. 2807-2819, 2000.
[24] J.U. Brackbill, D.B. Kothe & C. Zemach, A continuum method for modeling surface tension, *Journal of Computational Physics*, 100, pp. 335-354, 1992.
[25] D.L. Youngs, *Time-dependent multi-material flow with large fluid distortion, numerical methods for fluid dynamics*, Academic Press, pp. 273-285, 1982.

Hybrid density- and pressure-based splitting scheme for cavitating flows simulation

T. Alexandrikova, A. Pavlov & V. Streltsov
Mentor Graphics Corporation, Mechanical Analysis Division, Moscow, Russia

Abstract

A new numerical method to simulate steady-state isothermal liquid flows with hydraulic cavitation is proposed. The 3D averaged N-S equations with Lam-Bremhorst $\kappa - \varepsilon$ turbulence model are used. The barotropic state equation is developed basing on thermodynamic equilibrium relations.
Simulation of such flows faces a lot of numerical difficulties concerned with variations of density, speed of sound and time scale.
The method is a hybrid splitting scheme that is a mixture of "density-based" and "pressure-based" approaches. The splitting scheme is the "pressure-based" SIMPLE-type algorithm in the region of incompressible liquid flow without cavitation. The scheme degenerates to the "density-based" algorithm in a compressible region (2-phase state or pure gas). The proposed method differs from both "density-based" preconditioned algorithms and SIMPLE-type methods adapted to the case of cavitating flows.
The method has been tested on numerous typical 3D problems with hydraulic cavitation. Results of numerical simulation are in good agreement with experimental data. The algorithm shows high efficiency for the considered problems. The method has been implemented in FloEFD™.
Keywords: cavitating flows, numerical method, splitting scheme.

1 Introduction

A new numerical algorithm for simulation of cavitating flows is proposed in the paper. Cavitating flows are encountered in a wide range of applications playing mostly negative but sometimes positive role. We consider hydraulic cavitation

www.witpress.com, ISSN 1743-3533 (on-line)
doi:10.2495/MPF110041

that occurs in liquid flows when the pressure falls close to the saturated vapour pressure. In this case a two-phase flow takes place.

An approach when fluid is considered as a homogeneous and isothermal medium is used. The medium has such thermodynamic properties that it can be either an incompressible liquid or a highly compressible medium that corresponds to two-phase state or a compressible vapour that is treated as an ideal gas.

As cavitating flows are mostly turbulent, the averaged N-S equations are used in the paper. It is known that specific choice of turbulent model and its parameters influences on results of simulation (Senocak and Shyy [1]). However, this influence is not investigated in this paper because the problem requires a special research. To close the system of averaged N-S equations the Lam-Bremhorst $k-\varepsilon$ turbulent model and barotropic state equation are utilized.

Constructing numerical method to simulate cavitating flows, it is important to take into account the fact that regions of incompressible flow and highly compressible flow exist simultaneously in calculation domain. The speed of sound may drop from thousands in liquid flow to units in vaporized flow. It leads to supersonic flows with high Mach numbers, sometimes with shocks. Thereby the cavitation problems are characterized by wide range of Mach number from near zero to several tens in one calculation domain.

An important feature of cavitating flows is the presence of several time scales. It is common that the time scale is greater by several orders of magnitude in liquid than in vaporized region with a supersonic flow. The typical time step varies over 2 orders of magnitude over calculation domain in considered below problems.

Another important feature is that properties of the model medium are changing dramatically when crossing the boundary an incompressible flow region and a compressible one. The speed of sound is changing by 3 orders of magnitude in considered problems. As a result, nonlinearity of the equation system strongly manifests itself on the boundary of liquid and vaporized regions.

Difficulties in numerical simulation of flows with widely varied Mach number appear not only in multiphase flows but in many traditional gas dynamics problems as well. Nowadays nature of the difficulties has been investigated in details. The problems are related to change of character of flow parameters mutual influence while Mach number is varying. In a compressible flow all parameters (density, pressure, momentum and temperature) significantly influence on each others. The form of the mass conservation equation is such that it is naturally interpreted as a density equation. So, the density, momentum and energy are used as basic unknowns when considering compressible flows. Such methods are named "density-based".

While Mach number approaching to zero, the pressure-density dependence becomes weak. So, in the case of low Mach number and incompressible flows, the mass conservation equation is used to describe the pressure-momentum coupling. Such methods are named "pressure-based".

To date there are two main approaches to calculate all-speed compressible flows. The first one employs the "density-based" methods originally developed to simulate speed compressible flows (usually at M>0.3). The methods are adapted to low Mach number case by introducing artificial compressibility or using some preconditioning techniques (Kunz *et al.* [2], Li and Merkle [3], Li *et al.* [4, 5]). Without such adaptations the methods are almost useless for low-speed flows at $M \to 0$.

The second approach utilizes the “pressure-based” methods originally developed for incompressible flows. Usually these are SIMPLE-family splitting schemes (or “pressure-correction” methods) adapted to the case of compressible flows at high speed (Van der Heul *et al.* [6, 7]).

Presented in the paper approach differs from the both abovementioned approaches. At first glance the idea to apply the “pressure-based” in region of incompressible flow and the “density-based” approach in region of supersonic compressible flow looks quite natural. But it is not obvious how to couple these approaches. We propose a way of combining the approaches that is based on the following simple key idea. Employing the finite volume method, we suggest mixing fluxes and pressure approximations that correspond to “pressure-based” and “density-based” approaches on faces of control volumes. After that these mixed approximations are substituted in SIMPLE-type splitting scheme. Managing the mixing weight in fluxes, we can obtain either the original SIMPLE-type semi-implicit splitting scheme or the explicit “density-based” scheme or a mixture of these approaches.

In this paper only the steady-state regimes of cavitating flows are considered. Such flows (for averaged flow parameters) are realized in a large number of applications. It is too consuming in time to solve such problems by an algorithm developed for time-dependent problems.

Of course, unsteady effects, such as periodically coming off cavitating vortex, are of considerable interest, but we will not consider such flows in the paper.

2 Governing equations

Mass conservation and momentum equations:

$$\frac{\partial \rho}{\partial t} + \frac{\partial (\rho u_i)}{\partial x_i} = 0, \tag{1}$$

$$\frac{\partial \rho u_i}{\partial t} + \frac{\partial}{\partial x_j}(\rho u_i u_j) + \frac{\partial p}{\partial x_i} = \frac{\partial \tau_{ij}^{\Sigma}}{\partial x_j} + \rho \frac{\partial \varphi}{\partial x_i} - 2 \cdot e_{ijk} \Omega_j \rho u_k, \tag{2}$$

$$\tau_{ij}^{\Sigma} = (\mu + \mu_t) s_{ij}, \; \tau_{ij}^{R} = \mu_t \, s_{ij} - \frac{2}{3} \rho k \delta_{ij},$$

$$s_{ij} = \frac{\partial u_i}{\partial x_j} + \frac{\partial u_j}{\partial x_i} - \frac{2}{3} \delta_{ij} \frac{\partial u_k}{\partial x_k}, \; \varphi = \tfrac{1}{2} \Omega^2 r^2. \tag{3}$$

Here the Coriolis and centrifugal forces are taken into account.

Turbulence Lam-Bremhorst $k - \varepsilon$ model equations:

$$\frac{\partial \rho k}{\partial t}+\frac{\partial \rho k u_i}{\partial x_i}=\frac{\partial}{\partial x_i}\left(\left(\mu+\frac{\mu_t}{\sigma_k}\right)\frac{\partial k}{\partial x_i}\right)+\tau_{ij}^R\frac{\partial u_i}{\partial x_j}-\rho\varepsilon,$$

$$\frac{\partial \rho \varepsilon}{\partial t}+\frac{\partial \rho \varepsilon u_i}{\partial x_i}=\frac{\partial}{\partial x_i}\left(\left(\mu+\frac{\mu_t}{\sigma_\varepsilon}\right)\frac{\partial \varepsilon}{\partial x_i}\right)+f_1 C_{\varepsilon 1}\frac{\varepsilon}{k}\tau_{ij}^R\frac{\partial u_i}{\partial x_j}-f_2 C_{\varepsilon 2}\frac{\rho\varepsilon^2}{k},$$

$$\mu_t=C_\mu f_\mu\frac{\rho k^2}{\varepsilon},\ f_\mu=\left(1-\exp\left(-0.0165R_y\right)\right)^2\left(1+\frac{7.5}{\mathrm{Re}_t}\right),\ \mathrm{Re}_t=\frac{\rho k^2}{\mu\varepsilon},$$

$$f_1=1+\left(\frac{0.05}{f_\mu}\right)^3,\ f_2=1-\exp\left(-\mathrm{Re}_t^2\right),\ R_y=\frac{\rho\sqrt{k}y}{\mu},\ \mathrm{Pr}_t=0.9,$$

$$C_\mu=0.09,\ C_{\varepsilon 1}=1.44,\ C_{\varepsilon 2}=1.92,\ \sigma_k=1,\ \sigma_\varepsilon=1.3. \tag{4}$$

It is assumed that flow is isothermal. So, the energy equation is omitted. It is supposed that T is constant, $T=T_0$.

To close the system a state equation is needed. For practical purposes, it is important to be able to simulate flows of industrial liquids, such as liquid fuels, synthetic and mineral oils, freons and so on. But thermodynamic properties of such substances are incomplete or not widely available. Taking into account this fact, the state equation is derived in such a way that it requires just minimal and easy available parameters of a substance. We consider the barotropic state equation [27] for the homogeneous mixture consisting of liquid, its vapour and small amount of inert non-condensable gas (see Fig. 1). In practice, the most liquid fluids contain small amount of dissolved gases.

In Fig. 1 parameter p_E^L is pressure value below which vapour appears in the mixture, p_E^V is pressure value below which liquid disappears from the mixture. These two parameters p_E^L and p_E^V are derived from the condition of phases equilibrium employing Helmholtz thermodynamic potential [27]. In two-phase region the density-pressure dependency is

$$\rho=\frac{p-p_E}{RT}\frac{M_I}{Y_I},$$

where p_E is the saturated vapour pressure, M_I and Y_I is the molecular mass and mass fraction of the inert gas.

Thus, defining only the density of a liquid and the saturated vapour pressure at temperature T_0, the molecular masses of substance of this liquid and a dissolved inert gas, we obtain the state equation for the specific fluid:

$$\rho=\rho(p). \tag{5}$$

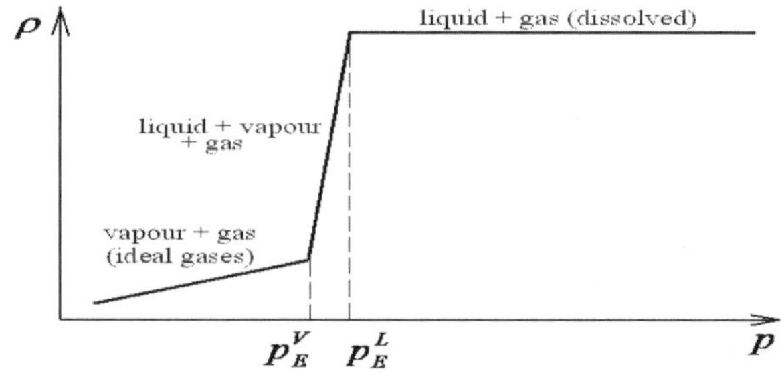

Figure 1: The barotropic state law $\rho(p)$.

3 Numerical method

Describing the numerical method, we pay the most attention to the operator-splitting technique. Details of spatial approximations are omitted. Being introduced by Yanenko [28], the operator-splitting technique has been further developed to treat SIMPLE-type methods in terms of this approach. Now this technique is well-accepted in CFD. Details of specific treating are presented in Churbanov *et al.* [29].

Following the finite volume method, the integral form of conservation equations (1)-(4) is approximated.

Let us describe the principal idea how to couple two approaches, the p-based semi-implicit SIMPLE-type approach and the ρ-based explicit approach, in one numerical method.

Assume that there are two well-established numerical methods. One is p-based method for incompressible (low-compressible) flows, for example SIMPLE splitting scheme with corresponding spatial approximations of fluxes and pressure:

$F_{inc}^{(\rho)}(m) = \sum a_k m_k$ is face mass flux approximation,

$F_{inc}^{(m)} = u_{i+1/2} \cdot \sum b_k m_k$ is face momentum flux approximation,

$p_{inc} = \sum c_k p_k$ is face pressure approximation.

Here m is momentum, u is velocity, a_k, b_k, c_k are approximation coefficients, and summation by index k is performed over the stencils of the approximations. Subscript "*inc*" is used to refer to incompressible (low-compressible) case.

The second well-established method is ρ-based one for high-speed compressible flows. For example well-known AUSM (Liou) with corresponding spatial approximations of fluxes and pressure:

$F_{comp}^{(\rho)} = (\rho u)_{AUSM}$ is the AUSM face mass flux approximation,

$F_{comp}^{(m)} = (u \cdot m)_{AUSM}$ is the AUSM face momentum flux approximation,

$p_{comp} = (p)_{AUSM}$ is the AUSM face pressure approximation.

Subscript "*comp*" is used to refer to compressible case.

It is proposed that the coupling is to be performed by combining spatial approximations of convective fluxes and pressure, and substituting these mixed approximations into any SIMPLE-family scheme.

Spatial approximations of fluxes and pressure are mixed on CV (control volume) faces with some weight w. The weight is defined on CV faces and depends on state of fluid in adjacent CVs. It is equal to zero in liquid (incompressible flow region) and 1 in containing vapor medium (compressible flow region).

Note that the combined fluxes and pressure approximations are treated differently in time, as it is shown below. The p-based approximations are treated implicitly where it is necessary, and the ρ-based approximations are treated explicitly in all cases.

Unsteady form of equations is used to obtain steady-state solutions. But time-step is treated as calculation parameter that is varying in space. Symbol "^" is used to mark parameters that refer to new iteration and are treated as unknowns. Symbol "~" is used to mark intermediate values of parameters. And parameters at old iteration do not have any "cup" symbol.

To simplify description, let us present deriving the method in 1d case.

The discrete mass equation is written as

$$\frac{V_i}{\Delta t_i}\delta\rho_i + \left(\hat{F}^{(\rho)}_{i+1/2}(\hat{m}) - \hat{F}^{(\rho)}_{i-1/2}(\hat{m})\right)S = 0, \qquad (6)$$

where $\delta\rho = \hat{\rho} - \rho$ and the weighted mass fluxes on CV faces are

$$\hat{F}^{(\rho)}_{i+1/2}(\hat{m}) = \left(1 - w_{i+1/2}\right)\cdot \hat{F}^{(\rho)}_{inc}(\hat{m}) + w_{i+1/2}\cdot F^{(\rho)}_{comp}(m),$$

$$\hat{F}^{(\rho)}_{inc} = \sum a_k \hat{m}_k, \quad F^{(\rho)}_{comp} = (\rho u)_{AUSM}.$$

Following the operator-splitting technique, the discrete momentum equation is written on two time levels, on intermediate and new ones. For "predictor", on intermediate time level, it is

$$\frac{V_i}{\Delta t_i}(\tilde{m}_i - m_i) + \left(\tilde{F}^{(m)}_{i+1/2} - \tilde{F}^{(m)}_{i-1/2}\right)S + \left(p_{i+1/2} - p_{i-1/2}\right)S = f_i, \qquad (7)$$

where the weighted momentum fluxes and pressure on CV faces are

$$\tilde{F}^{(m)}_{i+1/2} = \left(1 - w_{i+1/2}\right)\cdot \tilde{F}^{(m)}_{inc} + w_{i+1/2}\cdot F^{(m)}_{comp},$$

$$p_{i+1/2} = \left(1 - w_{i+1/2}\right)\cdot p_{inc} + w_{i+1/2}\cdot p_{comp},$$

$$\tilde{F}^{(m)}_{inc} = u_{i+1/2}\sum b_k \tilde{m}_k, \quad F^{(m)}_{comp} = (u\cdot m)_{AUSM},$$

$$p_{inc} = \sum c_k p_k, \quad p_{comp} = (p)_{AUSM}.$$

Here $\tilde{F}^{(m)}_{i+1/2}$ includes approximations of convective and diffusive fluxes of momentum. In 3d case it also includes Coriolis force, and f_i contains all other explicitly treated terms and forces.

The momentum equation is also written on new time level. The same $\tilde{F}^{(m)}_{i\pm1/2}$ but new pressure is employed:

$$\frac{V_i}{\Delta t_i}(\hat{m}_i - m_i) + \left(\widetilde{F}^{(m)}_{i+1/2} - \widetilde{F}^{(m)}_{i-1/2}\right)S + \left(\hat{p}_{i+1/2} - \hat{p}_{i-1/2}\right)S = f_i, \tag{8}$$

$$\hat{p}_{i+1/2} = \left(1 - w_{i+1/2}\right)\cdot \hat{p}_{inc} + w_{i+1/2}\cdot p_{comp},$$

$$\hat{p}_{inc} = \sum c_k \hat{p}_k, \quad p_{comp} = (p)_{AUSM}.$$

As usually for SIMPLE-type methods, an equation for pressure correction $\delta\hat{p}$ is derived to be used instead of the mass equation. For this purpose, eqn (7) is subtracted from eqn (8), and the result is substituted in the mass equation (6). This new equation replaces the mass equation in numerical method. Note that only equivalent algebraic transformations are employed, and the final system of discrete equations is fully equivalent to the original one.

Finally, the numerical method is written as the following equations that are solved consequentially.

The equation to calculate $\widetilde{m}$ (the "predictor" stage) is:

$$\frac{V_i}{\Delta t_i}(\widetilde{m}_i - m_i) + \left(\widetilde{F}^{(m)}_{i+1/2} - \widetilde{F}^{(m)}_{i-1/2}\right)S + \left(p_{i+1/2} - p_{i-1/2}\right)S = f_i, \tag{9}$$

The equation to calculate the pressure correction δp is:

$$\frac{V_i}{\Delta t_i}(D^{(n)}_p \cdot \delta p^{(n+1)}) + \Delta\left(\delta p^{(n+1)}\right) = -\left(\widetilde{F}^{(\rho)}_{i+1/2}(\widetilde{m}) - \widetilde{F}^{(\rho)}_{i-1/2}(\widetilde{m})\right)S, \tag{10}$$

where $D_p \cdot \delta p$ is approximation of density correction $\delta\rho$, and

$$\Delta(\delta p) = \left(\left(1 - w_{i+1/2}\right)F^{(\rho)}_{inc,\, i+1/2}\left(\frac{\Delta t}{V}G_{inc}(\delta p)\right) - \left(1 - w_{i-1/2}\right)F^{(\rho)}_{inc,\, i-1/2}\left(\frac{\Delta t}{V}G_{inc}(\delta p)\right)\right)S,$$

$$G_{inc}(\delta p)_i = \left(\left(1 - w_{i+1/2}\right)\delta p_{inc,\, i+1/2} - \left(1 - w_{i-1/2}\right)\delta p_{inc,\, i-1/2}\right)S, \tag{11}$$

$$\delta p_{inc,\, i+1/2} = \sum c_k \delta p_k .$$

The equations to calculate new momentum (the "corrector" stage) and pressure are:

$$\hat{m}^{(n+1)}_i = \widetilde{m}_i - \frac{\Delta t_i}{V_i} G_{inc}\left(\delta p^{(n+1)}\right)_i, \tag{12}$$

$$\hat{p}^{(n+1)}_i = p_i + \delta p^{(n+1)}_i . \tag{13}$$

The state equation to calculate new density is:

$$\rho^{(n+1)} = \rho\left(\hat{p}^{(n+1)}\right). \tag{14}$$

The coefficient D_p is approximated using $\left(\frac{\partial \rho}{\partial p}\right)_i$ in CV and relations $\left|\frac{\rho^{(n+1)}_{i\pm1} - \rho^{(n+1)}_i}{p^{(n+1)}_{i\pm1} - p^{(n+1)}_i}\right|$ on CV faces.

The local time step is defined differently in incompressible and compressible flow regions:

$$\Delta t_i = C_{inc} \frac{h_i}{u_i} \text{ in incompressible flow region,} \tag{15}$$

$$\Delta t_i = C_{comp} \frac{h_i}{u_i + c} \text{ in compressible flow region ,} \tag{16}$$

here $c = \left(\frac{\partial \rho}{\partial p}\right)^{-1/2}$ is the sound speed, $C_{inc} \approx 25$, $C_{comp} \approx 0.3$.

Such definition provide high rate of convergence to steady-state solution in both incompressible and compressible flow regions.

The internal iterations over index n are introduced. These are needed to take into account non linear dependence of the density increment on the pressure increment is in vicinity of two values of pressure, p_E^L and p_E^V. Three internal iterations were enough to provide robust calculations.

It is easy to see that setting w=0 leads to the standard SIMPLE (if only one internal iteration is performed).

Setting w=1 leads to the AUSM (with internal iterations).

The momentum equation:

$$\frac{V_i}{\Delta t_i}(\widetilde{m}_i - m_i) + \left((u \cdot m)_{AUSM,\, i+1/2} - (u \cdot m)_{AUSM,\, i-1/2}\right)S,$$

$$+ \left(p_{AUSM,\, i+1/2} - p_{AUSM,\, i-1/2}\right)S = f_i, \tag{17}$$

The mass equation:

$$\frac{V_i}{\Delta t_i}(D_p^{(n)} \cdot \delta p^{(n+1)}) + \left((\rho u)_{AUSM} - (\rho u)_{AUSM}\right)S = 0, \tag{18}$$

Other equations are as follows

$$\hat{m}_i^{(n+1)} = \widetilde{m}_i, \quad \hat{p}_i^{(n+1)} = p_i + \delta p_i^{(n+1)}, \quad \rho^{(n+1)} = \rho\left(\hat{p}^{(n+1)}\right).$$

In the original AUSM method eqn. (18) is used for direct calculation of $\delta\rho$ and $\hat{\rho} = \rho + \delta\rho$. New pressure $\hat{p}$ is defined from the inverse form of the state equation $\hat{p} = p(\hat{\rho})$.

Iterating abovementioned algorithm by n we obtain the original AUSM method.

Thus, we have obtained the method that couples p-based and ρ-based approaches.

To perform described below calculations, the following specific spatial approximations were used. For "*inc*" approximations of convective fluxes in incompressible flow regions, a combination of the central and the 2nd order upwind approximations was utilized. Limiters were used to provide monotonic properties of the approximation. For convective fluxes and pressure in compressible flow region well-known method AUSM [1] with some modifications as employed. For viscous fluxes central differences were used as usually.

The proposed numerical method has shown high efficiency and robustness solving steady-state problems. This is due to the fact that different well-established approaches are used in incompressible and compressible regions of flow with corresponding different local time steps (15), (16).

4 Results

In this section we present some results of cavitating flows simulation.

4.1 Centrifugal pump with two-dimensional curved blades

The geometry of the pump is presented in Fig. 2. In details it is described in Coutier-Delgosha *et al* [9] as well as results of experimental and numerical investigation. For calculations we use numerical mesh with 72500 control volumes. Similarly to [9] we specify the following boundary conditions – fixed mass flow-rate at the inlet and fixed static pressure at the outlet boundaries respectively. Presented numerical method utilizes the local time step, which values are chosen taking into account the stability condition in the whole computational domain. The results of flow simulation in cavitating and non-cavitating conditions were obtained.

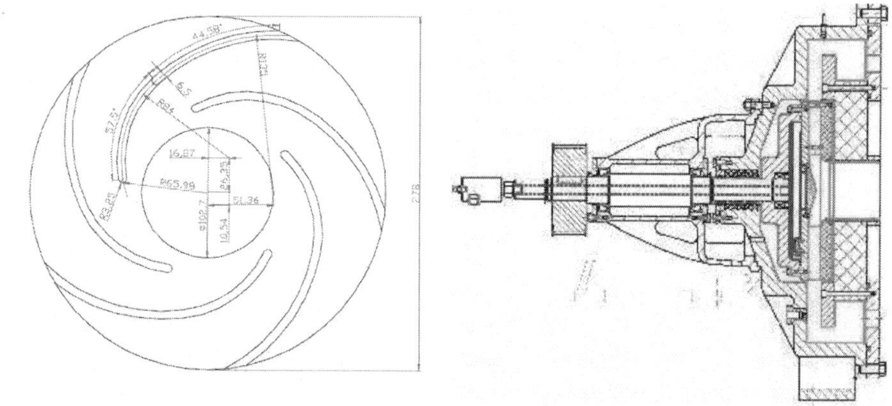

Figure 2: Centrifugal pump geometry, Coutier-Delgosha *et al.* [9].

The head drop chart H(NPSH) obtained by the calculation is drawn in Fig. 3. Calculations were performed considering various regimes in noncavitating and cavitating conditions, corresponding to various outlet pressure values and fixed mass flow-rate. The results were obtained for a large range of NPSH values, that correspond to different configurations of cavitation zones presented in the Fig. 4. The growth of cavitation zones and changes of their shape are observed with decrease of NPSH. Starting from NPSH=7 the cavitation zone attaching takes

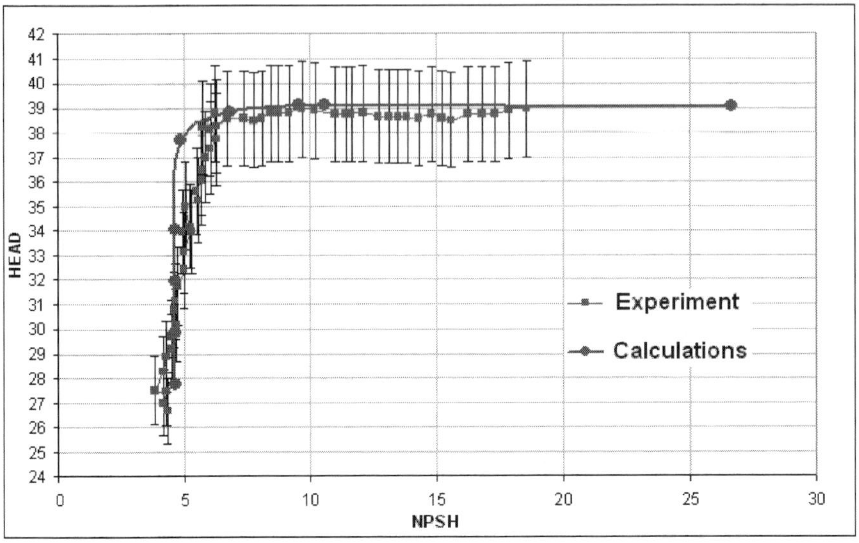

Figure 3: Head drop curves.

place on the outlet side of blades. Here H(NPSH) curve drop begins. With a further decreasing of NPSH the duct between blades is blocked by a cavity with a liquid-vapour mixture. Obtained numerical results are in a good agreement with experimental data.

4.2 Liquid fuel flow in a throttle nozzle

The results of liquid fuel flow in a throttle nozzle simulation are presented in this section. Model experiment for cavitating regimes of the flow investigation was described in detail in Winklhofer *et al.* [10]. In experiment the liquid fuel ran through a throttle made in the middle of a nozzle. A pressure drops were fixed and the authors measured mass flow-rates and visualized cavitation zones occurred in a narrow section. Detailed view of the throttle nozzle is shown in Fig. 5.

We simulated stationary 3D flows in a half of the geometry with a symmetry plane. As an inert attenuated gas air with mass fraction Y_{air}=0.1% was chosen.

Numerical results were obtained for 9 various regimes with pressure drop P_{in}-P_{out} from 10^6 to $9{\cdot}10^6$ Pa, where $P_{in} = 10^7$Pa. The numerical and experimental data of flow rate are presented in Fig. 6. Marked on the figure range of ratio error is 5%.

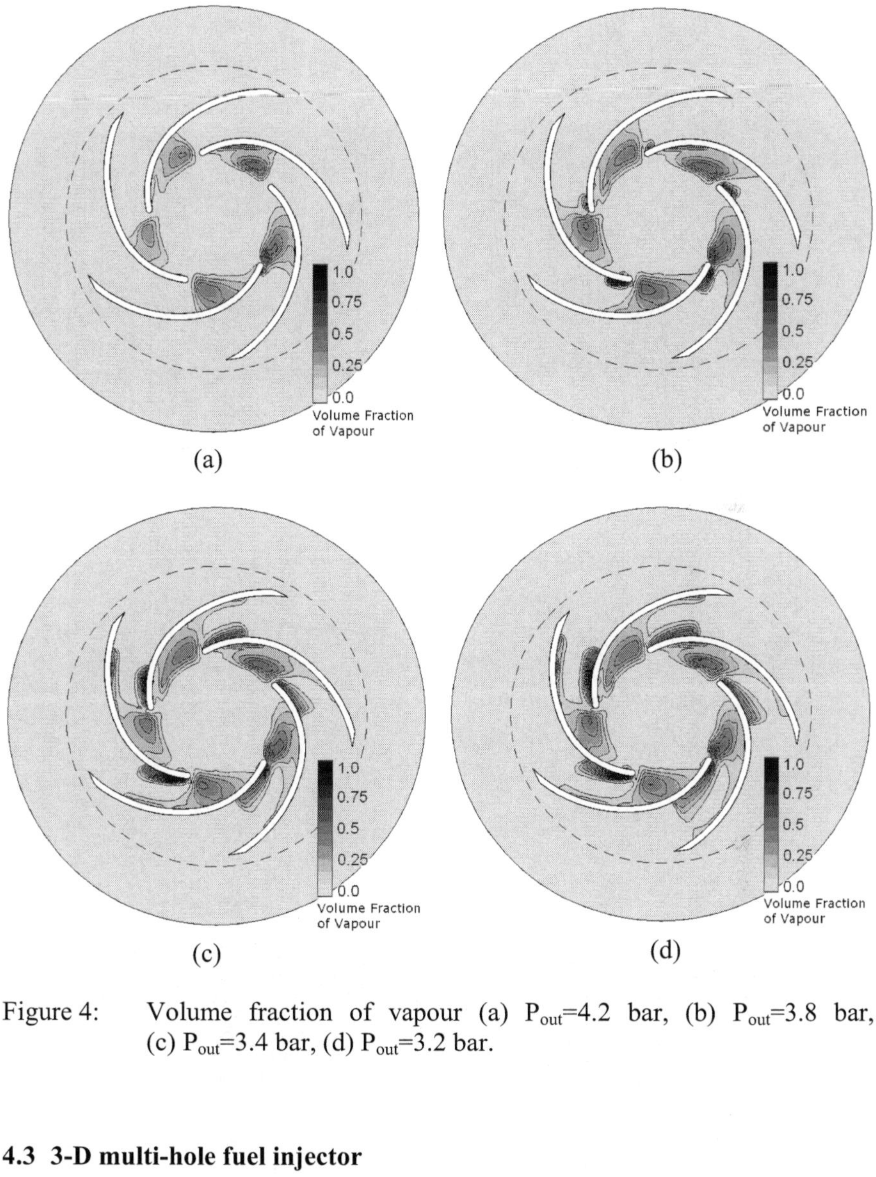

Figure 4: Volume fraction of vapour (a) P_{out}=4.2 bar, (b) P_{out}=3.8 bar, (c) P_{out}=3.4 bar, (d) P_{out}=3.2 bar.

4.3 3-D multi-hole fuel injector

We considered a model of car engine injector (Fig. 8) described in Schmidt *et al.* [11]. This model is axisymmetric so only 1/4 of the model can be considered. To exclude influence of outlet boundary condition on a flow in the injector we added enlarged domain with boundary condition $P=P_{out}$ to the model outlet, just as in [11]. For this computational domain we used the mesh with 110000 CVs

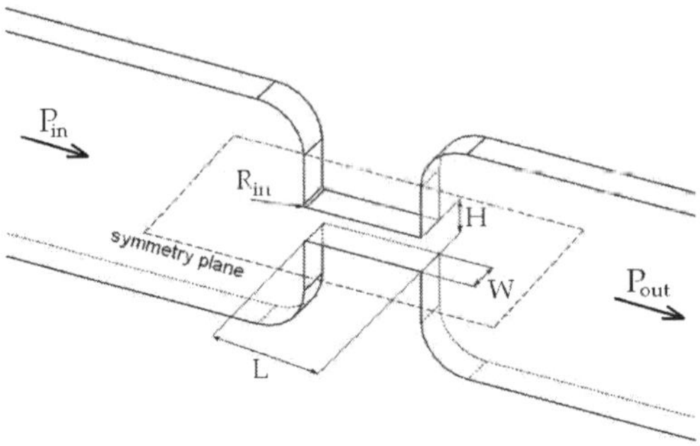

Figure 5: Throttle nozzle model. L=0.001 m., H=0.000299 m., W=0.0003 m., R_{in}=0.00002 m.

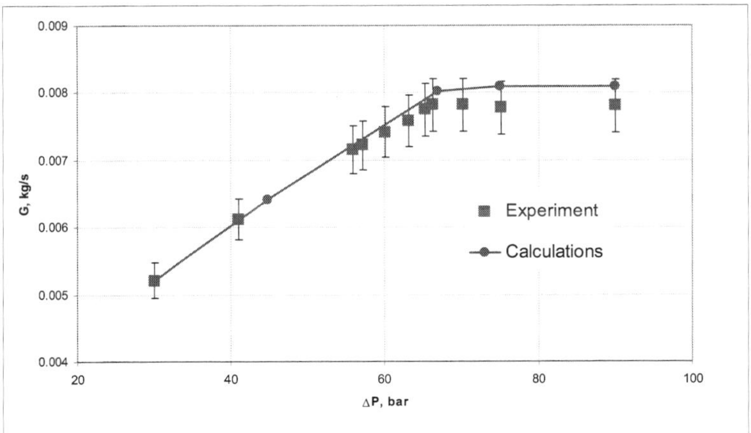

Figure 6: Flow rate characteristic of the throttle nozzle.

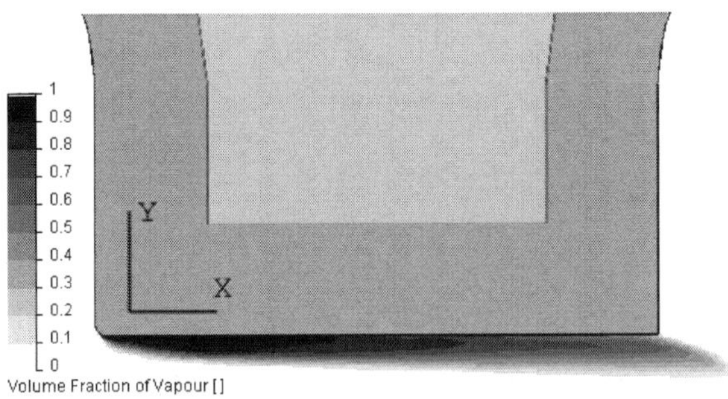

Figure 7: Volume fraction of vapour for pressure drop $\Delta P = 9 \cdot 10^6$ Pa.

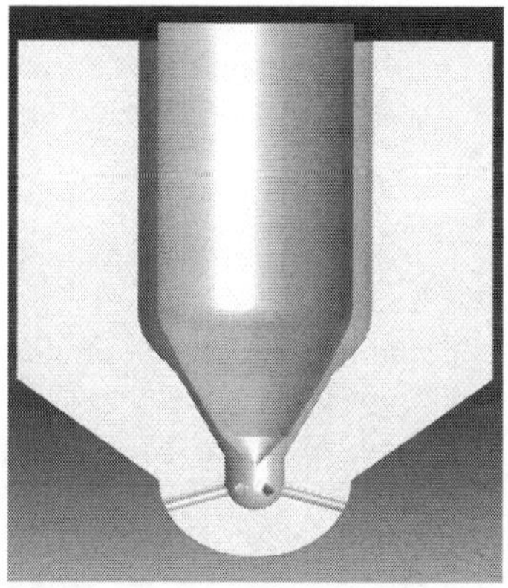

Figure 8: 180 degree section of the 3-D 6-hole injector geometry.

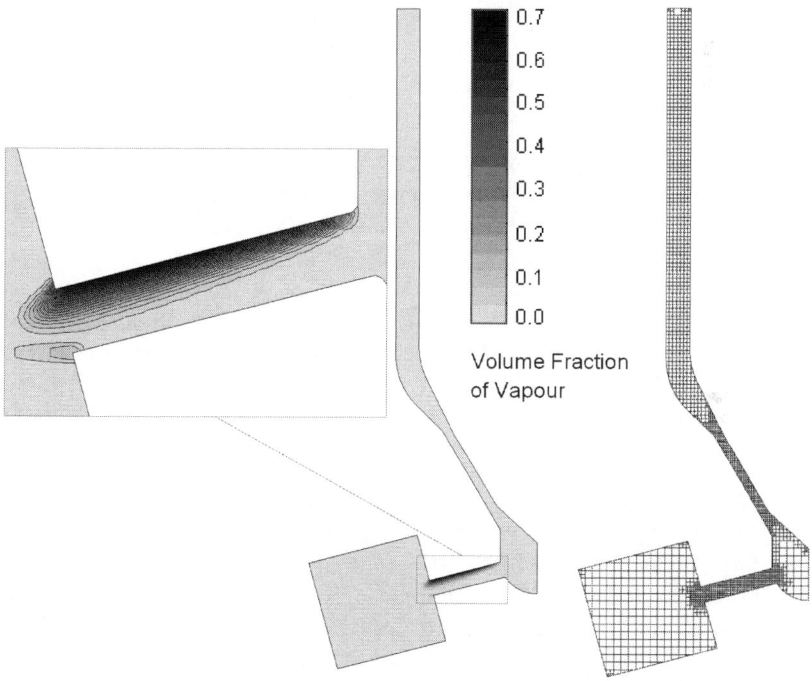

Figure 9: Mesh and steady-state cavitation zone. p_{in}=600 atm, p_{out}=26 atm, T_{init} = 333K.

(Fig. 9). Boundary conditions were defined by fixed values of pressure on the inlet and outlet boundaries, and different regimes were defined by different values of inlet pressure Pin, whereas outlet pressure was constant P_{out}=26 atm. Regime P_{in}=600 atm, P_{out}=26 atm, T_{init}=333 K was considered. The volume fraction of vapour for stationary solution of this task is presented in Fig. 9. Comparison of numerical result with external data shows good accordance of mass flow (60 g/s) for this regime. Cavitation zones also finely accord with calculations of Schmidt et al.

5 Conclusions

A hybrid method for simulating flows under hydraulic cavitation conditions is presented. It is based on coupling the well-established "pressure-based" and the "density-based" methods that allows efficient simulation of flows with both incompressible and highly compressible regions. Numerical experiments show that the new method is accurate, robust and efficient. It requires significantly less iterations compared to traditional "density-based" approaches to obtain a steady-state solutions.

Implemented in FloEFD™ method has been tested by solving a lot of typical 3D problems for cavitating flows with different liquids. Obtained results are in a good agreement with experimental data.

Acknowledgments

The authors thank Alexander V. Muslaev and Gennady E. Dumnov for helpful discussions and Alexander A. Sobachkin for continuous attention and support.

References

[1] Senocak I., Shyy W., A Pressure-Based Method for Turbulent Cavitating Flow Computations, Journal of Computational Physics 176, pp. 363–383 (2002)

[2] Robert F. Kunz, David A. Boger, David R. Stinebring, Thomas S. Chyczewski, Jules W. Lindau, Howard J. Gibeling, Sankaran Venkateswaran, T. R. Govindan, A preconditioned Navier-Stokes method for two-phase flows with application to cavitation prediction, Computers & Fluids, Volume 29, Issue 8, 31 August 2000, pp. 849–875

[3] Li DING, Charles L. MERKLE, A unified framework for incompressible and compressible fluid flows, Journal of Hydrodynamics, Ser. B, Volume 18, Issue 3, Supplement 1, Proceedings of the Conference of Global Chinese Scholars on Hydrodynamics, July 2006, pp. 113–119

[4] Li Ding, Xia Guoping, Merkle Charles L., Large-Scale Multidisciplinary Computational Physics Simulations Using Parallel Multi-Zone Methods, In: Anil Deane, Akin Ecer, James McDonough, Nobuyuki Satofuka, Gunther Brenner, David R. Emerson, Jacques Periaux and Damien Tromeur-Dervout, Editor(s), Parallel Computational Fluid Dynamics 2005, Elsevier, Amsterdam, 2006, pp. 227–234

[5] Li Ding, Xia Guoping, Merkle Charles L., Consistent properties reconstruction on adaptive Cartesian meshes for complex fluids

computations, Journal of Computational Physics, Volume 225, Issue 1, 1 July 2007, pp. 1175–1197
[6] Van der Heul D.R., Vuik C., Wesseling P., A conservative pressure-correction method for flow at all speeds. J.M. Burgers Center, Department of Applied Mathematical Analysis, Faculty of Information Technology and System, Delft University of Technology, Mekelweg 4, 2628 CD Delft, The Netherlands. 2002
[7] Van der Heul Duncan R., Vuik C. and Wesseling P., Efficient computation of flow with cavitation by compressible pressure, European Congress on Computational Methods in Applied Sciences and Engineering, ECCOMAS 2000
[8] Liou Meng-Sing, A Sequal to AUSM: AUSM+, Internal Fluid Mechanics Division, NASA Lewis Research Center, Cleveland, Ohio 44135, 1996
[9] Coutier-Delgosha O., Fortes-Patella R., Reboud J. L., Hofmann M., Stoffel B., Experimental and Numerical Studies in a Centrifugal Pump With Two-Dimensional Curved Blades in Cavitating Condition. Transaction of the ASME, Vol. 125, November 2003, pp. 970–978
[10] Winklhofer E., Kull E., Kelz E., Morozov A. Comprehensive hydraulic and flow field documentation in model throttle experiments under cavitation conditions// ILASS-Europe annual meeting, 2001. pp. 574–579.
[11] Schmidt S.J., Sezal I.H., Schnerr G.H., Thalhamer M., Forster M. Compressible Simulation of Liquid/Vapour Two-Phase Flows with Local Phase Transition, 6th International Conference on Multiphase Flow, ICMF 2007, Leipzig, Germany, July 9–13, 2007
[12] Tryggvason Gretar, Bunner Bernard, Esmaeeli Asghar, Al-Rawahi Nabeel, Computations of Multiphase Flows, Advances in Applied Mechanics, Elsevier, 2003, Volume 39, pp. 81–120
[13] Coutier-Delgosha O., Fortes-Patella R., Reboud J.L., Hakimi N., Hirsch C., Stability of preconditioned Navier-Stokes equations associated with a cavitation model, Computers & Fluids, Volume 34, Issue 3, March 2005, pp. 319–349
[14] Moukalled F., Darwish M., Sekar B., A pressure-based algorithm for multi-phase flow at all speeds, Journal of Computational Physics, Volume 190, Issue 2, 20 September 2003, pp. 550–571
[15] Tseng Chien-Chou, Shyy Wei, Modeling for isothermal and cryogenic cavitation, International Journal of Heat and Mass Transfer, Volume 53, Issues 1-3, 15 January 2010, pp. 513–525
[16] Saurel Richard, Petitpas Fabien, Ray A. Berry, Simple and efficient relaxation methods for interfaces separating compressible fluids, cavitating flows and shocks in multiphase mixtures, Journal of Computational Physics, Volume 228, Issue 5, 20 March 2009, pp. 1678–1712
[17] Abgrall Remi, Saurel Richard, Discrete equations for physical and numerical compressible multiphase mixtures, Journal of Computational Physics, Volume 186, Issue 2, 10 April 2003, pp. 361–396

[18] Metayer Le, Massoni J., Saurel R., Modelling evaporation fronts with reactive Riemann solvers, Journal of Computational Physics, Volume 205, Issue 2, 20 May 2005, pp. 567–610
[19] Saurel Richard, Abgrall Remi, A Multiphase Godunov Method for Compressible Multifluid and Multiphase Flows, Journal of Computational Physics, Volume 150, Issue 2, 10 April 1999, pp. 425–467
[20] Salvador F.J., Romero J.-V., Rosello M.-D., Martinez-Lopez J., Validation of a code for modeling cavitation phenomena in Diesel injector nozzles, Mathematical and Computer Modelling, Volume 52, Issues 7-8, Mathematical Models in Medicine, Business & Engineering 2009, October 2010, pp. 1123–1132
[21] Jenny Patrick, Tchelepi Hamdi A., Lee Seong H., Unconditionally convergent nonlinear solver for hyperbolic conservation laws with S-shaped flux functions, Journal of Computational Physics, Volume 228, Issue 20, 1 November 2009, pp. 7497–7512
[22] Lowe C.A., Two-phase shock-tube problems and numerical methods of solution, Journal of Computational Physics, Volume 204, Issue 2, 10 April 2005, pp. 598–632
[23] Tang H. S., Sotiropoulos F., A Second-Order Godunov Method for Wave Problems in Coupled Solid-Water-Gas Systems, Journal of Computational Physics, Volume 151, Issue 2, 20 May 1999, pp. 790–815
[24] Bell John B, Colella Phillip, Trangenstein John A, Higher order Godunov methods for general systems of hyperbolic conservation laws, Journal of Computational Physics, Volume 82, Issue 2, June 1989, pp. 362–397
[25] Schwendeman D.W., Wahle C.W., Kapila A.K., The Riemann problem and a high-resolution Godunov method for a model of compressible two-phase flow, Journal of Computational Physics, Volume 212, Issue 2, 1 March 2006, pp. 490–526
[26] De Vault K.J., Gremaud P.A., Jenssen H.K., Numerical Investigation of Cavitation in Multidimensional Compressible Flows. SIAM Journal of Applied Mathematics, 2007, v:67, n:6, pp. 1675–1692
[27] Volkov V.A., Dergunov I.M., Muslaev A.V., Streltsov V.Y., Numerical simulation of cavitating flows in injectors using FloEFDTM. Papers of 5th Russian National Heat Transfer Conference., Moscow, 25–29 October 2010, Volume 1, pp. 73–76 (in Russian)
[28] Yanenko N.N., The Method of Fractional Steps: The Solution of Problems of Mathematical Physics in Several Variables, English translation edited by M. Holr. Springer-Verlag, New York, 1971.
[29] Churbanov A.G., Pavlov A.N. and Vabishchevich P.N., Operator-splitting methods for the incompressible Navier-Stokes equations on non-staggered grids. Part. 1: First-order schemes, Int. J. Numer. Metods Fluids, Vol. 21, 1995, pp. 617–40.

Studies of different numerical models for a turbulent particulate flow in a square pipe with 90° bend

D. Schellander, D. Kahrimanovic & S. Pirker
Christian Doppler Laboratory on Particulate Flow Modelling, Johannes Kepler University, Austria

Abstract

In process industry simulating particulate flow is one of the actual top level research topics. Driven by the wish for readily available simulation results with predictive description of the processes in fluid and granular flow, many mathematical models have been proposed.

Among these models Eulerian and Lagrangian based particle models are most common. These models have been additionally augmented by sub-models in order to picture effects like particle rotation, rough wall treatment or inter-particle collisions. The last one includes in Lagrangian phase a very high increase of computational effort because of finding collision partners.

In this paper these models are tested at a very simple flow situation of a 90° bended square duct flow. From classical kinetic theory, Eulerian particle model is considered and also a discrete phase model, known as Lagrangian particle model. Finally, a hybrid Eulerian-Lagrangian model, EUgran+, is introduced using a new radial distribution coefficient, based on measurements of Gidaspow and Huilin. The presented hybrid model is based on an Eulerian model that is augmented by Lagrangian tracer particles. These tracer particles provide additional information on e.g. particle rotation and particle wall treatment that prepares information for the classical Eulerian phase, using coupling equations.

It was the aim to create a robust simulation model for simulating particulate flows with high and low loaded particle regions. The second aim was to create a model which gives comparable accuracy as a Lagrangian model with particle collision, without calculating collisions in Lagrangian phase, to decrease the computational time.
Keywords: hybrid model, particulate flow, transport system, tracer particles.

WIT Transactions on Engineering Sciences, Vol 70, © 2011 WIT Press
www.witpress.com, ISSN 1743-3533 (on-line)
doi:10.2495/MPF110051

1 Introduction

The pneumatic transport and separation of materials, like dust, sand, cement, coal and other particle based granular materials is in common use in the industry. It can be found in the chemical industry, the coal industry, cement industry and many other industries. Pulverised material is transported from one place in the company to another place using pneumatic conveying systems with pipes, containing elbows, bends and cyclones.

Since computer simulation is in our mind, scientists dream of simulating particulate flows, but there were and are already limitations. For particle interaction and flow simulation many equations are known, but nowadays it is not possible to calculate the flow of an arbitrary geometry exactly. Computers are nowadays more powerful than ten years ago but also today it is not possible to simulate realistic high loaded particle conveying systems. These high loaded regions are in a standard cyclone present and should be simulated.

For the simulation of particles in a fluid, two main models are now state of the art. Thus we firstly present the Discrete Phase Model (DPM) and the Eulerian Model (EM). The DPM describes single particles on their way through the simulation region. The EM describes the particles as a continuous phase, introducing a granular temperature. The granular temperature is a value for the particle velocity fluctuation (e.g. Dartevelle [1]). Both models have their advantages and disadvantages, especially in regions with low and high particle loadings.

Secondly our Hybrid-Model (HM) is presented and depicted in detail, called EUgran+ (e.g. Pirker et al. [2]). There are other hybrid models too, for example the Dense Discrete Phase Model (DDPM), described in e.g. Fluent [3] which is based on Lagrangian particle trajectories which get information from the Eulerian phase. The EUgran+ model is based on the Eulerian granular phase and is modified by information from Lagrangian tracer particles.

2 Fluid phase

The fluid phase is described by the Navier-Stokes equations. These equations are state of the art and used in Computational Fluid Dynamics (CFD) to describe a fluid flow. With additional models, like the Reynold Average Navier-Stokes equations, turbulence can be modelled (e.g. Fluent [3])

$$\frac{\partial}{\partial t}\left(\alpha_g \rho_g\right) + \nabla \cdot \left(\alpha_g \rho_g \mathbf{u}_g\right) = \mathbf{0}, \tag{1}$$

$$\frac{\partial}{\partial t}\left(\alpha_g \rho_g \mathbf{u}_g\right) + \nabla \cdot \left(\alpha_g \rho_g \mathbf{u}_g \mathbf{u}_g\right) = \\ -\alpha_g \nabla p_g + \nabla \cdot \alpha_g \tau_g + \mathbf{f}_{g,drag} + \mathbf{f}_{g,add}. \tag{2}$$

3 Discrete phase model

3.1 Introduction

The standard Lagrangian DPM is based on a translational force balance

$$\frac{\partial}{\partial t}\mathbf{u}_p = \frac{18\mu_g}{\rho_p d_p^2}\frac{C_D \mathrm{Re}_p}{24}\left(\mathbf{u}_g - \mathbf{u}_p\right) + \mathbf{g} + \mathbf{f}_{p,add}. \tag{3}$$

It can be seen in this equation that all forces in the Lagrangian phase are based on the mass of one particle. Additionally in this paper an angular momentum balance is used, to account for particle rotation and Magnus force calculations.

$$\frac{d\boldsymbol{\omega}_p}{dt} = \mathbf{t}_{p,g} + \mathbf{t}_{p,add} \tag{4}$$

3.2 Additional forces

3.2.1 Magnus force and torque

The standard DPM neglects the force created by particle rotation, the Magnus force, which was detected by Newton first. Here the acceleration of one particle, induced by the Magnus force is shown (e.g. Kahrimanovic [4]).

$$\mathbf{f}_{p,Magnus} = \frac{1}{2}\left|\mathbf{u}_g - \mathbf{u}_p\right| C_L r_p^2 \pi \left(\frac{\Omega_p \times \left(\mathbf{u}_g - \mathbf{u}_p\right)}{\Omega_p}\right)\frac{\rho_g}{m_p} \tag{5}$$

The Magnus force describes a force induced by rotation of a particle, because of different pressure regions within the boundary layer around the particle.

$$\Omega_p = \frac{1}{2}\nabla \times \mathbf{u}_g - \boldsymbol{\omega}_p \tag{6}$$

(6) describes the difference between the rotating fluid and the particle rotation, C_L is the lift coefficient (e.g. Lun and Lui [5]).

For the calculation of the angular momentum balance, the fluid-torque and additional torque from interparticle collisions is needed. The fluid-torque can be written as (e.g. Pirker et al. [2])

$$\mathbf{t}_{p,g} = \frac{\mu_g}{64 I_p} d_p^3 C_R \Omega_p \mathrm{Re}_R \tag{7}$$

with Re_R as rotating Reynolds-number and I_p as value in the main diagonal of the inertia tensor.

$$I_p = \frac{m_p d_p^2}{10}, \quad \mathrm{Re}_R = \frac{\rho_g d_p^2 \left|\boldsymbol{\Omega}_p\right|}{\mu_g}, \tag{8}$$

The rotation coefficient (e.g. Sommerfeld [6]) is written as

$$C_R = \begin{cases} \dfrac{64\pi}{\mathrm{Re}_R} & \ldots \quad \mathrm{Re}_R < 32 \\ \dfrac{12.9}{\sqrt{\mathrm{Re}_R}} + \dfrac{128.4}{\mathrm{Re}_R} & \ldots \quad 32 \le \mathrm{Re}_R < 1000 \end{cases} \tag{9}$$

Collisional forces between particles in a strand effects that the rotation of the particles decreases. The rotational energy is transformed via friction into temperature, also granular temperature which is neglected in this paper, because of the very small values. The following equation describes the interaction between collisions and angular velocity changing in one timestep.

$$\mathbf{t}_{p,add} = -C_t N_{s,p} \frac{\boldsymbol{\omega}_p}{t_{tp}} \tag{10}$$

The number of collisions in one timestep can be calculated as

$$N_{s,p} = n_s^2 d_p^2 g_0 \sqrt{\pi\Theta_s} \frac{t_{tp}}{V_c}, \tag{11}$$

with assumption that Eulerian phase and Lagrangian phase represents identical materials (e.g. Kahrimanovic [4], Fluent [3]). C_t is the torque factor which can be used to adjust the simulation. In this paper $C_t = 1$.

3.2.2 Particle-wall interaction

Particle-Wall interaction has a very high impact on the result of the simulation. So a well constructed model must be used. For the simulations in this paper a model depicted in Kahrimanovic [4] and Sommerfeld et al. [7] is taken. It can handle rough walls in a sense that the computational effort is in an acceptable range.

4 Eulerian phase model

The Eulerian granular phase model, in contrast to the Lagrangian model, considers the multitude of the particles as an artificial continuous solids phase. The individual values of particles are smeared out, the solid phase is based on mean values. The standard Eulerian granular model relies on the assumption that by inter-particle collisions energy is transferred from the mean flow towards chaotic fluctuations of the particles. To describe this behaviour an additional transport equation for the granular temperature must be solved (e.g. Fluent [3] and literature cited therein).

The main part of this paper is the hybrid model, this gives additional information for the momentum equation, which is written as

$$\frac{\partial}{\partial t}\left(\alpha_s \rho_s \mathbf{u}_s\right) + \nabla \cdot \left(\alpha_s \rho_s \mathbf{u}_s \mathbf{u}_s\right) =$$
$$-\alpha_s \nabla p_s + \nabla \cdot \alpha_s \tau_s + \alpha_s \rho_s \mathbf{g} + \mathbf{f}_{s,drag} + \mathbf{f}_{s,add} \tag{12}$$

with the drag force between fluid and solid as

$$\mathbf{f}_{s,drag} = K_{sg}\left(\mathbf{u}_g - \mathbf{u}_s\right) \tag{13}$$

4.1.1 Particle-wall interaction

The particle-wall interaction is based on the same method as mentioned in section 3.1.2.

5 Hybrid model: EUgran+

5.1 Additional forces for Eulerian granular phase and Lagrangian phases

For the exchange between the Eulerian and Lagrangian phase some additional equations are needed (e.g. Pirker et al. [2]). This must be done because the standard Eulerian granular model lacks some important physics. This includes particle rotation and therefore Magnus force and also the effects of wall roughness. A term describing the force between the two phase caused by different velocities of the Eulerian granular and Lagrangian phase is added. In Fig. 1 the hybrid model EUgran+ is shown in a modular way, so that the different modules and the interaction between them are displayed.

5.1.1 Dragforce between Eulerian and Lagrangian phase

Between the Eulerian and Lagrangian phase a dragforce is designed. This force is added to the Lagrangian phase and Eulerian phase.

$$\mathbf{f}_{p,drag} = \frac{3\left(1 + e_{p,s}\right)}{2d_p} \alpha_s g_0 \left|\mathbf{u}_s - \mathbf{u}_p\right| \left(\mathbf{u}_s - \mathbf{u}_p\right) \tag{14}$$

The equation is based on the force of solid-solid exchange for two different solids. The solid-solid exchange coefficient (e.g. Fluent [3]) for two different solids l and s is written as

$$K_{ls} = \frac{3\left(1 + e_{ls}\right)\left(\frac{\pi}{2} + C_{fr,ls}\frac{\pi^2}{8}\right)\alpha_s \rho_s \alpha_l \rho_l \left(d_l + d_s\right)^2 g_{0,ls}}{2\pi\left(\rho_l d_l^3 + \rho_s d_s^3\right)} \left|\mathbf{u}_s - \mathbf{u}_p\right|$$

In this paper only one diameter and density for particles is used, this changes the equation to the shown form (14). The radial distribution coefficient g_0 which was evaluated from Carnahan and Starling [13], Lun and Savage [14], Sinclair and Jackson [15] and Gidaspow [16], is calculated in a new way.

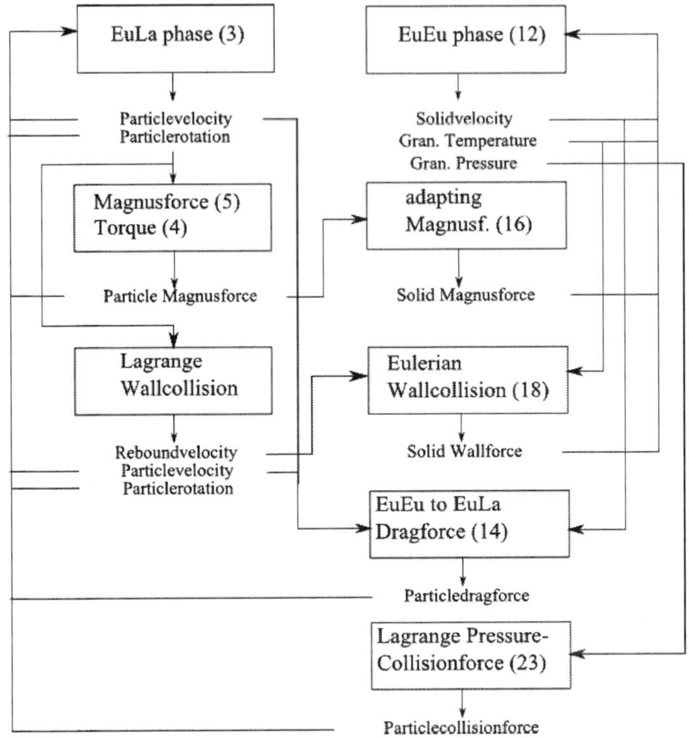

Figure 1: Simple flowchart of the hybrid model; the number in brackets refer to the corresponding equations in the text.

We propose to combine the equations from Gidaspow, Sinclair and Jackson. At the edges the results of both compared to experiments and computational developments, seem to be good but in the middle an enhancement seems to be possible.

$$g_0 = \begin{cases} 1 + \frac{23}{2}\alpha_s & \dots \quad \alpha_s < 0.07 \\ \frac{121}{100} + \frac{17}{2}\alpha_s & \dots \quad 0.07 \le \alpha_s < 0.46 \\ \frac{3}{5}\left[1 - \left(\frac{\alpha_s}{0.67}\right)^{1/3}\right]^{-1} & \dots \quad 0.46 \le \alpha_s \end{cases} \tag{15}$$

For this work, (15), was created using the measured-data from Gidaspow and Huilin and experiment-data from Alder and Wainright (e.g. van Wachem et al. [8]) and used for the HM simulations.

5.1.2 Magnus force

The Magnus force, calculated using Lagrangian tracer particles, can be adjusted for the Eulerian phase for each cell with equation

$$\mathbf{f}_{s,Magnus} = \alpha_s \rho_s \left\langle \mathbf{f}_{p,Magnus} \right\rangle \tag{16}$$

so that the same acceleration as of one representative particle is added to the solid phase. The average Magnus force in a cell is calculated at each timestep of each tracer particle. The particle timestep for Lagrangian tracer particles is not the same as the simulation timestep. For the n-th timestep of all particles hitting one cell

$$\left\langle \mathbf{f}_{p,Magnus} \right\rangle_n = \frac{(n-1)\left\langle \mathbf{f}_{p,Magnus} \right\rangle_{n-1} + \mathbf{f}_{p,Magnus,n}}{n} \tag{17}$$

is found.

5.1.3 Wallforce

For the interaction between wall and solid phase, as mentioned before, a restitution coefficient model is used. Additionally to them a wall force from Lagrangian part is added in the momentum equation (12). Based on the difference of velocities from Lagrangian and Eulerian phase after wall reflection

$$\mathbf{f}_{s,wall} = \alpha_s \frac{\rho_s V_{rebound}}{V_c t_{ts}} \left(\left\langle \mathbf{u}_{p,rebound} \right\rangle - \left[\mathbf{u}_s + \left\langle \mathbf{u'}_{s,rebound}(\Theta_s) \right\rangle \right] \right) \tag{18}$$

is found. Equation (18) represents a force from the Lagrangian phase to the Eulerian phase. It is assumed that all particles within the volume

$$V_{rebound} = u_{s,\perp} t_{ts} A_c \tag{19}$$

collide with the wall during the next timestep. This includes all particles with a velocity higher than

$$u_{s,\perp}(\Theta_s) = \left| \sqrt{\Theta_s} \mathbf{e}_n \right| \tag{20}$$

which is the velocity normal to the wall, and calculated with some assumptions from the granular temperature. It is known that (e.g. Sommerfeld et al. [7])

$$\Theta_s = \frac{u'^2_s + v'^2_s + w'^2_s}{3} \tag{21}$$

and additionally assumed $u' = v' = w'$. With this knowledge the rebound velocity (22) and the normal velocity (20) to the wall can be estimated.

$$\mathbf{u'}_{s,rebound}(\Theta_s) = \sqrt{\Theta_s} \mathbf{e}_{s,rebound} \tag{22}$$

5.1.4 Collisional force from granular pressure

If many particles are in one cell then additionally particles can be reflected or captured. This should be modelled with the additional collisional force based on the granular pressure gradient (e.g. Pirker et al. [2])

$$\mathbf{f}_{p,coll} = -C_p \frac{1}{\rho_p \alpha_s} \nabla p_s \tag{23}$$

with C_p as collisional pressure coefficient. C_p is used to adapt the strength of the force and in this paper chosen with 1.

6 Simulations and results

6.1.1 Simulation

For the simulations a simple geometry, a square duct with 90° bend as depicted in Fig. 2. is taken. The simulations conditions are presented in Tab 1.

Table 1: Simulation parameter.

Parameter	Value
Velocity at inlet v_{in}	10 m/s
Mass loading L	0.00206
Density fluid ρ_g	1.18 kg/m^3
Density particles $\rho_p = \rho_s$	2500 kg/m^3
Particle diameter $d_p = d_s$	77 µm
Mass-Flow rate $\dot{M}_p = \dot{M}_s$	0.000547 kg/s

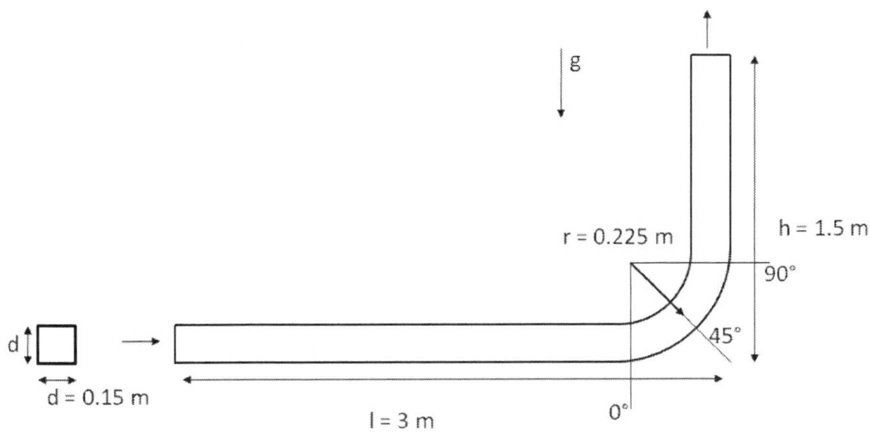

Figure 2: Geometry for simulation with approx. 350000 cells.

For this geometry, (e.g. Mohanarangam et al. [9, 10] and Kuan et al. [11]) simulations and experiments are done and described. For this work we decided to use this geometry, because there is known data for comparing our models with the experimental data. It should be recognized that the EUgran+ is designed for high loaded particle simulations, which is not the case here.

6.1.2 Results

The calculated velocities of the Lagrangian tracer particles and the granular phase are plotted at 3 positions in the duct bend, exactly at 0°, 45° and 90°, which is shown in Fig. 2. Additionally the measured data is depicted. In Fig. 3 the comparison of the different granular velocities can be seen. The Eulerian phase is only present at the cells which are nearest to the wall, which gives very bad results. The modelled Lagrangian phase (e.g. Kahrimanovic [4]) gives very good results, in regions which are reached from Lagrangian particles. The Lagrangian simulations are done with 3 different particle diameters 77 μm, 125 μm and 30μm and the average is plotted. The EUgran+ raises the granular phase away from the wall which gives better results and supports the building of particle strands at the Lagrangian part. The results are better as expected, the hybrid model is very close to the results of the Lagrangian model, regarding interparticle collisions.

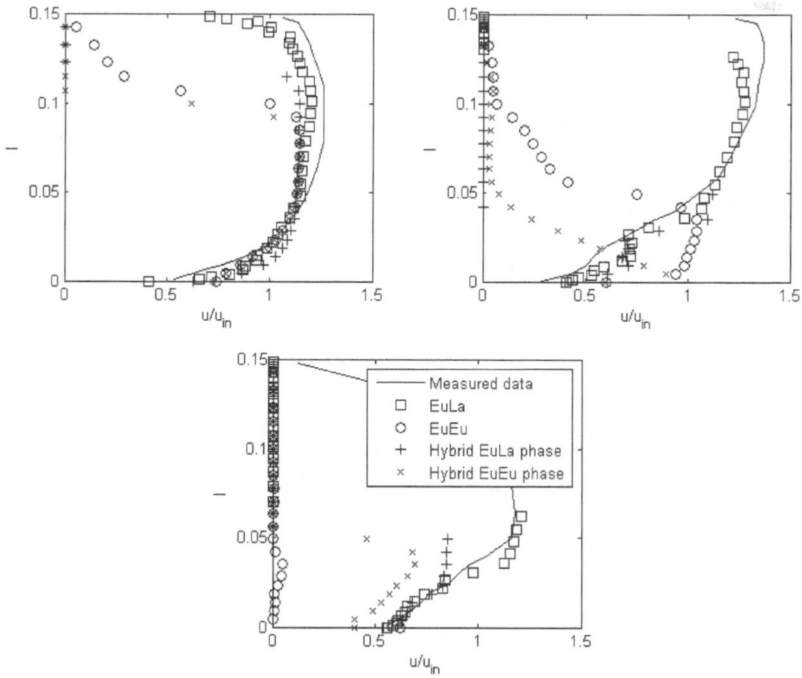

Figure 3: Particle and granular velocities at 0° (left upper), 45° (right upper) and 90° in bend.

Some values in Fig. 3 are zero, because there is no particle or granulate phase at this position for particles with diameter 77 μm. l is the distance from the outer wall in the bend and u/u_{in} is the average velocity at this point based on the velocity at the inlet.

7 Conclusion and outlook

The results with the Lagrangian model (e.g. Kahrimanovic [4]), which has a very high computational effort are very good, compared to the experimental data (e.g. Mohanarangam et al. [9, 10] and Kuan et al. [11]). The computational effort is increasing very strong, with additional particles. This runs into a computational limit. The hybrid model is useful as model for simulations with dilute and highly laden particle regimes, compared to the computational effort. The highly complex particle-particle interaction calculation is included in a shorter way, compared to calculation time, in the Eulerian phase. In regions where only a few particles are present the Lagrangian part has the main impact on the particle fluid interaction, in high loaded regimes the Eulerian part has the higher influence to the simulation result. For simulating cyclones and pneumatic conveying transport systems the EUgran+ is a good approach. For future work a wall treatment model for the Eulerian phase, which brings the high level wall treatment from the Lagrangian phase also to the Eulerian phase will be developed. Then the coupling of Eulerian and Lagrangian phase in wall treatment can be deleted, because both do the same.

For future, it would be additionally important to create a smooth exchange field between discrete phase and granular Eulerian phase (e.g. Pirker et al. [12]). In future work we will include this in the hybrid model. This reduces the computational time significantly (e.g. Pirker et al. [12]) and give a more stable simulation.

8 Abbreviations and notation

α	volume fraction	1
μ	dynamic viscosity	kg/(ms)
ρ	density	kg/m^3
Θ	granular temperature	m^2/s^2
τ	shear rate tensor	$kg/(ms^2)$
ω	angular velocity	1/s
Ω	particle fluid rotation	1/s
C	constant	1
d	diameter	m
e	restitution coefficient	1
$\mathbf{f}_p$	force, Lagrangian phase	m/s^2
$\mathbf{f}_s$	force, Eulerian phase	$kg/(m^2s^2)$

g	gravity	m/s^2
g_0	radial distribution	1
I	inertia tensor	kgm^2
m	mass	kg
p	pressure	Pa
r	radius	m
Re	Reynolds number	1
t	time	s
$\mathbf{t}$	angular momentum	$1/s^2$
$\mathbf{u}$	velocity	m/s

add	additional
$coll$	collision
g	gas phase
n	normal vector with length 1
p	particle, parcel of particles
$rebound$	rebound from particle wall collision
s	granular solid phase
tp	timestep
$\perp$	normal direction to wall

References

[1] Dartevelle, S., Numerical and granulometric approaches to geophysical granular flows, Ph.D. thesis, Michigan Technological University, Department of Geological and Mining Engineering, Houghton, Michigan, 2003

[2] Pirker, S., Kahrimanovic, D., Kloss, C., Popoff, B. & Braun, M., Simulating coarse particle conveying by a set of Eulerian, Lagrangian and hybrid particle models, Powder Technology 204, 2010

[3] Fluent 6.3 User's Guide, Fluent Inc., 2006

[4] Kahrimanovic D., Numerische Simulation und experimentelle Validierung von Gas-Partikel-Strömungen, Ph.D.-thesis, 2009

[5] Lun, C. K. K. & Lui, H. S., Numerical Simulation of dilute turbulent gas-solid flows in horizontal channels, International Journal of Multiphase Flow, 23, pp. 575 - 605, 1997

[6] Sommerfeld, M., Theoretical and Experimental Modelling of Particulate Flows, Lecture Series 2000 - 06, Karman Institute for Fluid Dynamics, 2000

[7] Sommerfeld, M., Crowe, C. & Tsuji, Y. Multiphase flows with droplet and particles, CRC Press, 1998

[8] van Wachem, B. G. M., Schouten, J. C. & van den Bleek, C. M., Comparative Analysis of CFD Models of Dense Gas-Solid Systems, AIChe Journal Vol 47, No. 5, 2001
[9] Mohanarangam. K., Yang, W., Zhang, H. J., and Tu, J. Y., Effect of particles in a turbulent flow within a 90° bend, *Seventh International Conference on CFD in the Minerals and Process Industries CSIRO*, Melbourne, Australia, 2009
[10] Mohanarangam, K., Tian, Z. F. & Tu, J. Y., Numerical simulation of turbulent gas-particle flow in a 90° bend: Eulerian-Eulerian approach, Computers and Chemical Engineering, 32, pp. 561–571, 2008
[11] Kuan, B., Yang, W., & Schwarz, M. P., Dilute gas-solid two-phase flows in a curved 90° duct bend: CFD simulation with experimental validation, Chemical Engineering Science, 2, pp. 2068–2088, 2007
[12] Pirker, S., Kahrimanovic, D. & Goniva, C., Improving the applicability of discrete phase simulations by smoothening their exchange fields, *Applied Mathematical Modelling*, 2010
[13] Carnahan, N. F., and K. E. Starling, Equations of State for Non-Attracting Rigid Spheres, J. Chem. Phys., 51, 635, 1969
[14] Lun, C. K. K., and S. B. Savage, The Effects of an Impact Velocity Dependent Coefficient of Restitution on Stresses Developed by Sheared Granular Materials, Acta Mech., 63, 15, 1986
[15] Sinclair, J. L., and R. Jackson, Gas-Particle Flow in a Vertical Pipe with Particle-Particle Interactions, AIChE J., 35, 1473, 1989
[16] Gidaspow, D., Multiphase Flow and Fluidization, Academic Press, San Diego, 1994

Height function interface reconstruction algorithm for the simulation of boiling flows

M. Magnini & B. Pulvirenti
Dipartimento di Ingegneria Energetica, Nucleare e del Controllo Ambientale, Università di Bologna, Italy

Abstract

In this paper we present our results on numerical study of vapor bubbles growing in quiescent superheated liquid, as effect of liquid evaporation at the interface. Height Function interface reconstruction algorithm is coupled with an evaporation model based on continuum field representation of source terms. The flow solver is a finite-volume CFD code. Interface is tracked within a Volume-Of-Fluid framework. Continuum-Surface-Force method accounts for surface tension effects. Vapor bubble heat-transfer-controlled growth is simulated for three different working fluids: water, HFE-7100 and R134a. Accuracy of interface reconstruction algorithm is of maximum importance. Unbalance between pressure gradients and surface tension forces at interface leads to the growth of an unphysical velocity field which switches original only diffusive heat transfer mechanism to combined diffusive-convective one. Height Function algorithm reduces the magnitude of this unreal velocity field. Standard test cases are considered to assess the performances of implemented version, through comparison with the widely used Youngs algorithm.
Keywords: volume of fluid, height function, evaporation, surface tension, bubbles.

1 Introduction

Modern numerical implementation of surface tension driven flows, with phase change, begins with the work of Juric and Tryggvason [1]. Earlier methods used a Lagrangian approach with a moving mesh following the interface. The limit of these methods was that they could not manage large interface deformation. Juric and Tryggvason developed a Front Tracking algorithm where Lagrangian evolution of interface is tracked on a fixed grid. Welch and Wilson [2] and Son

WIT Transactions on Engineering Sciences, Vol 70, © 2011 WIT Press
www.witpress.com, ISSN 1743-3533 (on-line)
doi:10.2495/MPF110061

and Dhir [3] reformulated two full Eulerian methods, called Volume-Of-Fluid (VOF) and Level-Set (LS), in order to take into account for phase change. Both methods solve a scalar field conservation equation, volume fraction for VOF and level-set function for LS. This scalar field is used to reconstruct interface shape and position. Level-set function varies smoothly through interface, giving an accurate interface shape approximation, but mass conservation is not preserved. On the contrary, Volume-Of-Fluid method preserves mass conservation, but volume fraction field varies sharply through interface, leading to poor quality interface representation. Several algorithms have been developed in order to give better interface representation on Volume-Of-Fluid background. An important issue in the simulation of boiling flows is the thermal boundary condition at the interface. Most of the methods impose interface temperature at equilibrium saturation temperature corresponding to system pressure, but different conditions may be taken at interface. A detailed review can be found in [1].

The aim of the present work is to model a two-phase flow with phase change by means of a finite-volume CFD code. VOF method is used to follow interface. Two different methods to reconstruct interface are compared: Youngs method [4] and Height Function (HF) method [5–7]. The comparison is carried out analyzing algorithms performances through standard test cases. An evaporation model based on continuum-field representation of source terms developed by Hardt and Wondra [8] is implemented. Finally, HF and evaporation models are employed to simulate the growth of a vapor bubble in quiescent superheated liquid.

2 Governing equations and numerical model

In the VOF method, a volume fraction α is defined for every domain cell. It represents the fraction of volume cell occupied by liquid, and it can be thought as the integral of an indicator function $I(\mathbf{x}, t)$ over the cell of volume V:

$$\alpha = \frac{1}{V} \int_V I(\mathbf{x}, t) dV \tag{1}$$

The indicator function $I(\mathbf{x}, t)$ is a multidimensional Heavyside step function with value 1 in liquid phase and 0 in vapor phase. Interface cells can be located by those cells with volume fraction included between 0 and 1. Generic fluid property Φ for every domain cell can be expressed in term of α as follows:

$$\Phi = \Phi_v + (\Phi_l - \Phi_v)\alpha \tag{2}$$

where Φ_v and Φ_l are vapor and liquid specific property. Since in this work both phases are always considered incompressible, the substitution of Eq. (2) in continuity equation yields the following volume fraction conservation equation:

$$\frac{\partial \alpha}{\partial t} + \nabla \cdot (\alpha \mathbf{u}) = \frac{S_\alpha}{\rho_l} \tag{3}$$

where S_α represents mass source term due to evaporation or condensation. Through Eq. (3) interface position is advected at every time step. Momentum conservation equation for a surface tension driven laminar flow, with constant surface tension σ, has the following form:

$$\frac{\partial(\rho\mathbf{u})}{\partial t} + \nabla \cdot (\rho\mathbf{u} \cdot \mathbf{u}) = -\nabla p + \nabla \cdot \left[\mu\left(\nabla\mathbf{u} + \nabla\mathbf{u}^T\right)\right] + \rho\mathbf{g} + \mathbf{F}_\sigma \qquad (4)$$

where $\mathbf{F}_\sigma$ represents surface tension force. It is modelled through Continuum-Surface-Force (CSF) method developed by Brackbill *et al.* [9]. By CSF method surface tension force $\mathbf{F}_\sigma$ can be expressed as:

$$\mathbf{F}_\sigma = \sigma\kappa\nabla\alpha \qquad (5)$$

where κ stands for interface curvature. Therefore surface tension, a surface force, is modeled as a volume force, localized on a finite thickness interface. Interface curvature κ has to be computed by the interface reconstruction algorithm, in this case Youngs or HF method. The flow equations set is completed by energy conservation equation:

$$\frac{\partial(\rho c_p T)}{\partial t} + \nabla \cdot (\rho c_p \mathbf{u} T) = \nabla \cdot (\lambda\nabla T) + S_E \qquad (6)$$

where S_E represents energy source term due to evaporation or condensation. Viscous heating is neglected. The task of the evaporation model is to compute correct mass and energy source terms, giving fields localized on interface cells.

Governing equations are discretized within a finite-volume framework. Volume fraction conservation equation is discretized first order in time through explicit scheme. PLIC technique by Youngs [4] is employed to compute mass transport across interface cell sides. Implicit first order scheme is used to solve momentum and energy conservation equations. Convective terms are discretized using a third order MUSCL scheme [10]. Diffusive terms are always discretized with central finite-difference scheme. For pressure-velocity coupling, PISO algorithm [11] is used.

3 Evaporation model

The Hardt and Wondra evaporation model [8] is implemented in order to evaluate mass and energy source terms on mass and energy equations. The model allows local interfacial temperature to deviate from saturation condition and local mass source is evaluated proportional to this interface superheating. According to Juric and Tryggvason [1], different interface temperature conditions may be taken, in order to account for entropy production due to phase change and for pressure rising due to curved interfaces. For the sake of simplicity, without loss of accuracy, entropy generation and non-equilibrium thermodynamics terms are neglected. Within these assumptions, a mass source equation was derived by Schrage study [12] of interface mass transfer. In order to express local interfacial mass flux $\dot{m}$, Schrage introduced the concept of local interfacial convective coefficient h_i, expressed as:

$$h_i = \frac{2\beta}{2-\beta} \frac{h_{lv}^2}{\sqrt{2\pi R}} \frac{\rho_v}{T_{sat}^{3/2}} \tag{7}$$

in which R is the gas constant, h_{lv} is vaporization latent heat and β is a constant called accommodation coefficient, for which details can be found in [12]. Local interfacial convective coefficient h_i is connected to local interfacial mass flux as follows:

$$\dot{m} = \frac{h_i}{h_{lv}}(T_i - T_{sat}) \tag{8}$$

where T_i is the interfacial temperature. An initial mass source field is evaluated on interfacial cells, where evaporation takes place, by Eq. (8). Such a localized source term may lead to numerical instabilities. To avoid instabilities, a diffusion equation in which initial field represents initial condition is solved. The effect is to smear the initial field over some cells. Diffused mass source field is then used to evaluate mass and energy source terms. Hardt and Wondra [8] demonstrated that this evaporation model leads to correct evaporation rate, by comparing their results with some benchmarks.

4 Evaluation of surface tension effects: HF and Youngs comparison

The task of the reconstruction algorithm is to reconstruct interface geometry, meaning normal vector $\mathbf{n}$ and curvature, which is implicit in volume fraction field. VOF based algorithms compute local curvature through derivatives of a chosen color function c, by the following relation:

$$\kappa = -\nabla \cdot \mathbf{n} = -\nabla \cdot \frac{\nabla c}{|\nabla c|} \tag{9}$$

where the first equation derives from geometrical considerations, see [9] for reference. Youngs [4] is one of the widest reconstruction algorithm used on VOF background, for this reason it was chosen as comparison. It works also with unstructured grids. It employs volume fraction α as color function to compute curvature, but volume fraction has a very sharp variation through interface, leading to poor accuracy computation. Height Function method derives a smoother color function field through local integration of volume fraction field, then derivatives are more accurate. It only works with Cartesian grids. The algorithm implemented is a merge of Malik *et al.* [6] and Hernandez *et al.* [5] schemes. It follows a brief summary of the algorithm:

1. For each interface cell, volume fraction field on a local stencil is considered. Stencil orientation, vertical or horizontal, depends on interface orientation, determined by normal vector direction, computed as $\mathbf{n} = \nabla\alpha/|\nabla\alpha|$;
2. Stencil extension, from 3×3 up to 7×3 cells, is chosen depending on volume fraction field variation, width is always 3 cells;

3. Stencil volume fraction field is adjusted to obtain a monotonic variation along height direction, in order to avoid errors when more than one interface cuts the stencil;
4. Volume fractions are summed columnwise, obtaining a local discrete height function H field;
5. If central column H value is not included on the cell for which curvature is being computed, curvature is taken equal to adjacent cell for which this condition is satisfied. Otherwise, curvature is computed through the extended form of Eq. (9):

$$\kappa = -\frac{H^{(2)}}{[1 + (H^{(1)})^2]^{3/2}} \tag{10}$$

where $H^{(1)}$ and $H^{(2)}$ denote first and second order derivatives of height function, computed through central difference scheme.

For axisymmetric simulations, curvature κ in Eq. (5) refers to total curvature, obtained summing principal curvatures *radii* κ_1 and κ_2. On a cylindrical coordinate reference frame (r, ϕ, z), κ_1 denotes interface curvature on $r - z$ plane, computed as reported from 1 to 5 steps. Curvature κ_2 refers to second principal curvature of the surface obtained through revolution of interface around z axis. It lies on a plane normal to interface and perpendicular to $r - z$ plane. Denoting interface profile in $r - z$ plane as $f(z)$, second curvature can be expressed as:

$$\kappa_2 = -\left(\frac{f'(z)}{|f'(z)|}\right)\frac{1}{f(z)[1 + (f'(z))^2]^{1/2}} \tag{11}$$

The discrete version of first order derivative of interface profile $f'(z)$ follows from HF algorithm, while discrete interface position $f(z)$ can be computed for each interface cell, given cell volume fraction and normal vector.

4.1 Test case 1: inviscid static droplet

A circular droplet of radius R is centered on a $L = 4R$ side square domain. Different uniform mesh sizes Δx are tested, in order to check convergence rate of the methods. The coarsest mesh has 20×20 elements, with $R/\Delta x = 5$. The most refined mesh has 160×160 elements, with $R/\Delta x = 40$. Surface tension and phases' densities are set as unity. Viscous and gravity effects are neglected. Pressure is given at all boundaries as boundary condition. For such a flow, momentum equation reduces to the form:

$$\frac{D(\rho \mathbf{u})}{Dt} = -\nabla p + \sigma\kappa\nabla c \tag{12}$$

If pressure gradient is balanced by surface tension force, left hand side term is zero at each time step. Any unbalance leads to the growth of an unphysical velocity field defined "spurious velocity" or "parasitic current". Due to the absence of external forces, the accuracy of the surface tension model and curvature calculation algorithms is attested by the comparison of unphysical velocity fields magnitude, in terms of the following non-dimensional error norms:

$$L_2(|\mathbf{u}^*|) = \frac{1}{U}\sqrt{\sum_{i=1}^{n}\frac{|\mathbf{u}_i|^2}{n}} \tag{13}$$

$$L_\infty(|\mathbf{u}^*|) = \frac{1}{U}\cdot \max(|\mathbf{u}_i|) \quad \text{for } i = 1,\ldots,n \tag{14}$$

where n is the number of domain cells and U is a velocity scale defined as $U = (\sigma/2\rho R)^{1/2}$. Velocity error norms convergence history is reported in Fig. 1. Youngs method results do not converge with mesh refinement. HF results show second order convergence for $R/\Delta x \leq 10$, then for higher resolutions convergence order is in the range $[1,2]$. Simulation time step is hold constant at $\Delta t = 5\cdot 10^{-7}$, while capillary time step constraint decreases as mesh is refined. For $R/\Delta x > 10$ it gets too close to simulation time step, yielding a worsening of convergence rate. The following non-dimensional pressure jump error norm is defined:

$$L_2(\Delta p^*) = \frac{1}{\Delta p_{ex}}\sqrt{\sum_{i=1}^{m}\frac{(\Delta p_i - \Delta p_{ex})^2}{m}} \tag{15}$$

where m is the number of interior droplet cells and $\Delta p_{ex} = \sigma/R$ is the exact value of pressure jump across the interface. Pressure convergence history is reported in Fig. 1. Youngs results show convergence (first order) only for $R/\Delta x \leq 10$. HF error norm shows convergence rate in the range $[1,2]$ for all mesh sizes.

4.2 Test case 2: isothermal bubble rising in viscous liquid

The second test case is the simulation of a gas bubble rising in a viscous liquid due to buoyancy forces. The effect of the combination of inertial, viscous and surface forces on bubble numerical terminal shape and velocity are compared with Bhaga and Weber experimental results [13]. Bhaga and Weber performed several test cases with air bubbles rising in a quiet water-sugar solution. Acting on sugar concentration, they could vary liquid density and viscosity to span a wide range of Eötvös (E) and Morton (M) numbers, obtaining different bubble shapes and terminal velocities, expressed as bubble Reynolds (Re) number. Surface tension variations were negligible. Non-dimensional numbers are defined as follows:

$$\mathrm{E} = \frac{\rho_l g D^2}{\sigma}, \quad \mathrm{M} = \frac{g\mu^4}{\rho_l \sigma^3}, \quad \mathrm{Re} = \frac{\rho_l U D}{\mu_l}$$

where D is bubble initial diameter and U is terminal bubble velocity, when the rise reaches a steady state. Numerical and experimental results are compared for 4 different cases characterized by the same Eötvös number. Variation of Morton number was possible in the simulations by changing only liquid viscosity. It was

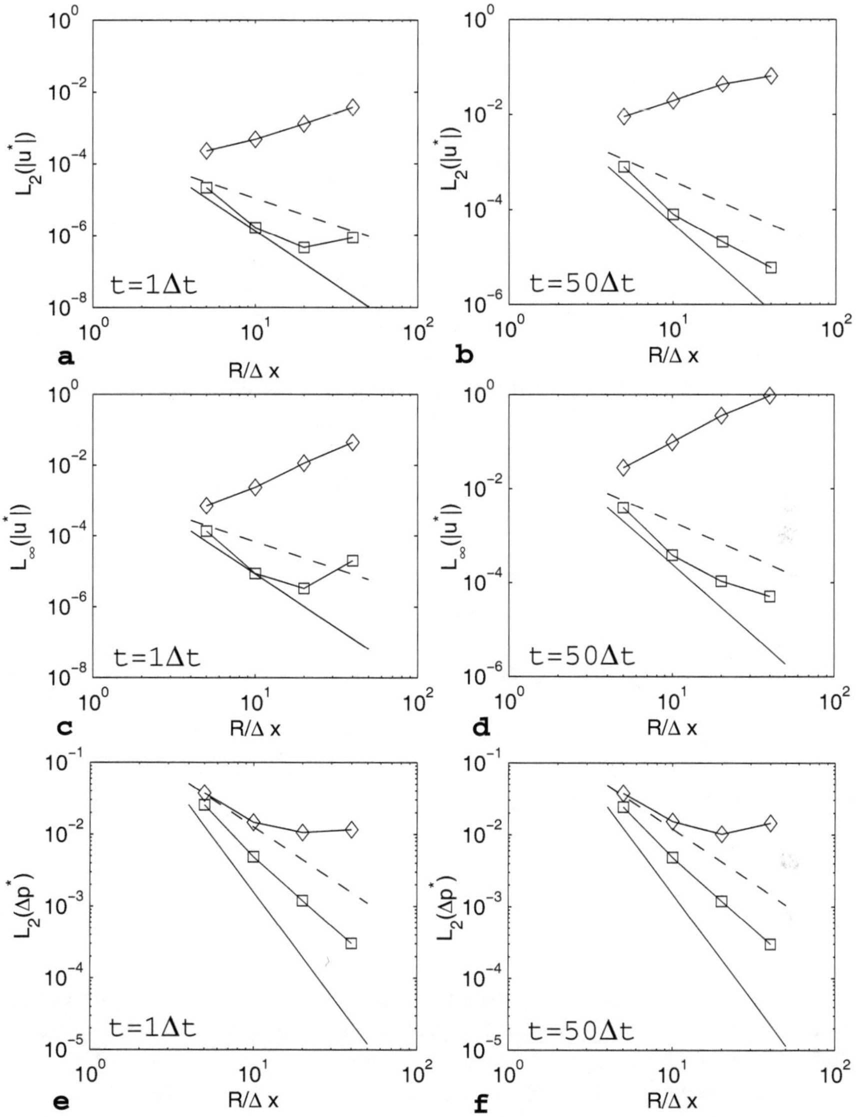

Figure 1: $L_2(|\mathbf{u}^*|)$ (a,b), $L_\infty(|\mathbf{u}^*|)$ (c,d) and $L_2(\Delta p^*)$ (e,f) error norms after one and fifty time steps. White squares represents HF and black diamonds are Youngs error norms. Dashed lines are first order and solid lines are second order curves.

proved that setting density ρ_l/ρ_g and viscosity μ_l/μ_g ratios as in the experiments leads to numerical errors. Since similitudes with experiments are guaranteed by Eötvös and Morton numbers, which involve only liquid properties, gas properties are set in order to fix aforementioned ratios to $\rho_l/\rho_g = 1000$ and $\mu_l/\mu_g = 100$. Bhaga and Weber tested that wakes behind the bubbles are closed and symmetric until $\mathrm{Re} < 110$. Since the simulations are performed on a 2D axisymmetrical domain, the cases chosen satisfy this condition.

A bubble of diameter D is centered at $(0, 2D)$ of a $[0, 4D] \times [0, 12D]$ axisymmetric rectangular domain, with $x = 0$ being the revolution axis. Domain sizes are set in order to avoid boundary influence, following Hua *et al.* sensitivity analysis [14]. Hua *et al.* show also that mesh resolution $D/\Delta x \geq 20$ ensures grid independence, then a 80×240 grid is chosen. Constant pressure is set on top and bottom boundaries. Free-slip condition is imposed on the domain side.

All the simulations are run until a steady state condition for the rising bubble is achieved, for both Height Function and Youngs algorithms. Then, bubble terminal velocity is computed and Reynolds number is evaluated for each case. Bubbles terminal shape for HF and Youngs cases are reported in Tab. 1, together with experimental and numerical Reynolds numbers. For cases I, II and III, both methods show similar performances, with errors in Reynolds number included below 5%. HF method performs slightly better than Youngs. Bubble shapes are very similar for both methods, close to experiments. Case IV is characterized by the highest bubble terminal velocity, then by the highest difference in gas-liquid phases velocity. Interface profile is highly deformed, the most in the bubble back, where interface changes its orientation in really a sharp way. Then, interface is poorly solved by the grid and HF algorithm computes wrong local geometry, leading to errors in capillary effects calculation. Wrong forces computation leads to the detachment of little parts of gas at bubble trail in the simulations. Then, the smaller bubble main body moves with a lower velocity with respect to experiments.

5 Vapor bubble growing in superheated liquid

Height Function interface reconstruction algorithm and evaporation model are implemented in order to simulate the growth of a spherical vapor bubble in an infinitely extended superheated liquid. In accord to Plesset and Zwick analysis [15], bubble growth process can be temporally split in two stages. A first stage, called inertia-controlled growth, which starts at bubble formation. At this stage, growth is governed by momentum interaction between bubble and surrounding liquid. Later, an asymptotic stage is reached, called heat-transfer-controlled growth. In this stage, characterized by a growth rate lower than first stage, growth is limited by heat transport to the interface. This asymptotic stage is the object of this study. Scriven [16] has derived an analytical solution for this stage, neglecting viscous and surface tension effects and considering the interface at saturation temperature. He obtained analytical bubble radius R as a function of time t:

$$R(t) = 2\beta\sqrt{\gamma t} \tag{16}$$

Table 1: Comparison of experimental and numerical Reynolds number. Figures represents bubble terminal shapes computed at $\alpha = 0.5$ contours. Solid line is HF shape, dashed line is Youngs shape. Errors between parenthesis are computed as err $= |\mathrm{Re} - \mathrm{Re}_{num}|/\mathrm{Re}$.

Test case	Experiments [13]	Simulations		
		Shapes	Re_{HF}	Re_{Y}
I	$\mathrm{E} = 116$ $\mathrm{M} = 848$ $\mathrm{Re} = 2.47$		2.37 (4%)	2.37 (4%)
II	$\mathrm{E} = 116$ $\mathrm{M} = 41.1$ $\mathrm{Re} = 7.16$		7 (2.2%)	6.94 (3.1%)
III	$\mathrm{E} = 116$ $\mathrm{M} = 1.31$ $\mathrm{Re} = 20.4$		19.66 (3.7%)	19.55 (4.2%)
IV	$\mathrm{E} = 116$ $\mathrm{M} = 0.103$ $\mathrm{Re} = 42.2$	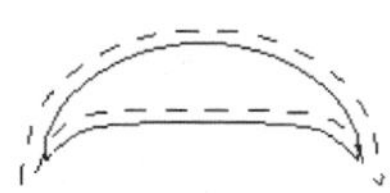	37.8 (10.4%)	39 (7.6%)

where β is a growth constant which details can be found in [16] and γ is liquid thermal diffusivity. This solution is used to validate numerical results.

The growth of a bubble of initial radius $R_0 = 0.1$ mm is simulated. The bubble is centered at $(4R_0, 0)$ of a $[0, 8R_0] \times [0, 4R_0]$ axisymmetric rectangular domain, with $y = 0$ being the revolution axis. A uniform mesh size is chosen, with 1μm element size. Such a fine grid is necessary in order to solve the thin

www.witpress.com, ISSN 1743-3533 (on-line)

thermal boundary layer surrounding bubble interface. As boundary conditions, pressure is fixed at all boundaries except for the axis. The fluid flow is laminar and gravity effects are neglected. All vapor and liquid properties are considered constant at saturation temperature. Initial bubble size is large enough to neglect vapor saturation temperature rising due to pressure jump across interface, then saturation temperature is equal for both phases. Initial pressure and velocity fields are zero. Initial temperature is saturation temperature for the bubble, while liquid is superheated at a temperature $T_\infty = T_{sat} + 5\,^\circ\mathrm{C}$.

A thin thermal boundary layer is placed around the interface on liquid side. Since the simulation starts at a $t = t_0$, when $R(t_0) = R_0$, a thermal boundary layer has been developing around the bubble since the beginning of heat-transfer-controlled growth stage. Temperature field around the bubble at $t = t_0$ can be extrapolated from Scriven solution as a function of spatial coordinate r and time. The thickness of initial thermal layer δ_T is defined as:

$$\delta_T = r(T = T_{sat} + 0.99(T_\infty - T_{sat})) - R_0 \tag{17}$$

Great attention has to be paid to initial thermal boundary layer position. Analytical initialization suggests thermal layer to begin at $r = R_0$. Nevertheless, in those cells which interface is less aligned with grid (close to $\pi/4$ and $3/4\,\pi$), cells centroids are located at $r > R_0$. Then, in those cells, thermal layer intersects bubble interface, leading to a faster initial growth rate than analytical. To avoid this effect, thermal boundary layer is initialized with a bit of misplacement, about 1-2 cells, outside bubble interface.

Three different fluids are tested. Water at atmospheric pressure, HFE-7100 at 0.52 bar, both with $\beta = 15.1$ and $\delta_T = 7\ \mu$m and R134a at 0.84 bar, with $\beta = 9.34$ and $\delta_T = 11\ \mu$m. The choice of each system pressure is done in order to have similar growth constants for the fluids.

During the whole simulation, bubble shape remains spherical. At initial stage of this work, this was proved not to happen using Youngs method for evaluating interface curvature. Moreover, growth was too fast due to high convective heat transfer led by unreal velocities related to errors in curvature calculation. Figure 2 shows bubble radius evolution obtained through HF method compared to analytical solutions, for all the fluids. Numerical data show good agreement with analytical results. For each fluid, bubble numerical growth rate follows a $\sqrt{t}$ proportional law, as it should be from Eq. (16). However, this does not happen during initial growth phase, in a more evident way for HFE-7100 and R134a. This initial phase can be meant as a settlement phase, in which thermal boundary layer arranges to fit interface position. This settlement is reflected on numerical growth rates lower than analytical ones at the beginning of the simulations. As detected by Kunkelmann and Stephan [17], liquid thermal conductivity is the parameter that rules the length of this thermal layer settlement phase. The higher is liquid thermal conductivity, the faster is thermal layer arrangement. For this reason numerical bubble growth rate deviation from analytical curve is more evident for refrigerant fluids and the most for HFE-7100, which has the lowest thermal conductivity.

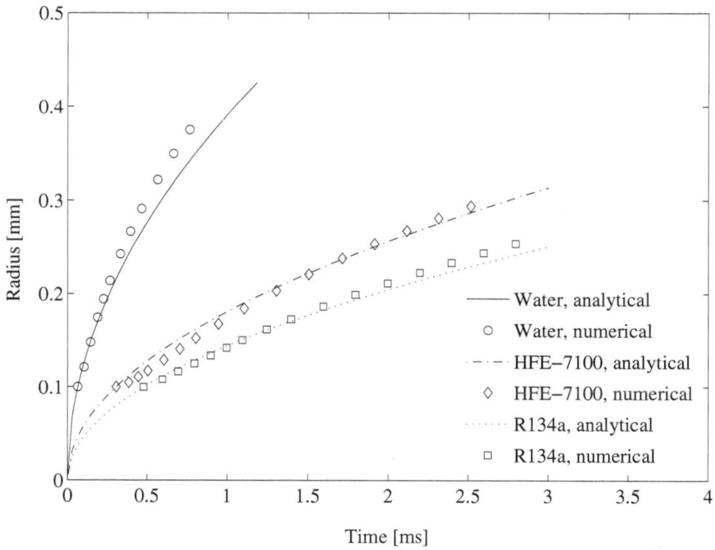

Figure 2: Vapor bubble radius over time for analytical and numerical solutions.

6 Conclusions

The main objective of this work was to implement the Hardt and Wondra [8] evaporation model in a finite-volume framework, in order to simulate a vapor bubble growing in superheated liquid for some working fluids. Bubble shape is computed in implicit way through VOF scheme and an interface reconstruction algorithm has the task to compute interface curvature. Accurate curvature computation is fundamental in order to correctly account for surface tension effects, leading to physical consistent interface shapes. This objective is reached implementing Height Function algorithm, which evaluate curvatures through derivatives of a color function obtained integrating volume fraction field. Inviscid static drop test case assesses HF performances compared to the widely used Youngs algorithm. Inaccuracies in terms of errors in velocity fields and pressure jump over interface scale with second order to mesh element size. Isothermal bubble rising test cases show good agreement of HF numerical results with experiments for $Re \leq 20$, with errors in numerical Reynolds number lower than 5%. HF algorithm coupled with evaporation model leads to excellent agreement of vapor bubble numerical growth rate compared to analytical solutions. The success of the simulations is related to correct treatment of initial thermal boundary layer. Fine grid resolution and proper boundary layer placement are necessary to obtain exact bubble growth rate.

References

[1] Juric, D. & Tryggvason, G., Computations of boiling flows. *Int J of Multiphase Flow*, **24**, pp. 387–410, 1998.
[2] Welch, S.W.J. & Wilson, J., A volume of fluid based method for fluid flows with phase change. *J of Computational Physics*, **160**, pp. 662–682, 2000.
[3] Son, G. & Dhir, V.K., Numerical simulation of film boiling near critical pressures with a level set method. *J of Heat Transfer*, **120**, pp. 183–192, 1998.
[4] Youngs, D.L., Time-dependent multi-material flow with large fluid distortion. *Numerical Methods for Fluid Dynamics*, eds. K.W. Morton & M.J. Baines, Academic Press, pp. 273–285, 1982.
[5] Hernandez, J., Lopez, J., Gomez, P., Zanzi, C. & Faura, F., A new volume of fluid method in three dimensions-Part I: Multidimensional advection method with face-matched flux polyhedra. *Int J for Numerical Methods in Fluids*, **58**, pp. 897–921, 2008.
[6] Malik, M., Sheung-Chi Fan, E. & Bussmann, M., Adaptive VOF with curvature-based refinement. *Int J for Numerical Methods in Fluids*, **55**, pp. 693–712, 2007.
[7] Cummins, S.J., Francois, M.M. & Kothe, D.B., Estimating curvature from volume fractions. *Computers and Structures*, **83**, pp. 425–434, 2005.
[8] Hardt, S. & Wondra, F., Evaporation model for interfacial flows based on a continuum-field representation of the source terms. *J of Computational Physics*, **227**, pp. 5871–5895, 2008.
[9] Brackbill, J.U., Kothe, D.B. & Zemach, C., A continuum method for modeling surface tension. *J of Computational Physics*, **100**, pp. 335–354, 1992.
[10] van Leer, B., Towards the ultimate conservative difference scheme. V. A second-order sequel to Godunov's method. *J of Computational Physics*, **32**, pp. 101–136, 1979.
[11] Issa, R.I., Solution of the implicitly discretized fluid flow equations by operator-splitting. *J of Computational Physics*, **62**, pp. 40–65, 1985.
[12] Schrage, R.W., *A theoretical study of interphase mass transfer*. Columbia University Press: New York, 1953.
[13] Bhaga, D. & Weber, M.E., Bubbles in viscous liquid: shapes, wakes and velocities. *Journal of Fluid Mechanics*, **105**, pp. 61–85, 1981.
[14] Hua, J., Stene, J.F. & Lin, P., Numerical simulation of 3D bubbles rising in viscous liquids using a front tracking method. *Journal of Computational Physics*, **227**, pp. 3358–3382, 2008.
[15] Plesset, M.S. & Zwick, S.A., The growth of vapor bubbles in superheated liquids. *J of Applied Physics*, **25**, pp. 493–500, 1954.
[16] Scriven, L.E., On the dynamics of phase growth. *Chemical Engineering Science*, **10**, pp. 1–13, 1959.
[17] Kunkelmann, C. & Stephan, P., CFD simulation of boiling flows using the Volume-of-Fluid method within OpenFOAM. *Numerical Heat Transfer, Part A*, **56**, pp. 631–646, 2009.

An experimental study of burnout and flow instability in sub-channels with subcooled void at low pressure

V. Chatoorgoon
Dept. of Mechanical Engineering, University of Manitoba, Canada

Abstract

An experimental study was undertaken to rationalize data obtained from previous experiments where the bundle was heated electrically until CHF occurred, destroying ultimately the test bundle. (We term this 'burnout' from here on). The bundle exit temperature was always subcooled; hence, the channels contained only subcooled void preceding 'burnout'. This study was undertaken to perform controlled experiments leading up to, but not including, the burnout power. Just prior to 'burnout' microscopic flow oscillations were exhibited between sub-channels. This was an interesting finding, as while the burnout power (especially for these bundles under sub-cooled boiling conditions) cannot be determined consistently with good accuracy through analytical methods, numerical methods for predicting the onset of flow oscillations are more accurate and well established. A stability analysis of the sub-channels of the single bundle, albeit without communication between the channels, was performed. Surprisingly, the predicted power at the onset of flow oscillations followed closely the experimental burnout power (of course, the latter was always slightly greater than the former, as would be expected).
Keywords: CHF, flow instability, subcooled boiling, fuel bundle, low pressure.

1 Introduction

Two-phase flow instabilities have been well researched, both experimentally and analytically. The analytical methods used for predicting them are also well established and have been abundantly benchmarked. However, in most reported experimental or analytical studies, the system considered was in a state of bulk boiling preceded by some small region of sub-cooled boiling. In all reported

WIT Transactions on Engineering Sciences, Vol 70, © 2011 WIT Press
www.witpress.com, ISSN 1743-3533 (on-line)
doi:10.2495/MPF110071

cases the sub-cooled boiling region had no more than a minor impact on the overall system stability.

It is rare when the entire system considered is subcooled and the instabilities are caused by subcooled void only. It is also rare when the system pressure is low and near atmospheric. Hainoun and Schaffrath [1] reported a numerical study of flow instabilities due to subcooled boiling. Their paper investigated primarily the steady-state pressure-drop versus flow-rate characteristic due to subcooled boiling. No results of actual flow oscillations, or instabilities, are presented. The cited reference does allude to subcooled boiling instability experiments done at Oak Ridge, but those results were not reported. The system pressure was 1.7MPa. Stoddard et al. [2] studied experimentally and analytically the onset of flow instability and critical heat flux in micro-channels with subcooled void at ~1 MPa and found that the onset of flow instability just preceded CHF. They also reported poor agreement between the measured CHF values and predicted values from recommended correlations.

Other than those referenced works, no experimental data on the instability of low-pressure, two-phase systems driven solely by sub-cooled boiling could be found. The information presented herein could be used to verify void models and/or validate thermal hydraulic codes. Three low-pressure, subcooled void driven instability experiments are presented.

The candidate bundle used in this study is labelled bundle ‘M’ for ease of communication. In previous tests the ‘M’ assembly was subjected to increasing powers at a constant forced inlet flow rate until burnout occurred. In those tests burnout invariably occurred before bulk boiling was attained at the bundle exit -- indicating that the exit flow temperature was less than the saturation temperature. The primary objective of those experiments was to determine the burnout power and not necessarily to understand the flow dynamics leading up to burnout.

Even though this generic study was conducted at low pressure and with simulated fuel bundles, the information gathered could provide useful insight into flow dynamics just before burnout occurs in fuel bundles.

2 Objectives

The objectives of these experiments were to:

(a) Gather data on parallel channel flow instability in a set of electrically-heated assemblies that would be useful for code validation
(b) Determine the power at the onset of flow instability (OFI)
(c) Determine whether the power at OFI, in a parallel arrangement with other similar assemblies, is related in any way to the burnout power obtained in single-assembly tests.

Unlike the burnout tests mentioned above, the ‘M’ assembly in these experiments was situated in parallel with two other assemblies (labelled the ‘H’ and ‘C’ assemblies). The total flow rate through all three assemblies was constant.

3 Test facility

The test loop used is shown in Fig. 1. It consisted of three parallel assemblies mounted on a common plenum (shown in Fig. 2) that was situated at the bottom of a large tank. Deionized light water was pumped from a holding reservoir into the plenum using a centrifugal pump. The coolant was driven upward through the heater assemblies after which it was returned to a holding reservoir located adjacent to the 40 hp pump. The flow exiting the heater assemblies was passed through a heat exchanger situated upstream of the holding reservoir to remove the added heat. The outlet of each heater assembly was connected to a common header, which in turn was connected to the heat exchanger inlet piping.

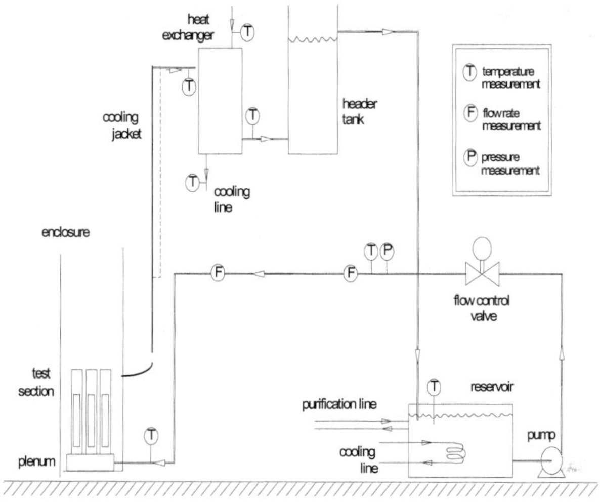

Figure 1: Schematic of flow loop.

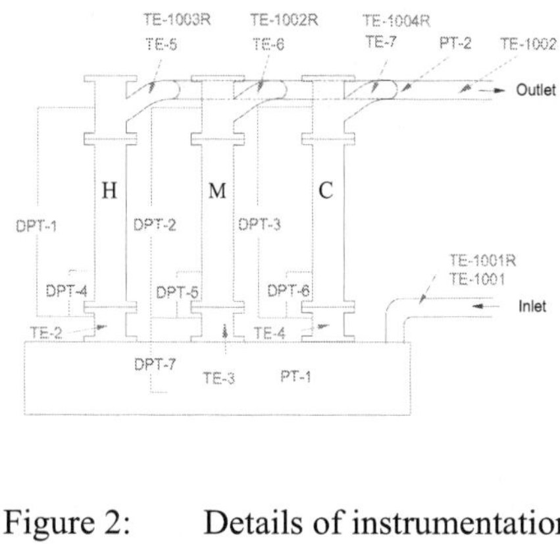

Figure 2: Details of instrumentation.

www.witpress.com, ISSN 1743-3533 (on-line)

4 Heater assemblies

Three different types of assemblies were used -- an H assembly, which comprised 7 finned elements; a C assembly, which comprised 6 finned elements, and the M assembly, which comprised 4 smooth elements. The tube around each heater assembly was made of a polycarbonate material and was fabricated specially for these assemblies. Each assembly was mounted on a spool piece, which was then mounted directly onto the plenum. All three assemblies were heated electrically using a DC power source with a total capacity of 1.7 MW.

The H assembly comprised 7 heater elements, each of length 0.6m. It had a cosine-shaped axial heat-flux profile. The C assembly comprised 6 heater elements of length 0.6 m. It also had a cosine-shaped axial heat-flux profile. The M assembly comprised 4 heater elements of length 0.47m. It had a uniform axial heat-flux profile.

5 Power connections

All three assemblies were connected electrically in a simple parallel arrangement. This delivered about 56% of the total power to the M assembly, 25% to the H assembly and 19% to the C assembly. This arrangement was chosen so as to introduce subcooled boiling in the M assembly while maintaining single-phase flow in the H and C assemblies. The H and C assemblies comprised indirect heaters whereas the M assembly was heated directly. The power distribution between the assemblies varied slightly with the applied power.

6 Instrumentation

The test facility was instrumented to measure coolant temperature, pressure, pressure-drop and assembly power. The flow rates through the H and C assemblies were deduced from DP cell pressure-drop measurements across the inlet of each assembly. These inlet DP cells were previously calibrated against known flow rates. An independent verification of the flow rate was also obtained from an energy balance applied across the assembly.

6.1 Steady-state flow rate

Steady-state flow rates in the three assemblies were determined as follows:

- Total loop flow rates were obtained from the FT-2 and FT-5 flow meters.
- Flow rates into the H and C assemblies were deduced from pressure-drop measurements obtained across the inlet end plates and also from an energy balance across each assembly.
- Flow rate through the M assembly was deduced from an energy balance across that assembly and also from subtracting the H and C assemblies flow rates from the total flow rate.

6.2 Pressure and pressure-drop

The following pressure and pressure-drop measurements were made:

- absolute pressure in the plenum; made with a Rosemount pressure cell.
- differential pressure across each inlet end-plate of the H and C assemblies; obtained with Sensotec DP cells.
- differential pressure across the inlet spool piece of the M assembly; obtained with a Sensotec DP cell.
- differential pressure across each assembly, made with a Rosemount DP cell.

6.3 Transient/oscillatory flow rate and pressure-drop

Transient or oscillatory flow rate was not measured directly because of the inherent difficulties in accurately measuring an oscillating flow rate. However, the transient/oscillatory pressure-drop across the H and C inlet end plates (DPT-4 and DPT-6), and the M inlet spool piece (DPT-5), were measured using fast-response Sensotec DP cells. These fast-acting Sensotec DP cells were sufficiently sensitive to capture details of the dynamics. The Sensotec DP cell placed across the inlet spool piece of the M assembly for recording oscillations in the inlet pressure-drop caused by oscillations in the inlet flow rate would not necessarily measure pressure oscillations between sub-channels of the M assembly. However, they were found to provide an indication of such oscillations. Oscillations in pressure-drop at any given location would indicate oscillations in the flow rate at that location. The frequency and phasing of the pressure-drop oscillations could be translated into a corresponding frequency and phasing of flow rate oscillations. The measured pressure-drop values could be compared directly with numerical predictions of pressure drop response. If they matched well, the corresponding calculated flow rates would be deemed realistic.

6.4 Temperature

The following temperatures were measured:

* inlet temperature to the plenum.
* inlet and outlet temperature of each bundle using thermocouples and RTDs.
* outlet temperature of all four inner channels of the M assembly using thermocouples.
* temperature in the outlet header using a thermocouple and RTD.

6.5 Assembly power

The assembly power was determined by measuring the current through each assembly and the voltage drop across each assembly. From the known resistance of the assembly and the measured voltage drop, a **direct** value of the applied power to each assembly was derived. An **indirect** value applied power was also deduced from an energy balance. The measured pressure-drop across each inlet end plate yielded a corresponding flow rate from the calibration curves. Combining this with the measured inlet and outlet temperatures from each assembly, the assembly power was calculated from the standard formula,

$$Power = \dot{m}\, C_p\, (T_{out} - T_{in}) \tag{1}$$

where $\dot{m}$ is the mass flow rate, C_p is the specific heat and $(T_{out} - T_{in})$ is the difference between the outlet and inlet temperatures. Of course, Eq. (1) is only strictly valid for single-phase flow. When two-phase flow occurs at the outlet, an error is introduced that is dependent on the energy content of the vapour.

7 Measurement uncertainty and errors

Precautions were taken to reduce experimental uncertainty and measurement errors. Due to brevity of space, this information is available, but omitted here.

8 Data acquisition system

A PC-based Data Acquisition System (DAS) and a Noise Analysis System (NAS) were used to collect, process and store the measured data. The DAS was used for recording steady or quasi-steady data while the NAS was used for recording unsteady, transient data. The NAS was developed for collecting data up to frequencies of 1kHz. The pressure-drop versus flow-rate characteristic of the inlet end plates of the H and C assemblies and the inlet spool piece of the M assembly were programmed into the DAS. Thus, direct values of flow rate were obtained from the DP cells connected across the H and C inlet end plates and the M inlet spool piece. Conversion of the raw data into the required engineering units was also performed automatically by the DAS. Further processing of the results into useful graphical displays was done using spreadsheet software.

9 Experimental procedure

9.1 Stability tests

The purpose of these tests was to examine the stability characteristics of the M assembly with sub-cooled void while the H and C assemblies were with single-phase conditions throughout. Specifically, the intent was to determine the M assembly power at the onset of flow instability (OFI). For this reason, the H and C assemblies were heated with only 25% and 19% respectively of the total applied power while the M assembly received the remaining 56%.

Three stability tests were done, all with a constant plenum temperature of 30°C. Commencing at zero power and a pre-set inlet flow velocity to the M assembly, the total power applied to all three assemblies was increased in increments of about 25kW until significant void (determined visually) was noticed in the outer channels of the M assembly. All the system parameters (temperature, pressure, pressure-drop, flow rate, current and voltage) were continually recorded by the DAS from the start of each test. The NAS data collection was activated and synchronized with the DAS when the total applied power reached 100 kW. The scan rate on the DAS was every 2 seconds. The scan rate on the NAS was every 0.01-second for the duration of the experiment.

The onset of nucleate boiling (ONB) was deduced (approximately) by listening for the change in sound emanating from the assemblies. (The onset of nucleate boiling on the inner channels of the M assembly could not be ascertained by this procedure as the sound from those channels was inaudible.) After each power increment the system was allowed time to settle down to a new steady state. This was determined by observing the M assembly outlet temperature on the DAS and waiting for it to become constant. After OSV was deduced to be attained in the M assembly, the total power increments were reduced to ~10 kW until the M inlet spool-piece pressure drop, monitored continually on an oscilloscope, showed small-scale oscillations. Thereafter the total power increments were reduced to about 5 kW, or less, until flow oscillations or instability was clearly evident on the oscilloscope.

The onset of flow oscillations (OFO) is distinctly different from the onset of flow instability (OFI). The former precedes the latter slightly. The OFI point is defined as that point where the oscillation amplitudes begin to diverge. Once the OFI power was determined, further increases in the total power were discontinued to protect the assemblies. After allowing sufficient time for data collection, the total power was reduced in preparation for the next test. The inlet velocity to the M was reset to a different value and the experimental process repeated. Three tests were done corresponding to average inlet velocities in the M assembly of 3.36, 2.43 and 1.09 m/s. The average velocity was determined by dividing the net flow rate through the M assembly by the product of the net flow area and the inlet flow density. For inlet velocities of 2.43 and 1.09 m/s, the repeatability of the experiments was assessed and confirmed. After flow instability was identified in the first power sweep, the total power was reduced briefly and increased again until OFI reoccurred. It was found that diverging flow oscillations appeared exactly at the same power (i.e. within 1–2 kW) as was obtained in the first sweep. This confirmed that the experiments were repeatable and the results were reproducible. During the tests, the test-section was also continuously filmed using two video cameras. The flow oscillations were, therefore, captured on film.

10 Experimental results

A summary of the test conditions at OFI in the M assembly, for the three tests performed, are given in Table 1. The onset of flow instability (OFI) is defined as that point where the oscillation amplitudes began to diverge with time, as inferred from the inlet pressure-drop oscillations. These oscillations may not have been noticeable with the naked eye, but they were measurable. The onset of flow instability for the M-assembly inlet spool piece for the three tests is shown in Figs. 3-6. Also shown is the inlet end-plate pressure drop for the H and C assemblies.

A cautionary note on comparing the magnitudes of the pressure-drop oscillations of the M-assembly inlet spool piece with the H and C assemblies' inlet end plate; the M-assembly inlet spool piece inherently had a low-pressure drop due to its open area. Thus, small-scale oscillations do not necessarily mean low amplitude flow oscillations. Flow oscillations between the inner and outer channels of the M assembly would induce oscillations in the inlet spool piece when the conditions and channel geometry are unequal. Since the inner and outer channels of the M were different, with different powers and flow conditions, any oscillations between the inner and outer channels would also induce oscillations in the flow within the inlet spool piece. Hence, the pressure oscillations across the M inlet spool piece recorded by the Sensotec DP cell are a result of the inner channel dynamics as well as the coupled dynamics between the M assembly and the H and C assemblies.

Figs. 3-5 show that the inlet flow to the H and C assemblies always oscillated in phase – an expected result since both assemblies were always with single-phase flow. Therefore, the H and C assemblies behaved, as expected, as a single assembly. With the apparent exception of the 3CH5 test, which is discussed subsequently, Figs. 3 and 4 show that, for the 3CH3 and 3CH4 tests, the M inlet spool piece pressure drop is almost, but not exactly, 180 degrees out of phase with the H and C assemblies' inlet end-plate pressure drop. However, the

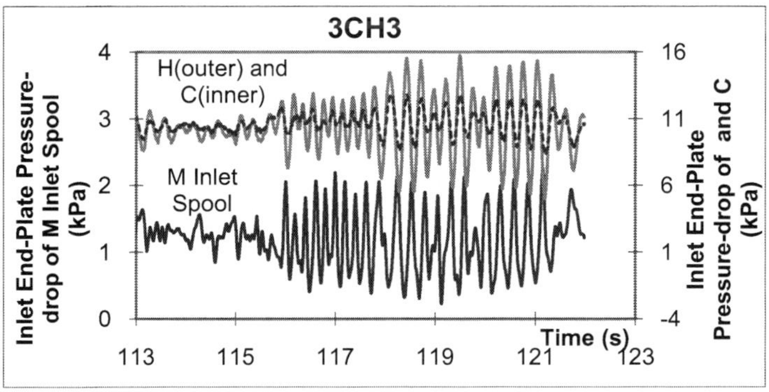

Figure 3: Onset of flow instability for test 3CH3.

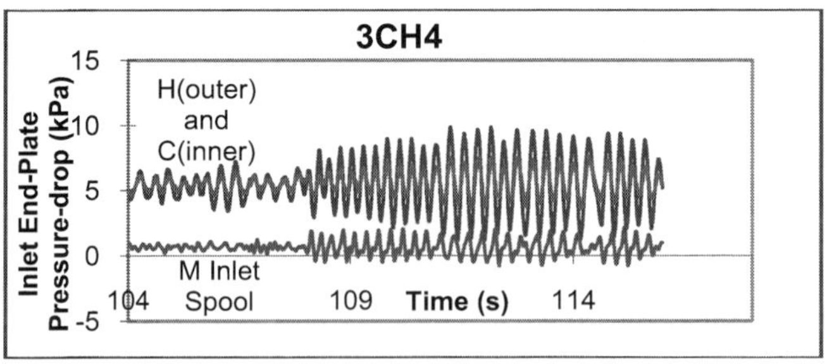

Figure 4: Onset of flow instability for test 3CH4.

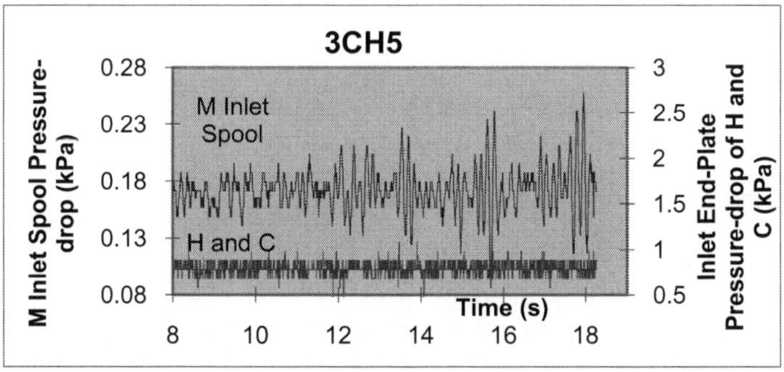

Figure 5: Onset of flow instability for test 3CH5 (130.7 kW).

periods of oscillation are equal. This indicates that the M assembly, for the 3CH3 and 3CH4 tests, behaved almost like a 'single' entity. Because the phase difference was slightly less than 180°, this indicates the possibility of some dynamic motions within the M assembly itself. If the M assembly behaved perfectly like a single entity, then its response would have been exactly 180° out of phase with the H and C assemblies' inlet pressure-drop oscillations. It was already established that the H and C assemblies behaved like a single entity. The lowest velocity test (3CH5) shows a different behaviour from the 3CH3 and 3CH4 tests (Fig. 5). Like before, Fig. 5 shows that the H and C assemblies' inlet pressure oscillations were always in phase with each other. However, the period of oscillation of the H and C assemblies has now doubled (~ 0.40 second) to about *twice* the period of oscillation of the M assembly. A plausible explanation is not obvious, but this result indicates that some dynamic motions may be occurring within the M assembly, especially at this lower flow rate.

The inlet flow oscillations in the M-assembly inlet spool piece (Fig. 5) at OFI clearly show two oscillation frequencies -- one with a period of about 0.2 second and the other with a period of about 2.0 seconds. There is no noticeable flow oscillation in the H and C assemblies in Fig. 5. It is possible that they were too small to be picked up at this power (130.7 kW in the M assembly).

Because the H and C assemblies always acted in phase and exhibited the same period, while the period of the oscillation in the M inlet spool piece was half the period of the oscillations in the H and C assemblies, this observation indicates that the M assembly behaved like multiple channels with phase differences. No other explanation is evident. Figs. 3-5 show evidence of two oscillation periods in the M assembly -- one of about 0.2 second and the other of about 2.0 seconds. However, the dominant mode seems to be the 0.2-second period. The period of oscillation in the H and C assemblies is about 0.4 seconds.

A plot of the M-assembly power at OFI versus average inlet velocity is shown in Fig. 6. Also shown is the burnout power obtained in previous burnout tests. It is noteworthy that the three OFI data points fall very close to the burnout data, suggesting a connection between burnout power and OFI power. It seems very plausible that in the single M-assembly burnout tests flow oscillations, and/or flow instability, were a precursor to burnout. Fig. 6 presents the OFI data in the non-dimensional form of $\frac{N_p}{N_s}$ versus N_s where N_p and N_s are the well-known Zuber and subcooling numbers respectively. The OFI data is approximated well by $\frac{N_p}{N_s} \approx 0.46$.

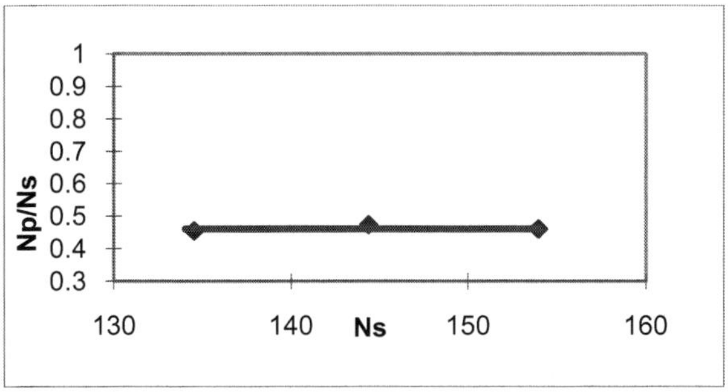

Figure 6: Np/Ns versus Ns for the M assembly.

A plot of the M-assembly power at OFI versus average inlet velocity is shown in Fig. 7. Also shown is the burnout power obtained in previous burnout tests. It is interesting that the three OFI data points fall very close to the burnout data, suggesting a connection between burnout power and OFI power. It seems very plausible that in the single M-assembly burnout tests flow oscillations, and/or flow instability, were a precursor to burnout.

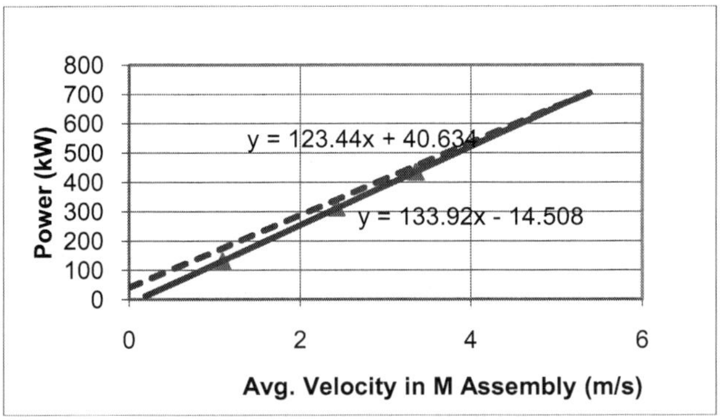

Figure 7: OFI power versus average inlet velocity for the M assembly.

Fig. 8 shows the comparison between burnout data and numerical prediction of incipient flow oscillations in the sub-M-assembly channels. The trend is similar, with the instability predictions slightly under the burnout data.

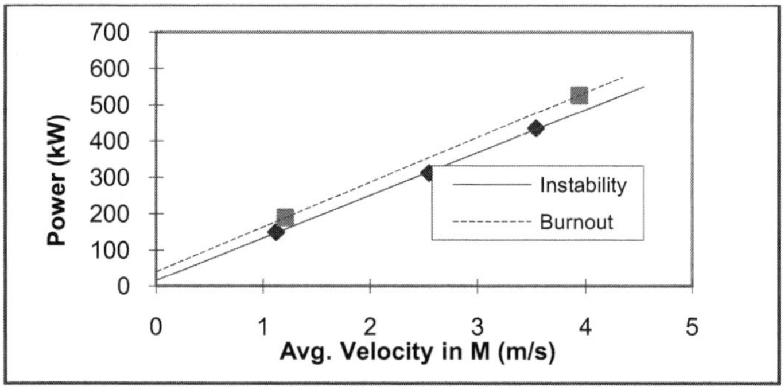

Figure 8: Comparison between burnout data in the M assembly vs. prediction of incipient flow oscillations using an instability numerical program (SPORTS).

11 Conclusions

M-assembly Stability Tests

The following were the findings of this study:

(a) Nucleate boiling precipitated flow oscillations. (b) Oscillations occurred between sub-channels of the M assembly, as well as between the M assembly and other assemblies.

(c) The onset of flow oscillations and instability (OFI) just preceded the single-assembly burnout, indicating a possible connection between burnout and OFI.

(d) If (c) is true, conditions just before burnout may be predictable through a numerical stability analysis of the assembly sub-channels.

(e) OFI occurred at an $\frac{N_p}{N_s} \approx 0.46$. This may be useful for scaling the data for pressure, velocity and sub-cooling effects.

(f) The close proximity of the OFI data to the single-assembly burnout data suggests that the burnout power is not significantly affected by the presence of other parallel assemblies.

It is unknown at this time whether points (b) and (c) would apply to CHF in fuel bundles, and whether flow oscillations between channels would occur and influence CHF. Further studies need to be undertaken to confirm whether this finding is generally applicable, or whether it is just a unique characteristic of this bundle geometry.

Acknowledgements

Sincere thanks are expressed to Dr. N. Popov for supporting this generic study and to Dr. H. Rummens for her invaluable technical expertise in designing the experiment. The experimental work was performed while at the Chalk River Laboratories of Atomic Energy of Canada, Ltd. (AECL-CRL). The experimental data presented in this paper are shared with permission from AECL-CRL. The interpretations of the data are those of the author.

References

[1] Hainoun, A., Schaffrath, A., 2001. Simulation of sub-cooled flow instability for high flux research reactors using the ATHLET code, Nucl. Eng. & Design, 207, p. 163-180.

[2] Stoddard, R.M., Blasick, A.M., et al., 2002. Onset of flow instability and critical heat flux in thin horizontal annuli, Experimental Thermal and Fluid Science, Vol. 26, Issue 1, April, p. 1-14.

Section 2
Turbulent flow

Transition length between water and air-water flows on stepped chutes

A. L. A. Simões[2], H. E. Schulz[1,2] & R. M. Porto[2]
[1]*Nucleus of Thermal Engineering and Fluids,*
School of Engineering at São Carlos, University of São Paulo, Brazil
[2]*Department of Hydraulics and Sanitary Engineering,*
School of Engineering at São Carlos, University of São Paulo, Brazil

Abstract

This study presents the steps followed to obtain mathematical models for the length of the transition region between the "full-water" and "full-mixed" flows in stepped spillways. This transition length is defined here as the distance along the flow, parallel to the pseudo-bottom, starting at the end of the "full-water" region and ending at the beginning of the "full-mixed" region. The definition is proposed based on experimental profiles of the surface obtained with an acoustic sensor in a stepped chute, which allows one to locate adequately the minima and maxima of the profile. A set of profiles obtained for different flow conditions is shown, and a comparison between predicted and calculated transition lengths is made. Experimental data and theoretical predictions superpose adequately for the present set of data.
Keywords: air-water flow, aeration, transition length, stepped spillways.

1 Introduction

For the design of stepped chutes it is necessary to know the behaviour of the air-water mixture along the flow. In many cases it is necessary to build bottom aerators that must be conveniently placed, for which the position of the inception point is needed (the section where the air begins to be captured by the water at the upper surface). It is also necessary to know the length of the black water region (upflow of the inception point), and the length along the flow needed for the air to travel until the bottom of the channel (downflow of the inception point). This length is here called "transition length", and is defined as the

WIT Transactions on Engineering Sciences, Vol 70,
www.witpress.com, ISSN 1743-3533 (on-line)
doi:10.2495/MPF110081

distance, parallel to the pseudo-bottom, which starts at the end of black water region and finishes at the beginning of the full-aerated region. The present analysis is valid for skimming flows.

It is generally accepted that skimming flows, as occur in smooth chutes or spillways, consists of three regions: (1) single-phase flow (water); (2) partially aerated flow, and (3) air-water two phase flow (Cain and Wood [5]). Similarly, these regions are also observed in stepped chutes, and their quantification is relevant, being motivated by the need of protection of the spillways from possible damages caused by cavitation. This study presents a mathematical model for the transition length in stepped spillways.

2 Theoretical formulation

Two formulations are presented to obtain predictions of the Transition length. The first one considers classical conservation principles, and the second considers an approximation based on analogies with basic transport equations.

2.1 Equation based on conservation principles (Model 1)

In this approximation, the principles of conservation were written in the Eulerian and integral forms. The control volume chosen has two input surfaces (section 1 and the free surface, through which air is captured), and one output surface (section 2), as shown in fig. 1. It is assumed that the main transport of mass, momentum and energy occurs through sections 1 and 2, as shown in the sequence. For steady state calculations, the integrations are performed over the areas of sections 1 and 2.

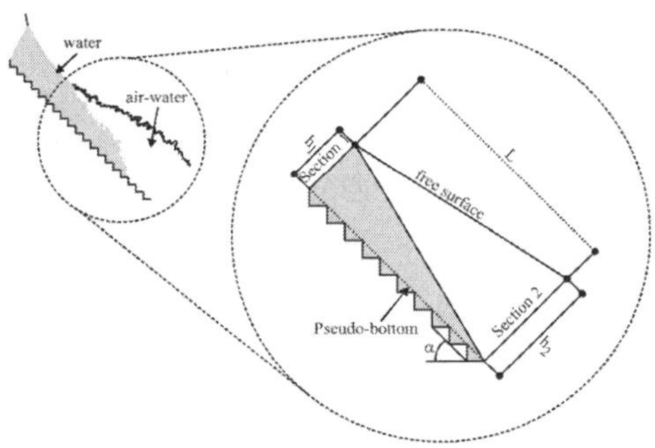

Figure 1: Control volume.

2.1.1 Conservation of mass

The mass conservation equation for steady flow is:

$$\iint_1 \rho_1 v_1 dA + m_{Surf} = \iint_2 \rho_2 v_2 dA \tag{1}$$

ρ, v, and A represent the densities, velocities and areas considered in the problem. The subscripts "*Surf*", 1 and 2 indicate the upper surface, and sections

1 and 2, respectively, while *m* represents the air mass flux. Because this flux is much lower than those given by the integrals on surfaces 1 and 2, its influence is considered here as composing the corrective factor of the flux on section 2.
As ρ_1, is constant, but ρ_2 varies along A_2, we have:

$$\rho_1 \iint_1 v_1 dA = \rho_1 V_1 A_1 \qquad \text{and} \qquad \frac{\iint_2 \rho_2 v_2 dA}{A_2} \neq \overline{\rho}_2 V_2 \tag{2}$$

The use of average values imposes a corrective factor (ω_0), which also accounts for m_{Surf} of eqn (1), furnishing

$$\omega_0 \overline{\rho}_2 V_2 A_2 = \rho_1 V_1 A_1 \tag{3}$$

Defining C as the volume fraction occupied by the air, the density of the mixture (ρ_2) is usually presented as:

$$\overline{\rho}_2 = \overline{C}\rho_g + (1-\overline{C})\rho_1 \;\Rightarrow\; \frac{\overline{\rho}_2}{\rho_1} \cong (1-\overline{C}) \;\text{ for }\; \frac{\overline{C}\rho_g}{\rho_1} << (1-\overline{C}) \tag{4}$$

Substituting 4 into 3, and using A=Bh, leads to:

$$\frac{V_2}{V_1} = \frac{1}{\omega_0 (1-\overline{C}) h^*} \tag{5}$$

where $h^* = h_2/h_1$.

2.1.2 Conservation of momentum

The momentum equation for steady flow is given by:

$$F = -\iint_1 \rho_1 v_1 v_1 dA + \iint_2 \rho_2 v_2 v_2 dA = -\beta_1 \rho_1 V_1^2 B h_1 + \omega_1 \overline{\rho}_2 V_2^2 B h_2 \tag{6}$$

β_1 is the Boussinesq coefficient, and ω_1 in section 2 is a weighting factor which takes into account the fact that $\rho \neq \overline{\rho}_2$. The control volume has a lateral trapezoidal form, as shown in fig. 1. ω_2 is defined as the fraction of this area occupied by water (ideally corresponding to the lower triangle in the figure). It is assumed that the mixture at the remaining fraction of the trapezoid (ideally the upper triangle) is $\overline{\rho}_2$, defined by eqn (5). The force F acting on the control volume is expressed as F=F_w+F_p+F_τ, where F_w is the weight, F_p is the force due to pressure and F_τ is the force due to the bottom shear stress. The three forces are given, respectively, by:

$$F_w = \frac{(h_1 + h_2)}{2} BL\rho_1 g\left[\omega_2 + (1-\omega_2)(1-\overline{C})\right]\sin\alpha \tag{7a}$$

$$F_{p1} - F_{p2} = \frac{\rho_1 g}{2} B\left[h_1^2 - (1-\overline{C})h_2^2\right]\cos\alpha \tag{7b}$$

$$F_\tau = -\tau_o BL \tag{7c}$$

The effects of shear stresses on the walls and the free surface were not considered. Combining eqns (5), (6) and (7), solving them for L/h_1, defining the nondimensional parameters $L^* = L/h_1$, and $Fr_1 = V/(gh_1)^{1/2}$, leads to:

$$L^* = \frac{-\beta_1 + \frac{\omega_1}{\omega_0^2}\frac{1}{(1-\overline{C})h^*} - \frac{1}{2Fr_1^2}\left[1-(1-\overline{C})h^{*2}\right]\cos\alpha}{\frac{(1+h^*)}{2Fr_1^2}\left[\omega_2 + (1-\omega_2)(1-\overline{C})\right]\sin\alpha - \frac{f}{8}} \tag{8}$$

The shear stress was obtained from the Darcy-Weisbach equation, and is related to the friction factor f at the bottom of the channel as:

$$f = \frac{\tau_o}{\frac{1}{8}\rho_1 V_1^2} \tag{9}$$

2.1.3 Conservation of energy

The energy equation for steady flow is given by:

$$\dot{Q} - \dot{W} = -\iint_1 \rho_1\left(e_1 + \frac{p_1}{\rho_1}\right)v_1 dA - \dot{E}_{Surf} + \iint_2 \rho_2\left(e_2 + \frac{p_2}{\rho_2}\right)v_2 dA \tag{10}$$

$\dot{Q}$ is the heat transfer across the control surfaces, $\dot{W}$ is the power transferred due to shear forces, $\dot{E}_{Surf}$ is the contribution due to the air flow through the surface (neglected here in relation to the other parcels), and $e = gz + v^2/2 + u$.
Eqn (10) is then represented as:

$$\begin{aligned}\dot{Q} - \dot{W} = -\rho_1 \iint_1 \left(gz_1 v_1 + \frac{v_1^3}{2} + u_1 v_1 + \frac{p_1 v_1}{\rho_1}\right)dA + \\ + \iint_2 \left(g\rho_2 z_2 v_2 + \rho_2\frac{v_2^3}{2} + \rho_2 u_2 v_2 + p_2 v_2\right)dA\end{aligned} \tag{11a}$$

The integrals at sections 1 and 2 are represented, in the next equations, by I_1 and I_2, respectively. Using mean values, a Coriolis coefficient α_{c1} must be used for section 1, and $p_1 = \rho_1 gh$. In section 2 a new weight coefficient ω_3 is used (similar to the Coriolis coefficient), $z_2 = 0$ and p_2 is assumed to be $p_2 = \omega_4\ \overline{\rho_2}, gh$, where $\overline{\rho_2}$ is the mean value of ρ_2 and ω_4 is an adjusted constant.

$$I_1 = g\left(L\sin\alpha + h_1\cos\alpha\right)V_1A_1 + \alpha_{c1}\frac{V_1^3}{2}A_1 + \bar{u}_1V_1A_1 \tag{11b}$$

$$I_2 = \frac{\omega_4\cos\alpha}{2}g\bar{\rho}_2h_2V_2A_2 + \omega_3\bar{\rho}_2\frac{V_2^3}{2}A_2 + \bar{\rho}_2\bar{u}_2V_2A_2 \tag{11c}$$

Combining eqns (3), (5), (10) and (11):

$$h_p = \omega_0\left[\left(Lsen\alpha + h_1\cos\alpha\right) + \alpha_{c1}\frac{V_1^2}{2g}\right] - \left(\omega_3\frac{V_2^2}{2g} + \omega_4\cos\alpha\frac{h_2}{2}\right) \tag{12}$$

Where h_p, known as "head loss", is given by:

$$h_p = -\frac{\dot{Q} - \dot{W}_r + \rho_1\bar{u}_1V_1A_1 - \bar{\rho}_2\bar{u}_2V_2A_2}{g\bar{\rho}_2V_2A_2} \tag{13}$$

The head loss may be simplified to a punctual loss, so that $h_p=KF_{r1}^2h_1/2$, where K is a constant. Writing then eqn (12) in nondimensional form, and solving it for $1-\bar{C}$, results:

$$(1-\bar{C}) = \frac{\sqrt{\omega_3}}{\omega_0}\left[h^*\sqrt{\omega_0\frac{2L^*sen\alpha}{Fr_1^2} + \frac{2}{Fr_1^2}\left(\omega_0 - \omega_4\frac{h^*}{2}\right)\cos\alpha + \omega_0\alpha_{c1} - K}\right]^{-1} \tag{14}$$

The set of eqns (8) and (14) allows to obtain L^*, and is here named "Model 1". In this study a least squares adjustment of the constants was made using a nonlinear procedure.

2.2 Analogy with basic transport equations (Model 2)

Schulz and Simões [3] presented an alternative analysis of the transition region also considering the volume of fig. 1. The authors considered that: 1) the rate of formation of voids in the water, indicated by $\dot{c}$, is proportional to the water flow rate that crosses the volume, q; and 2) $\dot{c}$ is also proportional to the slope of the surface, dh/dx. Joining both proportionalities, it leaded to:

$$\dot{c} = Kq\frac{dh}{dx} \tag{15}$$

K is a proportionality factor. The dimensions of the variables are: $|h|$=m, $|x|$=m, $|q|=\mathrm{m^2s^{-1}}$, $|\dot{c}|=\mathrm{s^{-1}}$, $|K|=\mathrm{m^{-2}}$. As mentioned by Schulz and Simões [3], eqn (15) is similar to the basic equations used in Transport Phenomena for mass and heat transfer, because it involves a first order derivative. Assuming $\dot{c}$, q and K as constants, the integration of this equation, from x=0 to x=L, produces:

$$L = \frac{Kq}{\dot{c}}(h_2 - h_1) \tag{16}$$

Although very simple, eqn (16) produced a correlation coefficient of about 0.73 (considered acceptable) when compared with the experimental data described in section 3. It was then "allowed" for one of the constants to vary. K was chosen, and the following general form was proposed:

$$K = \sum_i \phi_i (qh)^i \tag{17}$$

ϕ_i are constants. Eqns (15) and (17) were then rearranged to:

$$\dot{c} = \sum_i \phi_i (qh)^i \frac{d(qh)}{dx} \tag{18}$$

To obtain an adequate solution, it is necessary to conveniently truncate the series in eqn (18). For i=0, the solution is eqn (16) (acceptable). In this study we used i=1, which leads to:

$$L = \frac{\phi_0}{\dot{c}}(qh_2 - qh_1) + \frac{\phi_1}{\dot{c}}\left[(qh_2)^2 - (qh_1)^2\right] \tag{19}$$

This equation was compared with the measured data, together with an alternative semi-empirical form, in which all parcels have independent coefficients, that is,

$$L = \theta_1 qh_2 + \theta_2 qh_1 + \theta_3 (qh_2)^2 + \theta_4 (qh_1)^2 + \theta_5 \tag{20}$$

θ_i are constants. Eqn (20) has the same form of eqn (19), which is the reason of using it here. It is obviously expected that the last equation furnishes a better fit to experimental data, because it involves five adjustment constants (the former involves only two). Eqn (19) is named here Model 2a and eqn (20) is named Model 2b.

3 Experimental results

Experiments were conducted in a channel of the Laboratory of Environmental Hydraulics, in the School of Engineering at São Carlos (fig. 2a). The channel was 5 m long and 20 cm wide, allowing adjustments of the slope angle between 0° and 45°. In this study α = 45°. The height of the steps was s = 5 cm. The flow rate was controlled upstream by a sluice gate, and measured with an electromagnetic flowmeter. The profiles of the free surface were obtained positioning an ultrasonic sensor in 40 locations along 3.5 m of the channel. The sampling frequency adopted for the depth measurements was 50 Hz, and the sensor was maintained 120 s at each position to obtain representative data for the depths. A typical mean profile obtained in this study is shown in fig. 2b. This figure also shows the plot of the analytical solution for the single phase region (indicated by S_2, as obtained by Simões *et al.* [4]), and the minimum and maximum positions that limit the transition region.

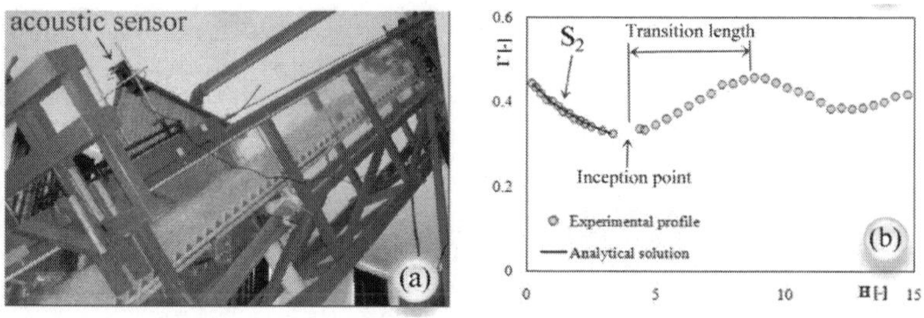

Figure 2: (a) Stepped chute and (b) experimental profile. The transition length is the distance between the minimum and maximum points of the surface profile.

3.1 Comparison with literature

The position of the inception point and the length of the transition region were here obtained from the measurements, as indicated in fig. 2b. A comparison was made with a prediction for the position of the inception point suggested by Boes and Hager [2]. Following a translation of the origin similar to that proposed by Boes [1], and using the measured points to adjust the theoretical curve, the difference between the positions of the measured and predicted inception points corresponds roughly to the transition length. The difference occurs because Boes and Hager [2] considered the full mixed region, while here the beginning of the transition region is considered.

3.2 Measured and predicted transition lengths

Figs 3, 4 and 5 show all the surface profiles measured in the present study. It is shown that the minima and maxima of the different profiles are well defined in most of the experiments, which points to the adequacy of this methodology. In some profiles the measurements are somewhat sparser, like shown in figs 5n (minimum) and 5o (maximum), which introduce errors in the evaluation of L, h_1 and h_2, but even so allowing to observe the transition region.

The proposed models for the transition length involve sets of constants that must be adjusted. For eqns (8) and (14) (Model 1), a nonlinear least squares adjustment using the set of experimental data furnished: $\alpha_{c1} = 3.61$; $\beta_1 = 1.44$; $\omega_3 = 0.042$; $\omega_1 = 0.086$; $K = 0.17$; $\omega_2 = 1$; $f = 0.0$; $\omega_0 = 0.056$; $\omega_4 = 0.91$. It must be said that other constants may also produce good adjustments, and that the physical principles used to derive the equations imposed the definition of this set of constants. The superposition between model and experimental data is shown

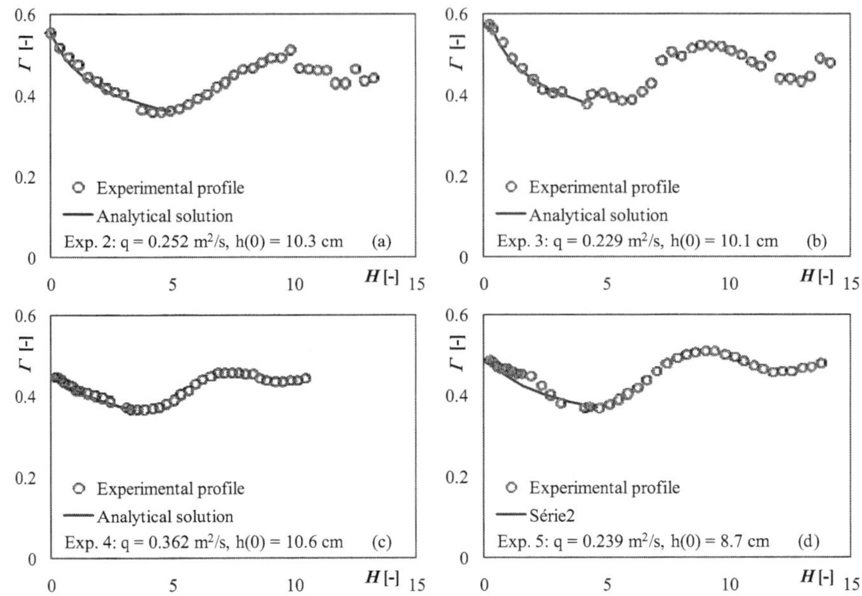

Figure 3: Experimental profiles and analytical solutions for runs 2 through 5.

in fig. 6a. As can be seen, the general trend of the experiments was captured by the model.

The convenience of a more restricted set of constants was one of the reasons for the search of an alternative quantification (Schulz and Simões [3]). Applying the least squares method to eqn (19) (Model 2a), the following constants were obtained: $\phi_0/\dot{c}=290.7$, $\phi_1/\dot{c}= -2290.1$. As can be seen in fig. 6b, also this equation follows the general trend of the experimental data.

Finally, applying the least squares method to eqn 20 (Model 2b), the following constants were obtained: $\theta_1=789.12$, $\theta_2=-976.5$, $\theta_3=-10610.5$, $\theta_4= 17265.15$, $\theta_5=0.1923$. Considering the number of constants, Model 2b (five constants) lies between Model 1 (nine constants) and Model 2a (two constants). Fig. 6c shows that it reproduces well the experimental data, and that it may be considered in further studies for the quantification of transitions lengths.

The correlation coefficients between measured and predicted values were: Model 1 = 0.94, Model 2a = 0.90, and Model 2b = 0.97.

4 Conclusions

Governing equations were derived for the calculation of transition lengths between full-water and full-mixed regions in stepped spillways. Two main procedures were followed: 1) using conservation principles, and 2) using an analogy with basic transport equations. The second procedure furnished an equation with two adjustment constants, which was modified, involving then five constants. The three predictions were compared with experimental data obtained

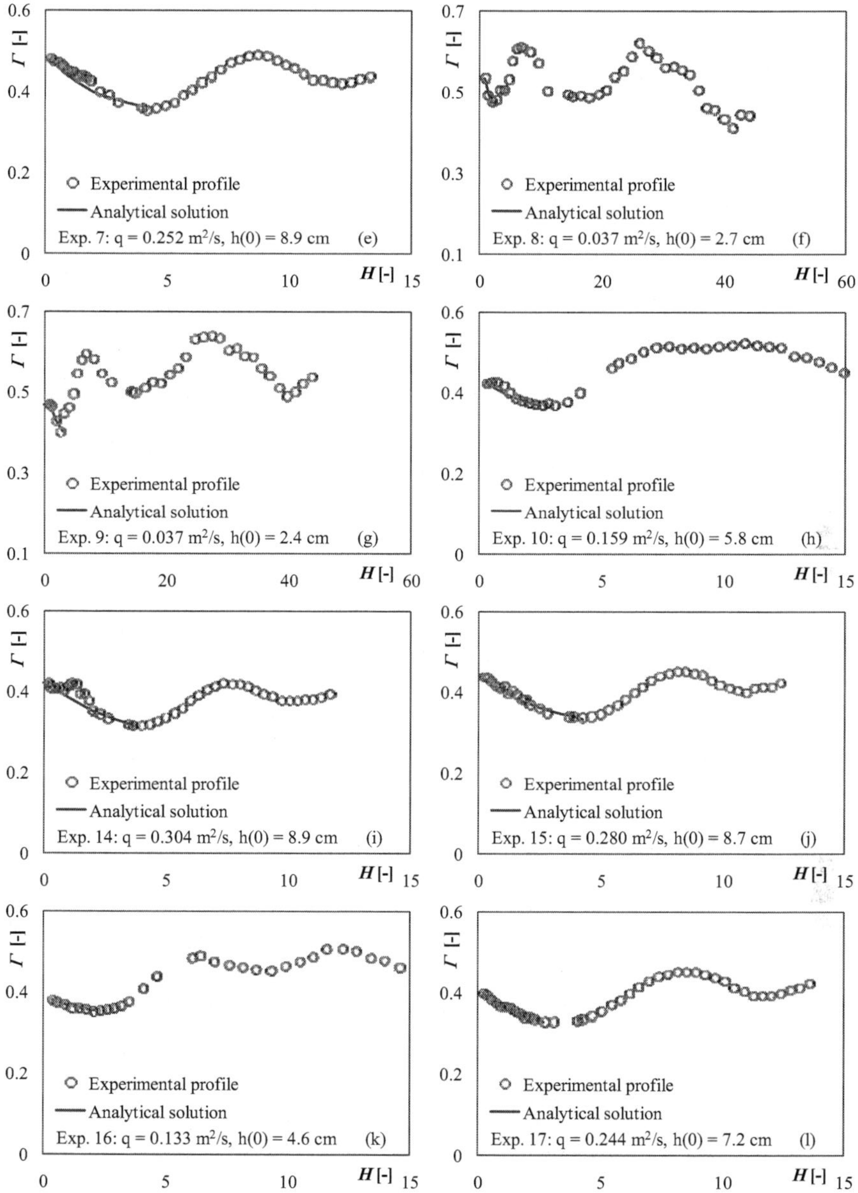

Figure 4: Experimental profiles and analytical solutions for runs 7 through 10, and 14 through 17.

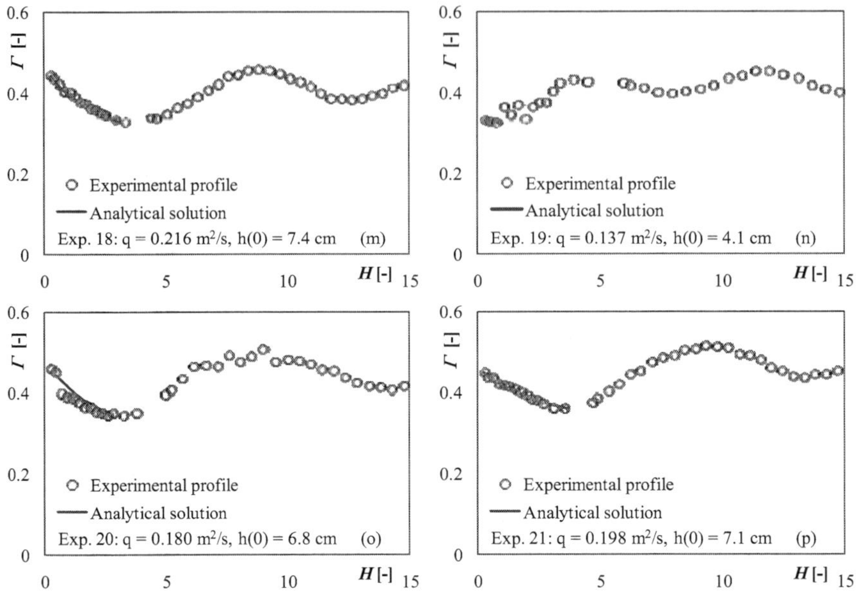

Figure 5: Experimental profiles and analytical solutions for runs 18 through 21.

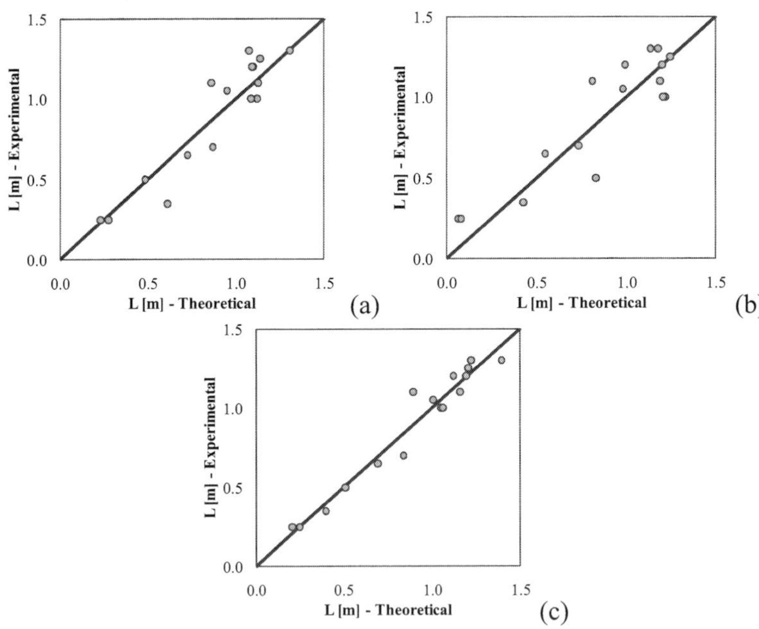

Figure 6: (a) Model 1 (eq. (8) and eq. (14)); (b) Model 2a (eq. (19)); (c) Model 2b (eq. (20)).

in a stepped chute inclined in 45°. The equations were adjusted using a set of data obtained for 16 different experimental conditions, in which the flow rate and the opening of the sluice gate were changed. Predicted and measured values showed adequate superposition for the three models. The best superposition was obtained for the second modified model.

Acknowledgements

The authors are indebted to FAPESP (Fundação de Amparo à Pesquisa do Estado de São Paulo), CAPES (Coordenação de Aperfeiçoamento de Pessoal de Nível Superior) and CNPq (Conselho Nacional de Pesquisa e Desenvolvimento Científico e Tecnológico), for financial support of this study.

References

[1] Boes, R.M. (2000). Two phase flows and energy conversion in large stepped channels. PhD Thesis. ETH Zurich, Zürich Switzerland [in German].

[2] Boes, R.M., Hager, W.H. (2003). Two-phase flow characteristics of stepped spillways. *J. Hydraulic Eng.* **129(9)**, pp. 661-670.

[3] Schulz, H.E.; Simões, A.L.A. (2011) "Alterative equations for transition lengths in aerated flows: analogies with basic equations of transport Phenomena", Lab. of Reology and Turbulence, School of Engg. at São Carlos, Univ. of São Paulo, LTR Report I/I/11, (vertedoresemdegraus. blogspot.com;stoa.usp.br/ltr/files/) [in Portuguese].

[4] Simões, A.L.A.; Schulz, H.E.; Porto, R.M. (2010) "Stepped and smooth spillways: resistance effects on stilling basin lengths" *Journal of Hydraulic Research* **48(3)**, pp.329-337.

[5] Cain, P.; Wood, I.R. (1981) Instrumentation for aerated flow on spillways. ASCE, *J. Hydraulic Eng.* **107**, No HY11.

Experimental measurements of macro-particle dispersion in grid turbulence and application to a stochastic numerical model for solid body turbulent diffusion

A. Joly[1], F. Moulin[2], S. Cazin[2], A. Astruc[2] & D. Violeau[1]
[1] *Saint-Venant Laboratory for Hydraulics, Université Paris-Est, France*
[2] *Institut de Mécanique des Fluides de Toulouse, Université de Toulouse INPT, France*

Abstract

The transport of solid bodies (e.g. algae, debris or sediment grains) along a coastline is a necessary consideration for the sustainable management of beaches and coasts, including any structure built along the shoreline. The use of a stochastic transport model allows to take into consideration a wide scale of physical processes, such as the current around a coastal structure, the turbulence generated by the flow and the effects of inertia and drag of each body. In order to validate the developed model, in view of industrial applications, a set of experiments was performed. The objective of the experiments was to validate to numerical model in isotropic turbulence. The experiment consisted of oscillating grid generated turbulence, inside which spherical particles were released. Measurements were done using PIV to quantify the turbulence and video particle tracking to measure displacement. The experimental result were then compared to various numerical simulations.
Keywords: particle diffusion, stochastic transport, isotropic turbulence, oscillating grid, particle image velocimetry, video particle tracking.

WIT Transactions on Engineering Sciences, Vol 70, © 2011 WIT Press
www.witpress.com, ISSN 1743-3533 (on-line)
doi:10.2495/MPF110091

1 Introduction

The transport of solid bodies (e.g. algae, debris or sediment grains) is a reoccurring problem for many industrial structures built along the shore line interested in the sustainable management of said coast. Currently most numerical models of solid body behaviour along a shoreline focus on the growth and evolution of an population (or ensemble) of bodies. These models require large time scales (days–months) and are relevant for large-scale areas (10–100 km), for example see [1] or [2] for algal blooms (a population of solid bodies). But these models prove ineffective when designing civil engineering works necessary for the sustainable management of coastal industrial structures. These problems require smaller scales of investigation (10 m–1 km and 1–24 h) as the motion of a body will be affected by tides, waves, and diffusion due to turbulence (see [3]). The model presented in [4] develops a stochastic approach for predicting the trajectory of individual particles. To validate this model, a set of experiments were done. In the first instance particles were released in static water of different densities; this was done to test the settling body velocities, as well as validate the particle tracking protocol. Finally particles were released into semi-isotropic turbulence, created using a pair of oscillating grids. Turbulent properties were quantified using PIV and LDV 2D measurements, and particle displacement statistics were measured using video particle tracking.

2 The particle transport model

The numerical particle transport model used in this paper is a stochastic model. It is divided into three stages. First the instantaneous local (at the emplacement of the particle) fluid velocity components (U_i) are calculated, using a stochastic model developed by [5]. It is then used to solve for the body velocity components (V_i), using the momentum equation. These are then integrated to give the position of the body (X_i), which is assumed to be isotropic (e.g. spheres):

$$dU_i(t) = -\frac{1}{T}U_i dt + C_i dt + B dW_i(t) \tag{1a}$$

$$dV_i(t) = \check{M} dU_i + \frac{1}{\tau_{part}}\left(U_i - V_i\right)dt + \check{g}_i dt \tag{1b}$$

$$dX_i(t) = V_i dt \tag{1c}$$

Where $dW_i(t)$ represents a Wiener process; the other coefficients are defined by the following equations:

$$T = \frac{1}{\frac{1}{2} + \frac{3}{4}C_0}\frac{k}{\varepsilon} \tag{2a}$$

$$\tau_{part} = \frac{2\overline{m}}{\rho_f S C_D(\mathrm{Re})\left|U_i - V_i\right|} \tag{2b}$$

$$\overline{m} = m + M \tag{2c}$$

$$\breve{M} = \frac{\rho_f \Omega + M}{\overline{m}} = \frac{\rho_f \Omega + M}{m + M} \tag{2d}$$

$$\breve{g}_i = \frac{m - \rho_f \Omega}{\overline{m}} g_i \tag{2e}$$

$$C_i = -\frac{1}{\rho_f} \frac{\partial \overline{P}}{\partial x_i} - \frac{1}{T} \overline{U_i} \tag{2f}$$

$$B = \sqrt{C_0 \varepsilon} \tag{2g}$$

For which ρ_f is the fluid density, k is the turbulent kinetic energy, ε is its dissipation rate, $\overline{U_i}$ are the mean fluid velocity components, $\overline{P}$ is the mean fluid pressure, m is the mass of the body, M is the magnitude of the added mass tensor, equal to $\frac{1}{2}\rho\Omega$ for a sphere, S is the cross-sectional area of the body, Ω is the area of the body, g_i are the components of gravity acceleration, C_D drag coefficient (for a sphere [6] provides an empirical solution as a function of the Reynolds number $\text{Re} = |U_i - V_i|\, D/\nu$), D is the sphere diameter, and finally $C_0 = 2.1$.

As can be seen from the coefficients in equation 1a this model requires characteristic flow variables ($\overline{P}$, $\overline{U_i}$, k and ε) to be input to solve for the instantaneous local fluid velocity. These values are usually solved using another numerical model. It should also be noted that the two characteristic time (T and τ_{part}) can vary greatly in time, space and from each other, making the resolution of equations 1 difficult (see [4]). Finally this model will be compared to two other numerical models: the model presented in [3], where the particles are transported using Brownian motion and none of the solid body dynamics are considered, and the model presented in [7] where the inertia of each particle is ignored. Focus will later be placed the impact of modeling solid body dynamics, and in particular inertia.

3 Settling velocities

This first experiment was undertaken as a mean of testing the particle tracking protocol, and to serve as a reference for particles released in semi-isotropic turbulence. To record the trajectories of the different settling particles, two cameras were placed perpendicular to each other, with a tank filled with water at their focus point. The tank was tall enough to ensure that particles reached their settling velocities before entering the cameras' field of vision. The effects of parallax were taken into account and calibrations were done before hand, allowing to compute the particle positions from pixels to millimeters.

$$\text{Camera 1}: \quad \frac{x(mm)}{x(pixels)} = \alpha_y y(mm) + \beta_x \quad \frac{z_1(mm)}{z_1(pixels)} = \alpha_y y(mm) + \beta_x \tag{3a}$$

Table 1: The diameters and densities of the Nylon Polyamide PA 6,6 particles released into the fluid.

Diameter D (mm)	Standard Deviation (%)	Mean body density ρ_s (kg/m^3)	Standard Deviation (%)
20	0.254	1129	0.0740
10	0.508	1128	0.301
5	1.02	1115	0.125
2	2.54	1062	2.42

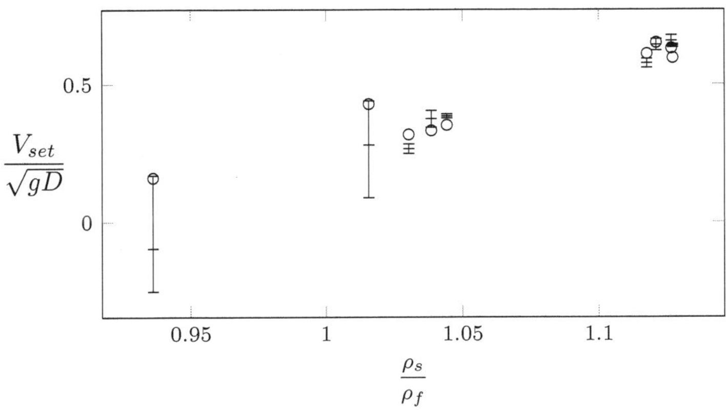

Figure 1: Dimensionless settling velocities against different particle fluid density ratios. "–" is the analytical solution (equation 4) plotted with its 95% confidence interval error bars and "○" are the experimental measurements.

$$\text{Camera 2}: \quad \frac{y(mm)}{y(pixels)} = \alpha_x x(mm) + \beta_y \quad \frac{z_2(mm)}{z_2(pixels)} = \alpha_x x(mm) + \beta_y \tag{3b}$$

The bodies used are spheres with diameters D and densities ρ_s, presented in table 1. They were released into two different fluid densities, $\rho_f = 1000$ and 1085 kg/m^3. Using equation (1b) it is possible to calculate the settling velocities of the bodies released in the fluid:

$$V_{set} = \frac{dX_z}{dt} = \frac{|m - \rho_f \Omega|}{(m - \rho_f \Omega)} \sqrt{\frac{2\,|m - \rho_f \Omega|}{\rho_f S C_D(\text{Re})} g_z} \tag{4}$$

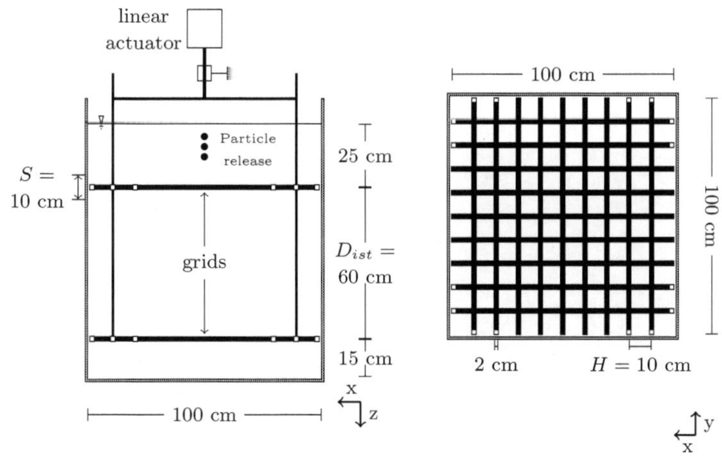

(a) Side view (b) Plan view

Figure 2: Dimensions for the turbulence generating system. The origin of the axis is located on the bottom right corner in the center of the lower grid.

Equation 4 is solved by iterations since Re depends on V_{set}. It should be noted that for this problem a rising velocity is assumed to be negative. The experimental values of V_{set} are then compared to the analytical results, for which the 95% confidence interval (calculated using the values in table 1) is presented in figure 1.

On figure 1 it is visible that the uncertainties are greater for low density ratios (ρ_s/ρ_f), but these correspond to the 2 mm diameter particles, for which the density had the greatest uncertainty (see table 1). In fact results show that for the smallest density ratio particles are settling, but the solution to the numerical model using the mean density gives a rising velocity. Nonetheless for large density ratios (with small particle densities uncertainties) the numerical model predicts the settling velocities accurately.

4 Turbulence quantification

The experimental device in figure 2 aimed to create near-isotropic turbulence by oscillating a pair of rectangular grids in a tank of still water. Several operating scenarios, in terms of oscillating of amplitudes (stroke S) and frequencies (f) were tested. The selected scenario, found to be most energetic, had a stroke of 10 cm and frequency of 1.67 Hz. The fluid velocities were then measured using Particle Image Velocimetry (PIV), in a 10×10 cm window of measurements located in between the two grids. Using these results the turbulent kinetic energy (k) and its rate of dissipation (ε) are estimated, see figure 3. Additional two-dimensional Laser Doppler Velocimetry (2D-LDV) was performed to validate the PIV measurements, and validate the horizontal isotropy.

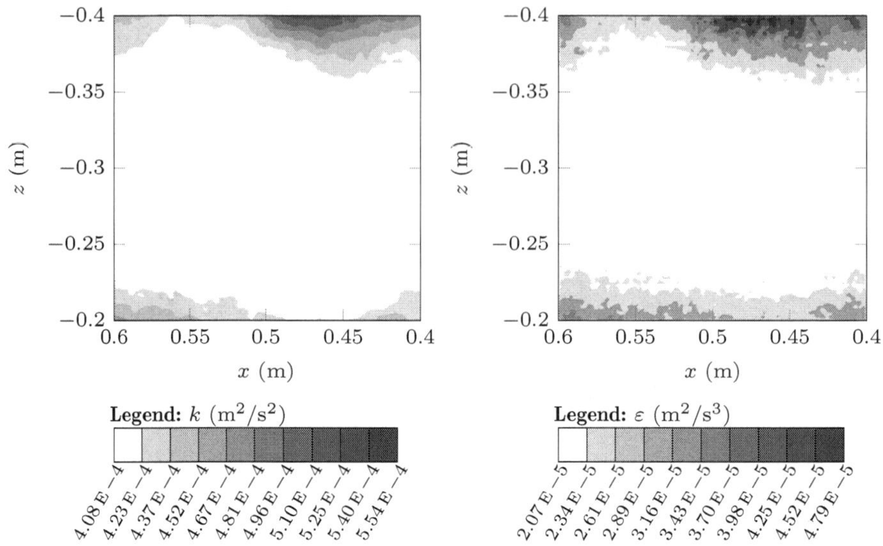

(a) Kinetic turbulent energy (k) (b) Turbulent energy dissipation rate (ε)

Figure 3: Turbulent quantification for double grid generated turbulence with an amplitude of 10 cm and frequency of 1.67 Hz.

Several papers give empirical formulae to model the turbulent characteristics of the flow resulting from a single oscillating grid (see [8–11]). Using these formulae, a crude hypothesis was stated; that the turbulent kinetic energy generated by each grid can be added. The semi-isotropic nature of the turbulence allows this assumption to give profile shapes for each turbulent characteristic in accordance to what can be expected. Constants were then adjusted to correspond to the constraints of the experiment. This gives the following equations:

$$\overline{U'_x} = \overline{U'_y} = \gamma_1 H^{\frac{1}{2}} S^{\frac{3}{2}} f \left[z^{-2} + (D_{ist} - z)^{-2} \right]^{\frac{1}{2}} \tag{5a}$$

$$\overline{U'_z} = \gamma_2 H^{\frac{1}{2}} S^{\frac{3}{2}} f \left[z^{-2} + (D_{ist} - z)^{-2} \right]^{\frac{1}{2}} \tag{5b}$$

$$k = \frac{1}{2} \left(2\gamma_1^2 + \gamma_2^2 \right) H S^3 f^2 \left[z^{-2} + (D_{ist} - z)^{-2} \right] \tag{5c}$$

$$\varepsilon = \frac{\gamma_3 \overline{U'_x}^3}{D_{ist}} \tag{5d}$$

Where H is the grid mesh size (10 cm), S is the stroke (10 cm), f is the frequency (1.67 Hz), D_{ist} is the distance between the two grids (60 cm), $\overline{U'_i}$ are the root mean square values of the velocity and γ_i are constants equal to 0.199, 0.252 and 5.05 respectively. 2D-LDV measurements have allowed to conclude

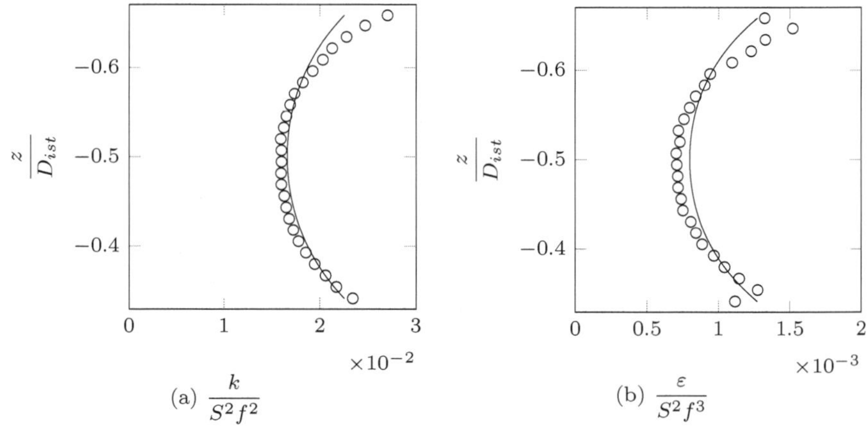

Figure 4: Vertical profiles for the turbulent kinetic energy and its dissipation. $S = 10$ cm and $f = 1.67$ Hz. "○" are the experimental values averaged along the horizontal axis and "——" is the empirical profile.

that $\overline{U'_x} = \overline{U'_y}$, but the constants γ_1 and γ_2 prove that the turbulence is not fully isotropic, but since their ratio is approximately 80%, the turbulence is semi-isotropic. In figure 4 the empirical and experimental profiles for the turbulent energy and its dissipation are plotted. These empirical values are in accordance with the measurements, and will therefore be used within the numerical model (equations 1 and 2) to calculate particle displacement statistics.

5 Particle displacement statistics

The particles described in table 1 were then released into the turbulence generated by the experimental setup presented in section 4, at location $(50, 50, 85)$ cm, for two different fluid densities, $\rho_f = 1000$ and 1084 kg/m^3. Three representative cases will be presented here: 20 mm particles in 1084 kg/m^3 fluid, 10 mm particles in 1000 kg/m^3 fluid and 5 mm particles in 1000 kg/m^3 fluid. Particle trajectories were recorded in a similar fashion to section 3. Statistics were then computed on the displacements over time of these particles (as the particle velocities were not measured). These statistics were also computed using the model presented in equations 1 and 2, for which the boundary conditions were provided by the empirical model in equations 5. Furthermore these numerical and experimental results where compared against two other numerical models; Brownian motion as described by [3] and an inertia free model presented by [7].

Figure 5 plots probability density functions for the horizontal and vertical displacements (X_{hor} and X_{vert}) of solid bodies in semi-isotropic turbulence. In this figure plots are associated with three characteristic numbers: N_r which is the number of experimental results, ρ_s/ρ_f which gives the density ratios and D/λ_l which shows the ratio of the particle diameter. The coefficient dt, which appears

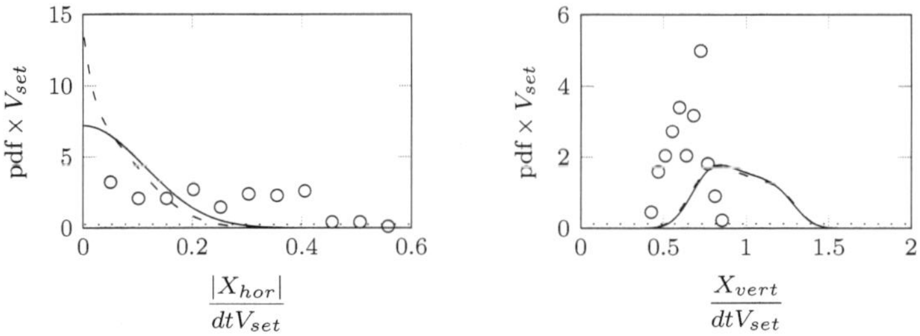

(a) $N_r = 105,\ \frac{\rho_s}{\rho_f} = 1.047,\ \frac{D}{\lambda_l} = 0.424$

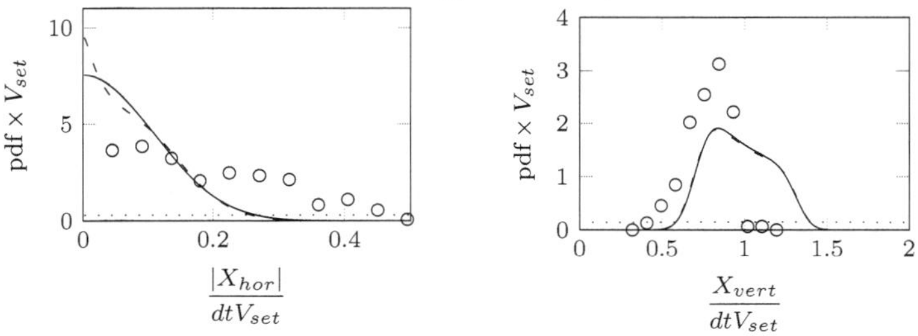

(b) $N_r = 178,\ \frac{\rho_s}{\rho_f} = 1.13,\ \frac{D}{\lambda_l} = 0.212$

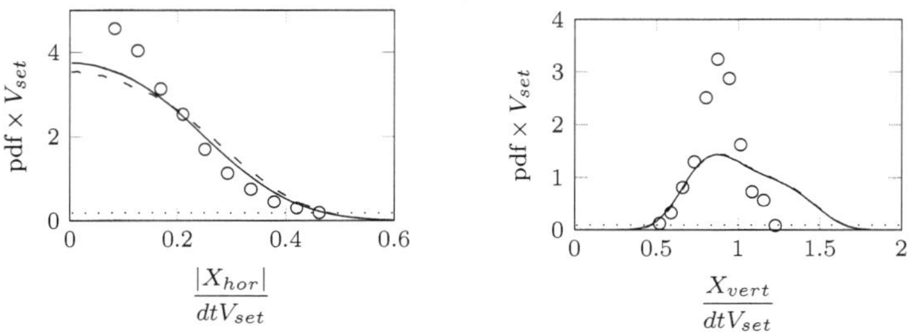

(c) $N_r = 354,\ \frac{\rho_s}{\rho_f} = 1.12,\ \frac{D}{\lambda_l} = 0.106$

Figure 5: Displacement statistics for bodies in grid turbulence of stroke 10 cm and a frequency of 1.67 Hz. "○" are the experimental values, "——" are the values for the numerical model (equations 1 and 2), "- -" are the values for the numerical model if particle inertia was ignored [7] and "· · · ·" are the values for the numerical model for particles following Brownian motion [3].

in this figure, represents the time step, whereas the large eddy characteristic length (λ_l) is calculated using this equation : $\lambda_l = C_\mu^{3/4} k^{3/2}/\varepsilon$, with $C_\mu = 0.09$. This figure shows that the horizontal particle displacement statistics are driven by the turbulent diffusion, whereas the vertical displacement statistics are driven by the buoyant forces. Furthermore it should be noted that the numerical model present in this figure have a few uncertainties associated with their resolutions, the main one being that particles are released above the window of measurement, and so the turbulent properties at the point of release haven't been validated. Nonetheless numerical results seem to be in accordance to the experimental results.

The first conclusion that can be derived from figure 5 is that a Brownian motion model (see [3]) is much too diffusive. [4] shows that the Brownian motion model would have been in accordance if the turbulence integral scale was much smaller. This implies that for turbulence of this magnitude, body properties (which are not considered in Brownian motion) have an impact on particle diffusion. The next conclusion is that for the largest particles (in figure 5(a)), the model from equations 1 and 2, seems to underestimate the horizontal diffusion, and overestimates the mean settling velocities. For smaller diameters (figures 5(b) and (c)) the model seems to give good results. When considering body properties ([4] and [7] models), it should be noted that particle inertia is most important in the cases where the particles are large and the density differences are small. For example, even though the model in equation 1 has difficulties sticking to the experimental results in figure 5(a), this model was developed under the assumption that the body was much smaller than the large scales of turbulence, adding inertial forces gives an improvement to just considering the drag forces (the model in [7]). These differences are less visible for the vertical displacement, where the displacement is driven by the buoyant forces. The assumption that the body was much smaller than the large scale of turbulence was done to keep the bodies passive; for the larger particles released in figure 5(a) it appears that its size cause it to affect the turbulence around it, requiring a more developed model such as Direct Numerical Simulations. When particles are smaller than the large scale of turbulence, but still of the same order (figure 5(b)), considering inertia gives an improvement, although considering the particle drag alone also gives a good estimate (the model in [7]).

6 Conclusion and further work

These experiments on the trajectories of solid bodies in semi-isotropic turbulence were undertaken to verify and observe the effect of solid body dynamics within an turbulent flow field. These experiment made it possible to conclude that the turbulent diffusion of these bodies is dependent on the forces acting upon it (see equation 1b). Therefore the first conclusion derived from these experiment is that a Brownian motion, which does not consider body properties, such as described in [3], might be limited when used in models developed to improve the sustainability of coastal structures. Furthermore considering the body's inertia gives an improvement to just considering drag forces (as was the case in [7]), but when these differences are most visible the particles' sizes will affect the turbulence around them, and results are only an approximation of the experimental results. Nonetheless for particle displacement statistics of smaller bodies, the model presented in this paper (equations 1 and 2) corresponds fairly well to the experimental results. It has therefore been shown that the model presented in [4] here can be representative of the displacement of solid bodies in a turbulent flow, provided there is at least an order of magnitude between the large turbulent eddies characteristic size and the particle diameter. Future validating work will focus on validating this solid body diffusion model for a more physical flow regime, in view of being later applied to engineering problems. The next step will then be to repeat this methodology for a deceptively simple problem of an open channel flow around a groyne. Finally more complex geometries of real bodies, such as algae, will be taken into consideration.

References

[1] Salomonsen, J., Flindt, M., Geertz-Hansen, O. & Johansen, C., Modelling advective transport of Ulva lactuca (L) in the sheltered bay, Møllekrogen, Roskilde Fjord, Denmark. *Hydrobiologia*, **397**, pp. 241–252, 1999.

[2] Donaghay, P. & Osborn, T., Toward a theory of biological-physical control of harmful algal bloom dynamics and impacts. *American Society of Limnology and Oceanography*, **42**, pp. 1283–1296, 1997.

[3] Issa, R., Rougé, D., Benoit, M., Violeau, D. & Joly, A., Modelling algae transport in coastal area with the shallow water equations. *Journal of Hydro-Environment Research*, **3**, pp. 215–223, 2009.

[4] Joly, A., Violeau, D. & Minier, J., Modelling of the turbulent diffusion of algae in a coastal environment through a stochastic method with an exact integrator. *1st European IAHR congress*, Edinburgh, 4–6 May, 2010, 2010.

[5] Pope, S., *Turbulent Flows*. Cambridge University Press: Cambridge, 2000.

[6] Almedeij, J., Drag coefficient of flow around a sphere: Matching asymptotically the wide trend. *Powder Technology*, **186**, pp. 218–223, 2008.

[7] Peirano, E., Chibbaro, S., Pozorski, J. & Minier, J.P., Mean-field/PDF numerical approach for polydispersed turbulent two-phase flows. *Progress in Energy and Combustion Science*, **32**, pp. 315–371, 2006.

[8] De Silva, I.P.D. & Fernando, H.J.S., Oscillating grids as a source of nearly isotropic turbulence. *Physics of Fluids*, **6(7)**, pp. 2455–2464, 1994.

[9] Cheng, N.S. & Law, A.W.K., Measurements of turbulence generated by oscillating grid. *Journal of Hydraulic Engineering*, **127(3)**, pp. 201–208, 2001.

[10] Al-Homoud, A. & Hondzo, M., Energy dissipation estimates in oscillating grid setup: LDV and PIV measurements. *Environmental Fluid Mechanics*, **7**, pp. 143–158, 2007.

[11] Holzner, M., Liberzon, A., Guala, M., Tsinober, A. & Kinzelbach, W., Generalized detection of a turbulent front generated by an oscillating grid. *Experiments in Fluids*, **41**, pp. 711–719, 2006.

Methodology to determine residence time distribution and small scale phenomena in settling tanks

T. Karches & K. Buzas
Department of Sanitary and Environmental Engineering, Budapest University of Technology and Economics, Hungary

Abstract

Hydraulic residence time is a crucial parameter of any wastewater treatment tank planning, operation and optimisation. Calculation of the residence time distribution at every point within the tank gives information about the dead-zones and short circuits and well-operating zones. Using the advective-diffusive transport equation supplemented with a scalar source term local mean age (LMA) can be detected if the flow field is given. Combining LMA with a tracer study the exact residence time can be determined. This method supports problem detection in a small wastewater treatment plant (2 dead zones and one short-circuiting path were detected) and also gave us better understanding of the operation of settling tanks.
Keywords: CFD, LMA, residence time, settling tank design.

1 Introduction

Computational fluid dynamics (CFD) in wastewater treatment processes has a significant role in recent years. Several studies demonstrate that CFD is an effective tool in design (Greenfield [8]), optimization (Bratchley et al. [3]) and operation (Brouckaert and Buckley [4]) Furthermore CFD studies revealed that with an extensive knowledge of fluid flow not only the hydraulic performance of e.g. an oxidation ditch could be improved (Yang et al. [17]) but we could have a closer look into the biological processes, too (Bartrand et al. [2]) For a few decades researchers used to focus only on the biology and chemistry in wastewater treatment, and hydraulics was mostly neglected.

WIT Transactions on Engineering Sciences, Vol 70, © 2011 WIT Press
www.witpress.com, ISSN 1743-3533 (on-line)
doi:10.2495/MPF1100101

According to ATV-A131 standard [1] the settling tanks have been handled by black box model, of which key parameter is hydraulic residence time. Most cases use a constant residence time – despite the fact that there is a distribution of residence time – calculated by $t = V/Q$, where V is the volume of the settling tank in m^3 and Q is the influent wastewater volume flux in m^3/s.

Residence time distribution (RTD) can be calculated using tracer study both experimentally and numerically (Gentric [7]). The field measurement based on the idea that a certain amount of LiCl is emitted to the wastewater tank and the concentration as the response function was measured as the function of time at the outlet. LiCl is a conservative material that never constituent of raw wastewater. The measured concentration data show how fast the particles go through the system, therefore this is useful in determining whether there are short –circuits or dead-zones and gives information how the wastewater tanks operate. If we know the flow, velocity and pressure fields, furthermore turbulence parameters, the tracer study can also be carried out numerically. If the tracer is injected at a short time interval, small concentrations will appear at the outflow where the rounding error's order of magnitude is close to the detected concentration. Eliminating this uncertainty the tracer is emitted to the inflow continuously and the response function is a cumulative function of concentration, which shows how fast the original fluid replaced by the tracer.

Numerical hydrodynamics should be completed with multiphase modelling in order to describe the behaviour of wastewater (and components) in the treatment process. Many efforts have been made, generally using particle tracking method (Saul and Stevin [14]), but when there is a strong coupling between fluid flow and the secondary phase, Eulerian-Eulerian multiphase model is more suitable (Ta et al. [16]).

Other approach to detect the spatial distribution of the fluid particles' age is the Local Mean Age (LMA) theory. LMA is used basically in ventilation techniques to describe mixing phenomenon of the air (Simons et al. [15]), but this method can be applied to water flow, too. Using the convection-diffusion scalar transport equation with zero-order kinetics the mean age of water can be calculated at each point in the tank and the computation time is less than in the RTD method. Furthermore, with LMA method reveal the problematic zones in the wastewater tanks and the geometry could be reconstruct or new boundary conditions could be set.

The theory section describes the mathematical background of this study. The results section presents a case study, where RTD and LMA methods were applied in a small wastewater plant in Sarisap, Hungary.

2 Theory of numerical methods

For the description of the fluid flow the conservation equations for mass and for the momentum has to be solved (Foias et al. [9]).

Conservation of mass

$$\frac{\partial \rho}{\partial t} + \nabla \cdot (\rho \vec{v}) = 0 \tag{1}$$

Conservation of momentum

$$\frac{\partial (\rho \vec{v})}{\partial t} + \nabla \cdot (\rho \vec{v}\vec{v}) = -\nabla p + \nabla \cdot (\bar{\bar{\tau}}) + \rho \vec{g} + \vec{F} \tag{2}$$

where p is the static pressure, $\rho\vec{g}$ is gravitational body force, $\vec{F}$ is external body force and $\bar{\bar{\tau}}$ is stress tensor, what can be written in form of:

$$\bar{\bar{\tau}} = \mu \left[\left(\nabla \vec{v} + \nabla \vec{v}^{T} \right) - \frac{2}{3} \nabla \cdot \vec{v} I \right], \tag{3}$$

where μ is molecular viscosity and I is the unit tensor.

Closure for the partial differential equation system is the k-ε turbulence model, where one equation is solved for the turbulence kinetic energy (k), see in eqn (4) and one for the turbulent dissipation (ε), see in eqn (5). This model assumes isotropic turbulence (Launder and Spalding [12]).

$$\frac{\partial (\rho k)}{\partial t} + \frac{\partial}{\partial x_j}(\rho k u_j) = \frac{\partial}{\partial x_j}\left[\left(\mu + \frac{\mu_t}{\sigma_k} \right) \frac{\partial k}{\partial x_j} \right] + G_k + G_b - \rho \varepsilon \tag{4}$$

$$\begin{aligned} \frac{\partial (\rho \varepsilon)}{\partial t} + \frac{\partial}{\partial x_j}(\rho \varepsilon u_j) &= \frac{\partial}{\partial x_j}\left[\left(\mu + \frac{\mu_t}{\sigma_\varepsilon} \right) \frac{\partial \varepsilon}{\partial x_j} \right] \\ &- \rho C_2 \frac{\varepsilon^2}{k + \sqrt{\upsilon \varepsilon}} + C_{1\varepsilon} \frac{\varepsilon}{k} C_{3\varepsilon} G_b \end{aligned} \tag{5}$$

where G_k is generation of the turbulent kinetic energy due to the mean velocity tensor and calculated as:

$$G_k = \mu_t S^2 \tag{6}$$

S is the deformation tensor

$$S \equiv \sqrt{2 S_{ij} S_{ij}} \tag{7}$$

μ_t is the eddy viscosity

$$\mu_t = \rho C_\mu \frac{k^2}{\varepsilon} \tag{8}$$

Here C_μ is a function of the mean strain and rotation rates, the angular velocity of the system rotation and the turbulence fields. G_b is generation of turbulent kinetic energy due to the buoyancy. Model constants are:

$C_{1\varepsilon}=1.44$, $C_{2\varepsilon}=1.9$, $\sigma_k=1$, $\sigma_\varepsilon=1.2$

For the LMA calculation an additional convection–diffusion scalar transport equation needs to be solved:

$$\frac{\partial \rho \varphi}{\partial t} + \nabla \cdot \left(\rho \vec{v} \varphi - \Gamma \nabla \varphi \right) = S \tag{9}$$

where ρ is the fluid density, φ is the local mean age of the fluid v is the fluid velocity, Γ is the diffusion coefficient, S is the source term of the scalar. S is taken as equal to 1 and for the diffusion term in turbulent flows:

$$\Gamma = \frac{\mu_{eff}}{\sigma_{SC}}, \tag{10}$$

where μ_{eff} is the turbulent viscosity and σ_{SC} is the Schmidt number. In the simulation it has a constant value of 0.7.

Assuming steady state condition eqn (9) can be simplified as:

$$\nabla \cdot \left(\rho \vec{v} \varphi - \frac{\mu_{eff}}{\sigma_{SC}} \nabla \varphi \right) = 1 \tag{11}$$

The boundary condition for eqn (11) is a zero value at the inlet and a zero gradient at the outflow boundary (Chanteloup and Mirade [6]).

For describing the different phases' behaviour Fluent Mixture model [10] was used. Continuity equation takes into account a mass averaged velocity, momentum equation sums up the individual momentum for all phases and a volume fraction as a scalar variable is introduced which can be calculated by dividing the secondary phase volume in respect with the total volume. Mixture model uses a so-called slip velocity, which is the velocity difference between the secondary and primary phase (Manninen et al. [13]).

The above mentioned equations – eqns. (1), (2), (4), (5), (11) and multiphase model – should be discretized and solved on a predefined mesh. For that second order discretisation scheme was used on all equations to reduce numerical diffusion and for the pressure-velocity coupling, the SIMPLE (Semi-Implicit Method for Pressure Linked Equations) scheme was required.

3 Results and discussion

Study site was a small municipal wastewater treatment plant about 2000 PE in Sarisap, Hungary. The wastewater tank consists of two parts, one is the aeration tank with cylindrical form and a cone shaped settling tank as the figure 1 shows. The applied mesh is built with tetrahedral elements with approx. 550 000 cell number and 107 000 node number. Size function was used where significant gradients in flow field variables may occur, at the inflow, outflow and at the inner pipes. Boundary condition was a constant volume flux (185 m^3/d due to the 85% cumulative frequency of the inflow water) at the inlet, an atmospherical pressure at the outlet. In the aeration tank air source term is built in the model to assume the effect of the aeration on the mixing. At the surface was the outflow of the air due to a negative source term to satisfy the continuity.

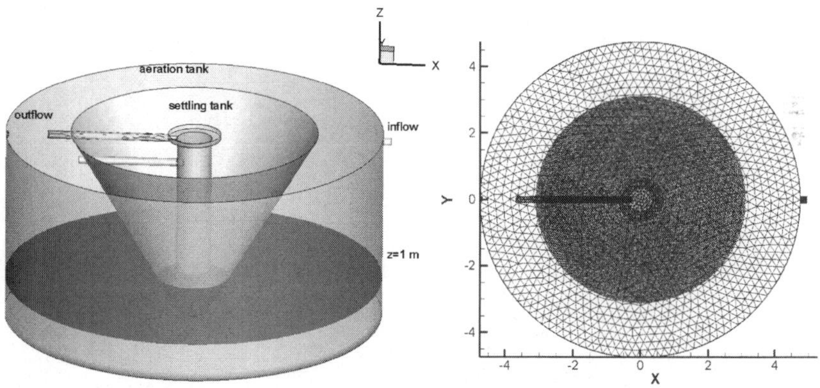

Figure 1: Aeration and settling tank geometry 3D Cartesian coordinates (left) and applied mesh 2D Cartesian coordinates (right).

As the solution converged, tracer was injected to the tank, replacing the original fluid and the outflow volume fraction of the tracer was detected. In the field study LiCl was discharged, which concentration was measured in the effluent water during five days. According to the literature (Buttz and Daigger [5]) five days the 95 percent of the LiCl added must be measured in the samples. The measured data can be transformed to a variable, which can be compared to the numerical results.

$$E(t) = \frac{C(t)}{\int_0^\infty C(t)dt} \tag{12}$$

$$F(t) = \int_0^t E(t)dt \tag{13}$$

E(t) is the normalised concentration response function and *F*(t) is the cumulative response function, which comes from the numerical model. Figure 2 shows the measured and calculated data (Karches and Melicz [11]).

From the *F*(t) the average residence time is:

$$\bar{t} = \int_0^{\infty} \left(1 - F(t)\right) dt \tag{14}$$

Solving eqn (14), the average residence time from the measured data is around 40 h, from the numerically calculated data it is 45,3 h. In a previous study (Karches [11]) it is revealed that there was a hole between the aeration and the settling tank, therefore a small amount of fluid could escape faster than it was planned and diminish the residence time. The V/Q ratio is 48,7 h, which is close to the model result.

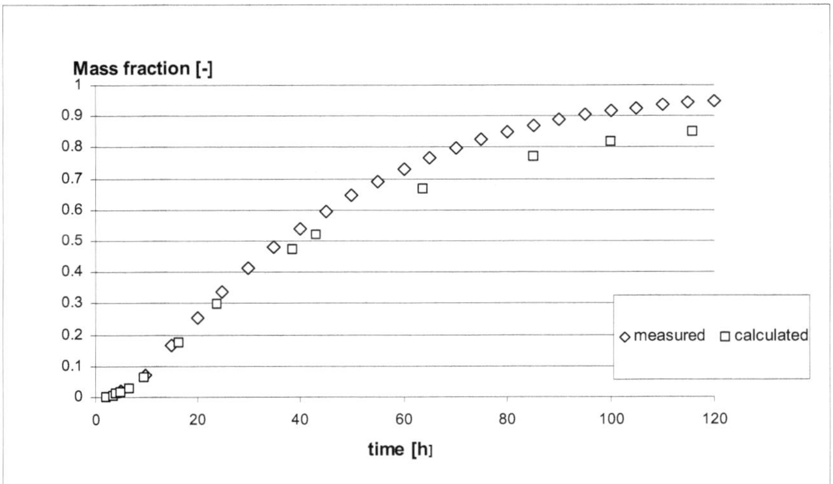

Figure 2: RTD study measured and calculated data which represents the mass fraction of the tracer at the outlet boundary.

When the flow field is given, LMA calculation (solving the eqn. 11) can be made separately. LMA predicts the average time which is needed to reach that point in the wastewater tank. Figures 3 and 4 show the distribution of the local fluid age at different sections. It can be seen that there are several zones where the fluid age is much higher than in the other zones. Because of the high fluid age the dissolved oxygen level decreases, anaerobic zones could appear in the aeration tank, which has a direct negative effect on the wastewater effluent quality. If a particle reaches the high fluid age zone, remains there for a long time, but not forever due to the turbulence. In the settling tank, because of the bad design, the fluid uses only a small volume, the residence time is not sufficient for the proper settling and the fact was confirmed by the LMA calculation. Another interesting fact is that aeration does not cease the dead-zones, probably because of the low intensity of the air inflow.

While RTD calculation predicts the occurrence of dead-zones, LMA shows the location, spatial dimensions of them. The problem of the dead-zone could be solved if the flow direction would not be the normal direction of the inlet boundary, but under an angle e.g. 60°. The settling problem could be solved if the outflow would be at the margin of the settling cone, because the particles path would be longer and there will be more time for the separation.

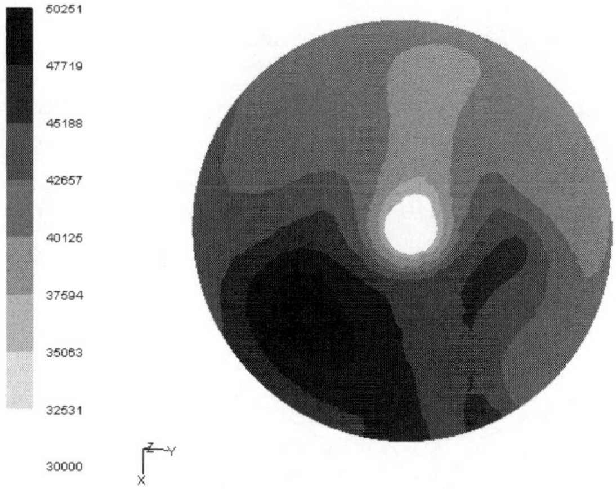

Figure 3: Local mean age of the fluid at z=1m plane. The dark colour represents the high water age. Scale is in second.

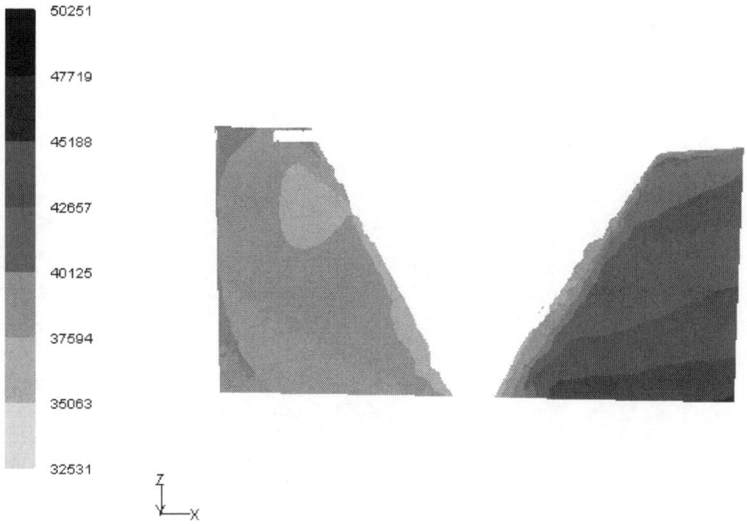

Figure 4: Local mean age of the fluid at XZ plane. The scale is in second.

www.witpress.com, ISSN 1743-3533 (on-line)

4 Conclusion

This study focused on the hydraulic aspects of a wastewater tanks. With the help of CFD the turbulence flow dynamics could have been predicted in a small wastewater treatment plant. Multiphase modelling was required on describing the aeration and carry out the RTD study with which could predict the average residence time in the wastewater tank and have impressions whether there are short-circuits and dead-zones. CFD flow field evaluation using velocity vectors, streamline imaging helped us to find location of the hydraulic phenomena but these effects on the residence time was not calculated. Local mean age of fluid particle is the link between the hydraulics and the biological processes. We find zones in the aeration tank, where the fluid's age was extreme high which could cause high ammonium concentration in the effluent. In the settling tank the residence time was not enough for the sedimentation process which results high suspended solid concentration in the effluent. With numerical simulation not only the problem detection is possible, but extensive knowledge of the fluid flow and residence time could help us to understand better the biological processes, too.

Acknowledgements

This work is connected to the scientific program of the "Development of quality-oriented and harmonized R+D+I strategy and functional model at BME" project. This project is supported by the New Hungary Development Plan (Project ID: TÁMOP-4.2.1/B-09/1/KMR-2010-0002

References

[1] ATV-DVWK Arbeitsblatt-A-131. GFA, Hennef, 2000.

[2] Bartrand, T.A., Farouk, B., Greene D.J., Haas C.N., Liberti L., Notarnicola M. and Santoro D., Use of CFD for wastewater disinfection process analysis: E.coli inactivation with peroxyacetic acid (PAA). *International Journal of Chemical Reactor Engineering* Vol. 3:A46, 2005.

[3] Bratchley E., Do-Quang Z., Janex M.L., Laine J.M. and Savoye P., Impact of water quality and reactor hydrodynamics on wastewater disinfection by UV, use of CFD modeling for performance optimization. *Water Science and Technology* Vol. 38 No. 6 pp. 71-78, 1998.

[4] Brouckaert C.J. and Buckley C.A., The use of Computational Fluid Dynamics for Improving the design and operation of water and wastewater treatment plants. *Water Science and Technology* Vol. 40 No. 4-5 pp. 81-90, 1999.

[5] Buttz J.A. and Daigger G.T. Upgrading wastewater treatment plants. Technomic Pub Co Vol 2. pp. 13-53, 1998.

[6] Chanteloup V. and Mirade P-S. Computational fluid dynamics (CFD) modelling of local mean age of air distribution in forced-ventilation food plants. *Journal of Food Engineering* Vol. 90 pp. 90-103, 2009.

[7] Gentric C., Leclerc J.P., Moullec Y.L. and Potier O. Flow field and residence time distribution simulation of a cross-flow gas-liquid wastewater treatment reactor using CFD. *Chemical Engineering Science* Vol. 63 pp. 2436-2449, 2008.

[8] Greenfield P.F., Howes T., Johns M.R., Keller J. and Wood M.G. Computational fluid dynamic modelling of wastewater ponds to improve design. *Water Science and Technology* Vol. 31 No. 12 pp. 111–118, 1995.

[9] Foias C., Manley O. Rosa R. and Temam R. (2001) *Navier-Stokes equations and turbulence* Cambridge Books Online pp. 1-3.

[10] Fluent 6.3 Users Guide. Ansys Inc., USA

[11] Karches T. and Melicz Z., Problem detection in a small wastewater treatment tank with the tool of computational fluid dynamics. 9th international Congress on Advances in Civil Engineering.KTU, Trabzon, Turkey, 27-30 Sept. 2010.

[12] Launder B.E. and Spalding D.B. The numerical computation of turbulent flow. *Computer methods in applied mechanics and Engineering*, Vol. 3 Issue 2. pp. 269-289, 1974.

[13] Manninen M., Taivassalo V. and Kallio S. *On the mixture model for multiphase flow*. VTT Publications 288, Technical Research Centre of Finland, 1996.

[14] Saul J. A. and Stovin R.V. A computational fluid dynamics (CFD) particle tracking approach to efficiency prediction. *Water Science and Technology* Vol.37. Issue 1 pp. 285-293, 1998.

[15] Simons MW., Waters J. R. and White P. R. S., Local mean age of air: predictive techniques compared. *Building Services Engineering Research and Technology* Vol. 20. No. 4211-218, 1999.

[16] Ta CT, Beckley J., and Eades A. A multiphase CFD model of DAF process. *Water Science and Technology* Vol. 43 No 8 pp. 153-157, 2001.

[17] Yang Y., Wu Y., Yang X., Zhang K. and Yang J. Flow field prediction in full-scale Carrousel oxidation ditch by using computational fluid dynamics *Water Science and Technology* Vol. 62 No 2 pp. 256-265, 2010.

Section 3
Bubble and drop dynamics

Heat and mass transfer in evaporating turbulent drop-laden flow

R. Groll
Center of Applied Space Technology and Microgravity, University of Bremen, Germany

Abstract

This work deals with the computational modelling of the mass transition of evaporating liquid drop-laden gas flows. In the present study the evaporation model due to Abramzon and Sirignano (1989) has been extended by introducing an additional transport equation for a newly defined quantity $\overline{a}$, defined as the phase-interface surface fraction. This allows the change in the drop diameter to be quantified in terms of a probability density function. The source term in the equation describing the dynamics of the volumetric fraction of the dispersed phase α^D is related to the evaporation time scale τ_Γ.
Keywords: Euler/Euler approach, Euler/Lagrange approach, gas/liquid flow, evaporation model, heat and mass transfer, volume and surface fraction.

1 Introduction

The dimensionless numbers characterizing the heat and mass exchange process are Nusselt number Nu and Sherwood number Sh. Several models (e.g. Renksizbulut and Yuen [1] and Park et al. [2]) are based on the correlation model of Ranz and Marshall [3] and have been developed to model the parameters. The evaporation rate model of Abramzon and Sirignano [4] considers additionally the latent heat flux of the evaporated liquid leaving the droplet. The correct capturing of the gas phase humidity requires the liquid vapor mass ratio Y, influenced by convective, conductive, turbulent and thermal diffusive effects, to be computed from an appropriate transport equation in addition to the equation governing the temperature field T.

In an evaporating process the mass transfer rate on the drop surface depends on the drop size. The polydispersed spray consists of drops with different diameters.

WIT Transactions on Engineering Sciences, Vol 70, © 2011 WIT Press
www.witpress.com, ISSN 1743-3533 (on-line)
doi:10.2495/MPF110111

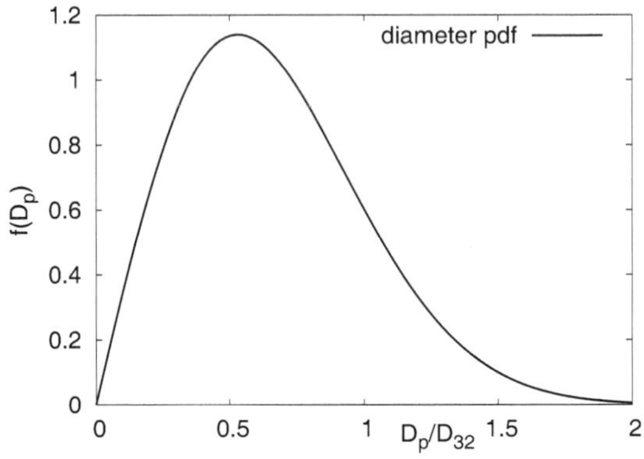

Figure 1: PDF of a normalized particle drop diameter in a polydispersed spray.

To determine the mass transfer, i.e. the evaporation rate for such a case, the drop diameter distribution of the spray is necessary. Using the following approach based on a particle diameter probability density function (PDF) (see Fig. 1 for its graphical representation):

$$f\left(\frac{D_p}{\mathrm{E}(D_p)}\right) = \frac{\pi}{2}\frac{D_p}{\mathrm{E}(D_p)} e^{-\frac{\pi}{4}\frac{D_p^2}{\mathrm{E}(D_p)^2}} \quad \text{with } f(D_p) = \frac{9}{8}\pi\frac{D_p}{D_{32}^2} e^{-\frac{9}{16}\pi\frac{D_p^2}{D_{32}^2}} \tag{1}$$

depending on the Sauter mean diameter,

$$D_{32} = \frac{\mathrm{E}\left(D_p^3\right)}{\mathrm{E}\left(D_p^2\right)} = \frac{6\bar{\alpha}^D}{\bar{a}} \quad \Leftrightarrow \quad \frac{\bar{a}}{\pi\mathrm{E}\left(D_p^2\right)} = \frac{\bar{\alpha}}{\frac{\pi}{6}\mathrm{E}\left(D_p^3\right)} \tag{2}$$

with the volumetric fraction α and the newly defined variable $\bar{a}$ being the phase-interface surface fraction,

$$\alpha = \frac{V_{\text{liquid}}}{(V_{\text{liquid}} + V_{\text{gas}})} \qquad a = \frac{A_{\text{phase interface}}}{(V_{\text{liquid}} + V_{\text{gas}})} \tag{3}$$

the mass transfer rate are calculated as shown in the further work. The expectation values of the squared and cubic diameters in terms of the Sauter mean diameter only, result from this modelled PDF (eq. 1), e.g. Groll et al. [5]:

$$\mathrm{E}(D_p) = \frac{2}{3}D_{32} \qquad \mathrm{E}\left(D_p^2\right) = \frac{16}{9\pi}D_{32}^2 \qquad \mathrm{E}\left(D_p^3\right) = \frac{16}{9\pi}D_{32}^3 \tag{4}$$

A time dependent relation has to be defined to calculate the change of the drop diameter when simulating motion of a polydispersed phase. The change of the expectation value of the diameter squared is presumed to be constant in accordance to the d^2-law (Kastner [6]):

$$\frac{\mathrm{d}}{\mathrm{d}t}\mathrm{E}\left(D_p^2\right) = -\Gamma. \tag{5}$$

Integration of this equation results in the time-dependent solution for the drop diameter:

$$D_p(t) = \sqrt{D_p^2(0) - \Gamma t}. \tag{6}$$

Substitution of the expectation value with the given density function definition

$$-\Gamma = \frac{\mathrm{d}}{\mathrm{d}t}\mathrm{E}\left(D_p^2\right) = \frac{16}{9\pi}\frac{\mathrm{d}}{\mathrm{d}t}D_{32}^2 = \frac{32}{9\pi}\dot{D}_{32}\frac{\mathrm{d}}{\mathrm{d}t}D_{32} \Rightarrow \dot{D}_{32} = -\frac{9\pi}{32}\frac{\Gamma}{D_{32}} \tag{7}$$

reveals that the deviation of the Sauter mean diameter depends on the evaporation constant of the d^2-law. Based on this formulation, the following equation has been derived:

$$\frac{\mathrm{d}}{\mathrm{d}t}\mathrm{E}\left(D_p^3\right) = \frac{16}{9\pi}\frac{\mathrm{d}}{\mathrm{d}t}D_{32}^3 = \frac{16}{9\pi}\left(\dot{D}_{32}D_{32}^2 + D_{32}\frac{\mathrm{d}}{\mathrm{d}t}D_{32}^2\right) = -\frac{3}{2}\Gamma D_{32}. \tag{8}$$

With this expression, the time-dependent change of an expected drop volume is defined. The outcome of the last equation is used for the determination of the mass transfer of a drop with the expected mass $\bar{m}_p$:

$$\frac{\mathrm{d}\bar{m}_p}{\mathrm{d}t} = \frac{\pi}{6}\rho^D\frac{\mathrm{d}}{\mathrm{d}t}\mathrm{E}\left(D_p^3\right) = -\frac{\pi}{4}\rho^D D_{32}\Gamma. \tag{9}$$

Based on the evaporation model due to Abramzon and Sirignano [4], the mass transfer rate at the surface of a drops cloud representing a function of the modified Sherwood number Sh^* and the mass transfer coefficient B_M is to be determined by using the following equations:

$$\dot{m} = \pi D_{21}\rho^C D_{\alpha\beta}\mathrm{Sh}^*\ln\left(1 + B_M\right) \tag{10}$$

The definitions of the Sherwood number and its modification and the mass transfer coefficient are given in Section 3. By equalizing the PDF-dependent mass transfer rate and the modelled mass transfer rate, the evaporation factor $\mathrm{Sh}^*\ln\left(1 + B_M\right)$ in the model of Abramzon and Sirignano [4] is defined as:

$$\frac{\pi}{4}\rho^D D_{32}\Gamma = -\dot{m}_p = \dot{m} = \pi D_{21}\rho^C D_{\alpha\beta}\mathrm{Sh}^*\ln\left(1 + B_M\right)$$

$$\Rightarrow \Gamma = 4\frac{D_{21}}{D_{32}}\frac{\rho^C}{\rho^D}D_{\alpha\beta}\mathrm{Sh}^*\ln\left(1 + B_M\right). \tag{11}$$

In accordance to the d^2-law, the evaporation constant should be independent of the particle diameters. This fact brings an additional constraint to the expectation values of the probability function: D_{21}/D_{32} =const. The given PDF fulfills this condition as follows:

$$\frac{D_{21}}{D_{32}} = \frac{\mathrm{E}\left(D^2\right)\mathrm{E}\left(D^2\right)}{\mathrm{E}\left(D\right)\mathrm{E}\left(D^3\right)} = \frac{8}{3\pi}. \tag{12}$$

With the proposal of Abramzon and Sirignano, the time-dependent modelling of the drop diameter probability function is closed. Combining this information with the relation of the change of the specific volume α and the specific surface a, described in section 2, the transport equations are developed.

2 Evaporation progress

The model developed serves for calculation of the evaporation rate of spherical water drops. Water is a liquid dispersed phase, which satisfies the d^2-law (e.g. Kastner [6]). Keeping in mind the definition of the life time of a drop T and its diameter loss rate, it is known, that each drop with a diameter

$$D_p < (\Gamma T)^{\frac{1}{2}} \tag{13}$$

is evaporated completely. To determine the number of the evaporated drops in a cloud, the probability density function has to be integrated in the following way:

$$\int_0^{(\Gamma T)^{\frac{1}{2}}} f\left(D_p\right) \mathrm{d}D_p. \tag{14}$$

Consequently, the time change of the particle number is obtained as follows:

$$\frac{\mathrm{d}\bar{n}}{\mathrm{d}t} = -\lim_{T\to 0}\left[\frac{\bar{n}}{T}\int_0^{(\Gamma T)^{\frac{1}{2}}} f\left(D_p\right)\mathrm{d}D_p\right] = \lim_{T\to 0}\left[\frac{\bar{n}}{T}\left(e^{-\frac{9\pi}{16}\frac{\Gamma T}{D_{32}^2}} - 1\right)\right] \tag{15}$$

$$= \lim_{T\to 0}\left[\frac{\bar{n}}{T}\left(e^{mT} - 1\right)\right] \quad ; \quad m = -\frac{9\pi}{16}\frac{\Gamma}{D_{32}^2} \tag{16}$$

$$= \lim_{T\to 0}\left[\frac{\bar{n}}{T}\left((1 + mT)^{\frac{1}{mT}\cdot mT} - 1\right)\right]$$

$$= \lim_{T\to 0}\left[\frac{\bar{n}}{T}mT\right] = \bar{n}m = -\frac{9\pi}{16}\frac{\Gamma}{D_{32}^2}\bar{n}.$$

The time change of the volumetric fraction $\bar{\alpha}^D$ is calculated by using the results represented by Eqs. (8) and (15):

$$\frac{\mathrm{d}\bar{\alpha}^D}{\mathrm{d}t} = \frac{\pi}{6}\mathrm{E}\left(D_p^3\right)\frac{\mathrm{d}}{\mathrm{d}t}\bar{n} + \bar{n}\frac{\pi}{6}\frac{\mathrm{d}}{\mathrm{d}t}\mathrm{E}\left(D_p^3\right)$$

$$= -\frac{\pi}{6}\frac{16}{9\pi}D_{32}^3 \cdot \frac{9\pi}{16}\frac{\Gamma}{D_{32}^2}\bar{n} - \bar{n}\frac{\pi}{6}\cdot\frac{3}{2}\Gamma D_{32} = -\frac{45}{32}\pi\frac{\Gamma\bar{\alpha}^D}{D_{32}^2} \tag{17}$$

The time change of the Sauter mean diameter (Eq. 2) consists of the deviations of the volumetric fraction and the surface fraction:

$$\dot{D}_{32} = \frac{6}{\bar{a}^2}\left(\bar{a}\frac{\mathrm{d}\bar{\alpha}^D}{\mathrm{d}t} - \bar{\alpha}^D\frac{\mathrm{d}\bar{a}}{\mathrm{d}t}\right) = D_{32}\left(\frac{1}{\bar{\alpha}^D}\frac{\mathrm{d}\bar{\alpha}^D}{\mathrm{d}t} - \frac{1}{\bar{a}}\frac{\mathrm{d}\bar{a}}{\mathrm{d}t}\right) \tag{18}$$

Utilizing the results following from the Eqs. (7) and (17), the change of the surface fraction can also be formulated in terms of the evaporation constant and the Sauter mean diameter:

$$\frac{1}{\bar{a}}\frac{\mathrm{d}\bar{a}}{\mathrm{d}t} = \frac{1}{\bar{\alpha}^D}\frac{\mathrm{d}\bar{\alpha}^D}{\mathrm{d}t} - \frac{1}{D_{32}}\frac{\mathrm{d}D_{32}}{\mathrm{d}t} = -\frac{9\pi}{8}\frac{\Gamma}{D_{32}^2} \quad \Rightarrow \quad \frac{\mathrm{d}\bar{a}}{\mathrm{d}t} = -\frac{9\pi}{8}\frac{\Gamma}{D_{32}^2}\bar{a}. \tag{19}$$

The source terms of the α-equation (Eq. 17) and the a-equation (Eq. 19) stay in following relationship:

$$\frac{1}{\bar{\alpha}^D}\frac{\mathrm{d}\bar{\alpha}^D}{\mathrm{d}t} = \frac{5}{4}\cdot\frac{1}{\bar{a}}\frac{\mathrm{d}\bar{a}}{\mathrm{d}t}. \tag{20}$$

Obviously, the source terms in both transport equations can be formulated in terms of the same parameter: τ_Γ (Eq. 23). With the definition of the particle diameter probability density function as the starting point, the evaporation process is finally modelled by the following two transport equations:

$$\partial_t\left(\rho^D\bar{\alpha}^D\right) + \partial_j\left(\rho^D\bar{\alpha}^D < u_j^D >^D\right) = -\frac{5}{4}\rho^D\tau_\Gamma^{-1}\bar{\alpha}^D \tag{21}$$

$$\partial_t\left(\rho^D\bar{a}\right) + \partial_j\left(\rho^D\bar{a} < u_j^D >^D\right) = -\rho^D\tau_\Gamma^{-1}\bar{a} \tag{22}$$

The convective transport in both equations is defined by the volume-fraction-weighted averaged particle velocity $< u_i^D >^D$ (e.g. Politis [7]). The first equation (Eq. 21) originates from the well-known mass balance of the dispersed phase with a mass transfer defining source term. The second equation (Eq. 22) governs the surface fraction of the dispersed phase, being the synonym for the cloud surface per volume. The evaporation time scale τ_Γ, the source terms of this evaporation-describing two-equation model depend upon, reads:

$$\tau_\Gamma = -\bar{a}\left(\frac{\mathrm{d}\bar{a}}{\mathrm{d}t}\right)^{-1} = \frac{8}{9\pi}\frac{D_{32}^2}{\Gamma} \tag{23}$$

Accordingly, the transport of both quantities α^D and a depends on the evaporation constant Γ, which is given by the d^2-law. Introducing the definition of the evaporation constant, the final expression serving for the determination of the evaporation time scale is given by:

$$\tau_\Gamma = \frac{D_{32}^2}{12}\frac{\rho^D}{\rho^C}\left[D_{\alpha\beta}\mathrm{Sh}^*\ln\left(1 + B_M\right)\right]^{-1}. \tag{24}$$

By solving the equations of the α-a-model the evaporation rate of a spray stream can be quantified. In such a way, the mass balance of a two-phase flow is completely satisfied.

3 Radiation, heat and mass transfer

In accordance to the transport of the absolute humidity Y the density of the continuous phase consisting of the vapor and the carrier gas phase (p.e. air) changes depending on that value:

$$\rho^C = \frac{\rho^{\text{vapor}}}{Y} = \frac{\rho^{\text{air}}}{1-Y}. \tag{25}$$

The radiative heat flux per volume $\dot{q}^D_{\text{rad}}$, absorbed by a particle, is defined in terms of its surface area per volume a, the absorption coefficient of the dispersed phase ϵ^D_{abs} and the difference of the fourth powers of the absolute temperatures.

$$\dot{q}^D_{\text{rad}} = a\epsilon^D_{\text{abs}}\sigma\left[(T^C)^4 - (T^D)^4\right], \quad \epsilon^D_{\text{abs}} = \epsilon^{\text{water}}_{\text{abs}} = 0,92 \tag{26}$$

with the Stephan-Boltzmann constant $\sigma = 5,67051 \cdot 10^{-8} W/(m^2 K)$. This term represents an additional source term in the equation of thermal transport. Due to the equivalence of emission and absorption coefficients, the radiative flux can take negative values, if the dispersed phase is on the higher temperature level compared to the continuous one.

As defined in the evaporation model of Abramzon and Sirignano [4], the heat transfer at the surface of an evaporating drop depends on the latent heat $L(T^D)$, the temperature difference and the mass transfer $\dot{m}$

$$Q_L = \dot{m}\left[\frac{c^D_p\left(T^C - T^D\right)}{B_T} - L\left(T^D\right)\right]. \tag{27}$$

The heat transfer coefficient B_T is to be computed from the "Abramzon and Sirignano"-model algorithm. By utilizing Eq. (21), this heat source term can be expressed in terms of τ_Γ in a volume specific way.

Using the following definition of the evaporation enthalpy

$$\Delta h_v = \frac{R_0}{M^{\text{vapor}}}\left(\frac{1}{T^D} - \frac{1}{T_{\text{ref}}}\right)^{-1} \ln\frac{p_{\text{sat}}\left(T_{\text{ref}}\right)}{p_{\text{sat}}\left(T^D\right)} \tag{28}$$

the heat, which is used for the evaporation process, can be subtracted from the heat of the continuous phase. The temperature T_{ref} denoting the gas temperature near the drop surface is formulated by the $1/3$-law (Hubbard et al. [8]):

$$T_{\text{ref}} = T^D + A_r\left(T^C - T^D\right) \quad \text{with} \quad A_r = \frac{1}{3}. \tag{29}$$

The final two equations completing the Euler/Euler computational scheme are those governing the thermal transport between the phases being represented by the particular set of temperature equations. The liquid mass leaving the droplet in the evaporation process influences strongly both the heat and mass transfer between the phases. This process is accounted for by the Spalding heat and mass transfer coefficients B_T (Eq. 27) and B_M (Eq. 23). This algorithm for their determination is pointed out by the evaporation model proposed by Abramzon and Sirignano [4].

The relative humidity $0 \leq \phi \leq 1$ is defined by the ratio of the vapor mass ratio to its maximum: $\phi = Y/Y_{\text{sat}}$. The mass transfer per surface of a cloud of spherical particles

$$\frac{\dot{m}}{A_O} = \rho^C D_{\alpha\beta} B_M \frac{\text{Sh}}{\text{E}(D_p)} \quad \text{with } A_O = \pi\,\text{E}\left(D_p^2\right)$$
$$\Rightarrow \qquad \dot{m} = \rho^C \pi D_{21} D_{\alpha\beta} \text{Sh} B_M \tag{30}$$

depends on the Sherwood Number Sh and the mass transfer coefficient B_M:

$$B_M = \frac{Y_{\text{sat}} - Y_\infty}{1 - Y_{\text{sat}}} = \frac{Y_{\text{sat}}}{1 - Y_{\text{sat}}} - \frac{Y_\infty}{1 - Y_{\text{sat}}} \tag{31}$$

The heat flux

$$Q_L = \text{Nu}\pi D_{21} \lambda^C \left(T^C - T^D\right) - \dot{m} L\left(T^D\right) \tag{32}$$

depending on the Nusselt number Nu can also be defined in terms of the heat transfer coefficient B_T and the liquid temperature dependent latent heat $L(T^D)$:

$$B_T = \frac{c_p^D\left(T^C - T^D\right)}{Q_L/\dot{m} + L\left(T^D\right)} \Rightarrow Q_L = \dot{m}\left(\frac{c_p^D\left(T^C - T^D\right)}{B_T} - L\left(T^D\right)\right). \tag{33}$$

Utilizing the heat flux definition given by Eq. (32), the mass transfer rate formulae can be finally written as a function of the Nusselt number

$$\dot{m} = \pi D_{21} \frac{\lambda^C}{c_p^D} \text{Nu} B_T. \tag{34}$$

In such a way, the mass transfer from the liquid to the gas phase can be determined modelling the *Nusselt number* Nu and the *Spalding Heat transfer coefficient* B_T with the evaporation model of Abramzon and Sirignano [4].

4 Computational results and discussion

The flow configuration simulated of a turbulent, gas/liquid channel flow, being discretized for the quarter channel, is considered. Similar as in the particle-laden cases, a three-dimensional solution domain was adopted. The results obtained by the Euler/Euler computational model were compared with the results obtained by an Euler/Lagrange method (e.g. Sommerfeld et al. [9]) solving the transport equations of the carrier phase (Euler framework) and individual tracking of the droplet parcels (14.800 trajectories were introduced into the flow field displayed in Figs. 2–4.

The boundary conditions of the flow configurations considered (channel dimensions, bulk Reynolds number, turbulence intensities,...) correspond to the geometry used for the experimental investigation of the particle-laden flow by Kulick et al. [10]. Instead of solid particles, the flow was laden with droplets characterized by their uniform distribution along the entire inlet cross-section. This flow configuration (no experimental data for the evaporation process exist) with the mass loading of the liquid phase being set to $Z = 1\%$ is simulated, corresponding to a dilute liquid/gas two-phase flow. As already noted, the present evaporation model was assessed by contrasting the Euler/Euler results to the results of the Euler/Lagrange method exclusively.

Figs. 2–4 show some selected results obtained by both approaches. In order to provide the fully-developed flow and turbulence conditions, a completely saturated gas phase (relative humidity was taken to be 100%, outside the left boundary of Fig. 2) was computed over the duct length of $260h = 5.2m$ (h-channel half-width), prior to the onset of the evaporation process. The liquid phase evaporation was introduced by heating the duct walls (in such a way the temperature derivative corresponding to the difference between $353K$ prescribed at the inlet cross-section - $x = 0$ - and the constant wall temperature of $368K$ was imposed, Fig. 2) causing a decrease in the relative humidity, Fig. 2. The temperature range ($T < 100^oC$) was chosen to prevent complete evaporation. The work focused primarily on the evaporation process due to concentration gradient. The temperature range prescribed corresponds to the high saturation pressure derivative dp_{sat}/dT (i.e. dY_{sat}/dT).

A somewhat slower penetration of the heat flux into the flow core (Fig. 2, lower), obtained by the Euler/Lagrange method in conjunction with original evaporation model of Abramzon and Sirignano [4] implemented in the AVL SWIFT commercial flow solver, causes weaker temperature gradients in this region, leading to a less intensive evaporation process, a fact represented by a slightly higher vapor concentration (weaker decrease from the initial 100%, Fig. 2 lower). Fig. 3a shows the isolines of the relative Reynolds number (important for the determination of the drag force relaxation time scale τ_p in the volume fraction weighted momentum equation for the continuous phase) based on the

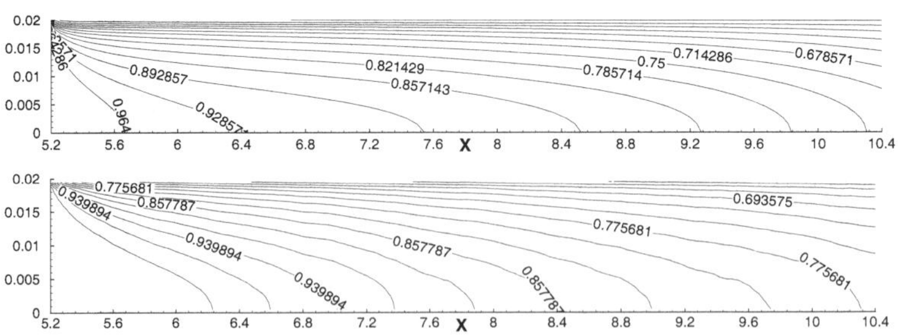

Figure 2: Comparison of the relative humidities obtained by the present Euler/Euler scheme (upper) and the Euler/Lagrange method (lower).

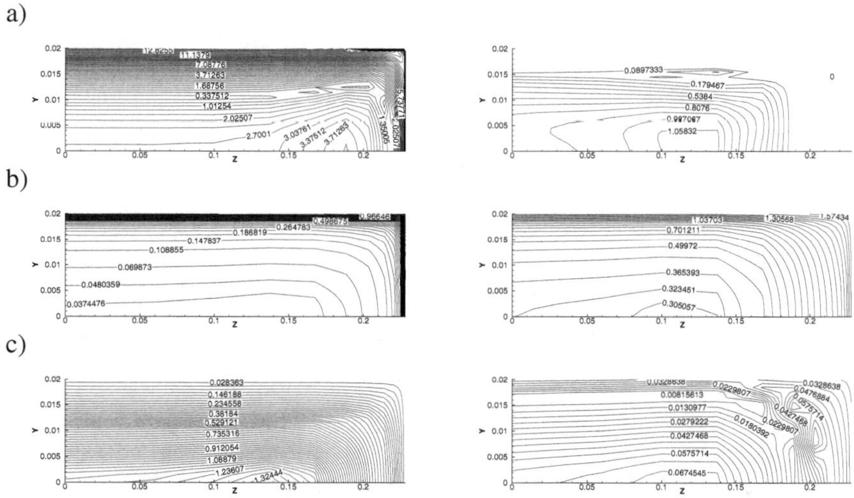

Figure 3: Isolines of a) Reynolds number Re_{rel}, b) mass transfer coefficients B_M and c) time scale of evaporation τ_Γ across the duct at two selected longitudinal locations $x/h = 20$ (left) and 140 (right) obtained by the present Euler/Euler scheme.

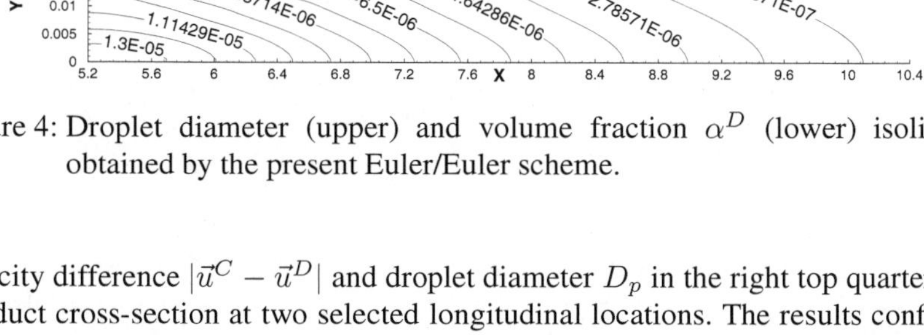

Figure 4: Droplet diameter (upper) and volume fraction α^D (lower) isolines obtained by the present Euler/Euler scheme.

velocity difference $|\vec{u}^C - \vec{u}^D|$ and droplet diameter D_p in the right top quarter of the duct cross-section at two selected longitudinal locations. The results confirm the general reduction of the Reynolds number due to the droplet evaporation (diameter reduction and consequently the volume fraction α^D reduction, Fig. 4). This tendency is particularly pronounced in the near-wall flow regions. Fig. 3b displays the evolution of the mass transfer coefficients B_M in the Abramzon and Sirignano evaporation model. This coefficient, representing indeed a measure of the vapor fraction being absorbed by the surrounding gas phase, increases due to warm up of the gas phase. The ratio of the droplet surface to the droplet

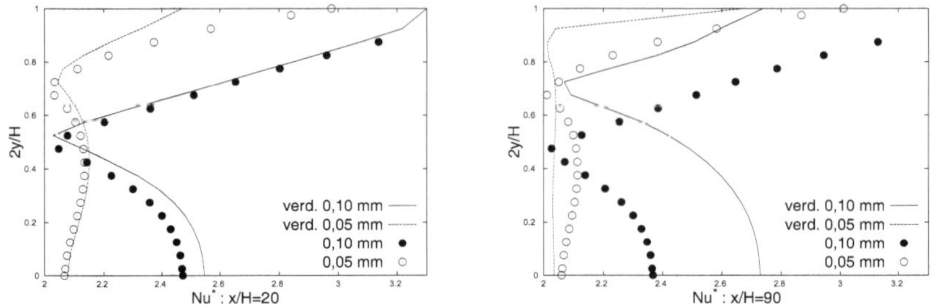

Figure 5: Modified Nusselt-number at two selected locations: $x/h = 20$ and 90.

volume increases by the droplet diameter reduction. Due to the temperature raise, resulting in the intensification of the evaporating process, the time scale of the evaporation is decreasing, Fig. 3c. Fig. 5 illustrates the influence of the evaporation (non-evaporating liquid phase was also computationally simulated) and droplet diameter (two different D_p-values were analyzed: 50 and 100 μm) on the Nusselt-number redistribution across the channel. Whereas no significant changes in the Nusselt-number behavior in the case without evaporation (symbols) are noticed, the modified Nusselt number exhibits a decrease in the region with the dominating influence of the small droplets (near wall) if the evaporation is accounted for (lines). In contrast, the increasing effect of the modification (in terms of the Reynolds number, Fig. 3a) is pronounced in the flow core.

References

[1] Renksizbulut, M. & Yuen, M.C., Experimental study of droplet evaporation in a high-temperature air stream. *ASME Journal of Heat Transfer*, **105**:384–388, 1983.

[2] Park, T.W., Aggarwal, S.K. & Katta, V.R., Gravity effects on the dynamics of evaporating droplets in a heated jet. *Journal of Propulsion and Power*, **11**(3):519–528, 1995.

[3] Ranz, W.E. & Marshall W.R., Evaporation from drops: parts I & II. *Chemical Engineering Progress*, **48**:141–146, 173–180, 1952.

[4] Abramzon, B. & Sirignano, W.A., Droplet vaporization model for spray combustion calculations. *Int. J. Heat Mass Transf.*, **32**:1605–1618, 1989.

[5] Groll, R., Jakirlić, S. & Tropea, C., Comparative study of Euler/Euler and Euler/Lagrange approaches simulating evaporation in a turbulent gas liquid flow. *Int. J. for Num. Meth. in Fluids*; **59**:873–906, 2009.

[6] Kastner, O., Theoretische und experimentelle Untersuchungen zum Stoffübergang von Einzeltropfen in einem akustischen Rohrlevitator. *Dissertation*, Technische Fakultät der Universität Erlangen-Nürnberg, 2001.

[7] Politis, S., Prediction of Two-Phase Solid-Liquid Turbulent Flow in Stirred Vessels. *PhD Thesis, Imperial College London*, 1989.

[8] Hubbard, G.L., Denny, V.E. & Mills, S.F., Droplet vaporization - effects on transient and variable properties. *Int. J. Heat Mass Transfer*, **18**:1003–1008, 1975.

[9] Sommerfeld, M., Kohnen, G. Qiu & H.H., Spray evaporation in turbulent flow: numerical calculations and detailed experiments by phase-doppler anemometry. *Revue de Institut Francais du Petrole*, **48**(6):677–695, 1993.

[10] Kulick, J.D., Fessler, J. & Eaton, J.K., Particle response and turbulence modification in fully developed channel flow. *J. Fluid Mech.*, **277**:109–134, 1994.

Investigating the influence of fines in fluidized bed reactors using 3D ECT images

C. Rautenbach, M. C. Melaaen & B. M. Halvorsen
Institute for Process, Energy and Environmental Technology, Telemark University College, Norway

Abstract

Electrical Capacitance Tomography (ECT) has become a useful measurement tool in process technology applications, especially in fluidized bed research. The ECT system is neither intrusive nor invasive which make the system practically viable for monitoring the internal flow behaviour in a fluidized bed. The sensor is placed on the outside of the non-conductive experimental reactor thus making implementation very convenient. ECT also presents user friendly equipment that is safe and easy to use compared to some of the other tomographic modalities currently available.

There has been postulated that the insertion of fines into a powder will give more uniform flow behaviour in a fluidized bed. Smaller and more evenly distributed bubbles are observed. These conditions lead to better mixing of gas and solids in a fluidized bed and thus improving the reactions in the reactor. These phenomena have been investigated in the present study using ECT and a reconstruction program developed in the present study. This reconstruction program created three dimensional images of the fluidized bed reactor under consideration. The reconstructions allow the user to get a three dimensional visual image of the flow behaviour inside the experimental reactor without disturbing the flow. Bubble- and bed characteristics of several powders with different percentages of fines were investigated. The volume, location and shape of individual bubbles were studied and thus the average bubble size, volume and frequency of a particular experimental set up was calculated. These quantities are of great importance for numerous industrial applications. Applications of these results are in almost all fields involving fluidization. The research is part of ongoing global research in optimizing and understanding fluidized beds better.

WIT Transactions on Engineering Sciences, Vol 70, © 2011 WIT Press
www.witpress.com, ISSN 1743-3533 (on-line)
doi:10.2495/MPF110121

The results are given, and its concluded that the insertion of fines into a powder would have the effect of smaller bubbles and cause Geldart type **D** powders to be less prone to slugging behaviour.
Keywords: fluidization, ECT, fines, bubble characteristics.

1 Introduction

Process tomography has been used to visualise internal flow behaviour of numerous industrial processes [1]. In the past intrusive techniques were used in an attempt to visualise flow processes that would otherwise not lend it self to be studied visually. The problem with such intrusive measuring techniques is that they alter the normal flow behaviour of the process. Doubts thus arises whether the recorded data is an accurate reflection of how that process would otherwise behave, react or change. In fluidized beds the flow behaviour is of great importance. The better a rector can be understood the better predictions can be made of its behaviour. Optimizing these reactors' mixing capabilities increases production and thus cost effectiveness. Better modeling and non-intrusive measurements of bubbling behaviour, mixing and circulation properties can have a drastic impact on optimizing fluidized bed reactors. With this in mind the influence of fine in a fluidized bed reactor was investigated in the present study using the ECT system and a reconstruction program developed for the present study.

1.1 The ECT system

A two plane ECT system was used in the present study. It consists of two arrays of electrodes and each array contains twelve electrodes. In Figure 1(a) a schematic drawing of the ECT sensor is given. The location and size of the electrodes were designed by Process Tomography Ltd. [2]. The sensor is covered by a grounded screen to protect the electrodes from external noise because the sensor operates with a soft field and is very susceptible to external interference. The non-invasive neither intrusive nature of the sensor can be observed in Figure 1(b). The electrodes are placed on the circumference of the experimental tower and thus does not influence the internal flow behaviour. The ECT system produces a cross-sectional image showing the distribution of electrical permittivities of the content of the experimental tower from measurements taken at the boundary of the vessel [3]. The capacitance reading is taken between each set of electrodes and thus $n/2(n-1)$ different capacitance values are recorded in one measurement, where n is the number of electrodes. These measurements are interpreted and illustrated as a colourful image using a reconstruction algorithm provided by Process Tomography Ltd. [2]. An example of an image depicting the solid fraction distribution in a fluidized bed is given in Figure 2(b). The resolution of the image is usually relatively low but can be sampled at high sample rates. Off-line image processing can also improve the quality of the image dramatically [4].

The measuring planes are situated at two different locations. One at height of 156.5 mm and the other at a height of 286.5 mm above the gas distributor. The

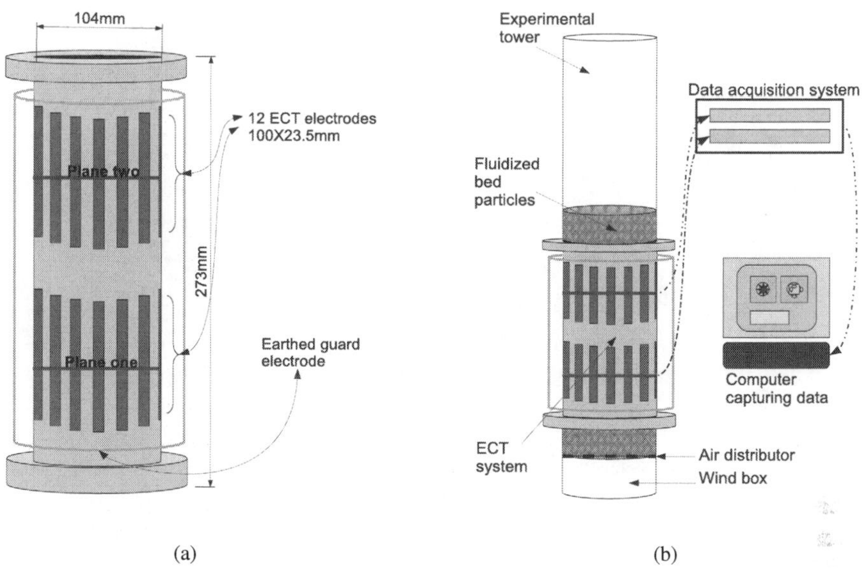

Figure 1: A not to scale drawing of the two plane ECT system utilised in the present study. (a) The two measuring planes of the system are indicated together with the electrodes and earthed guard screen. (b) The ECT system together with the experimental tower, acquisition system and the computer recording the data.

lower plane will be called plane one and the upper plane, plane two (refer to Figure 1(a)). Even though the ECT system calculates averages, the data that are obtained are viewed as a slice through the bed at the center of each sensor. Plane one and plane two is thus located at the center position of the electrodes (refer to Figure 1(a)).

The obtained image consists of pixels and each pixel represents an average solid fraction value. The average is taken over a rectangular volume equal to $9.77\times10^{-7}m^3$ [3]. A 32×32 pixel image is produced and the pixels that falls outside the circular tower will assume zero values (refer to Figure 2).

Theoretically, the more electrodes one uses the smaller the electrodes become and the more dominant the background noise can become. Thus a trade of has to be made so that the electrode is not too small but also not too large. When the electrodes are too large the resolution of the produce image will be very low. The system with twelve electrodes can capture up to a hundred 32×32 matrix maps of solid fractions per second and increasing to two hundred frames per second for an eight electrode sensor [3]. Practically between six and sixteen electrodes are normally used [2].

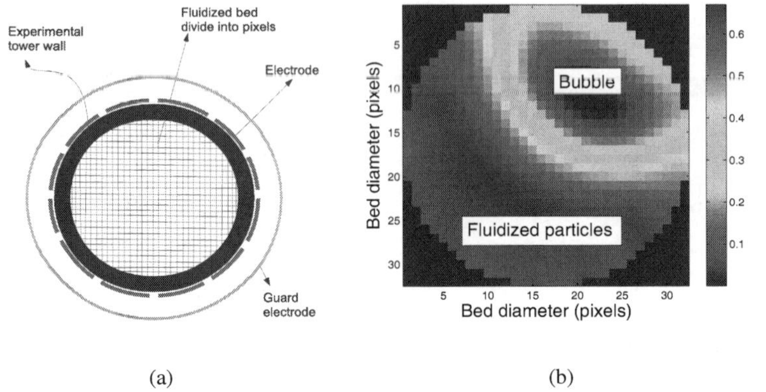

Figure 2: (a) A cross-sectional view of the ECT sensor together with the 1024 pixels created by the reconstruction program and (b) a cross-sectional image of the experimental tower indicating the solid fraction values inside the tower at a particular plane. Red indicates particles at minimum fluidization conditions and blue indicates air.

1.2 Experimental set-up and procedure

This system also requires calibration. First the tower is left empty so that just air is present. The ECT software then calibrates this as the low permittivity material (the blue colour in Figure 2(b)). Then the tower is filled with the powder that is investigated and the ECT software calibrates this as the high permittivity material (the red colour in Figure 2(b)) [2]. This provides the reconstruction algorithm the necessary boundary conditions to accurately represent the permittivities in-between these two extreme values.

A series of four powder and powder mixes were used during the present study. The characteristics of the investigated powders are presented in Table 1. For convenience abbreviations will be used in the figures and they are as follows: the mixture containing fifty percent 100–200μm powder and fifty percent 400–600μm will be called '*mix 1*' and the mixed powder containing eighty three percent of the 750–1000μm powder and eight and a half percent 400–600μm and 100–200μm powder respectively will be called '*mix 2*'. The mean particle diameter was calculated using the *surface-volume mean* diameter and is defined as

$$\overline{d}_{sv} = \frac{1}{\sum_i x_i/d_i}, \tag{1}$$

with x_i the mass fraction of the particular particle size and d_i the particular particle size. The minimum fluidization velocity (u_{mf}) was determined using either pressure drop measurements or the standard deviation of the average solid

Table 1: Relevant parameters of the powders used in the present study.

Powders used	Mean particle size ($\bar{d}_{sv}$) [μm]	Solid fraction (ϵ_s) [-]	u_{mf} [m/s]
100–200μm	153 [5]	0.68	0.022
400–600μm	482.9	0.68	0.21 [6]
750–1000μm	899.15	0.67	0.42 [6]
mix 1 : 50% 100–200μm, 50% 400–600μm	265.58	0.66	0.037
mix 2 : 8.5% 100–200μm, 8.5% 400–600μm, 83% 750–1000μm	-	0.7	0.27

fraction measurements or both [4]. Both methods are acceptable and the values obtained agrees significantly.

Gas was fed into the 10.4cm experimental tower through a porous plate distributor. Air was used as fluidizing fluid and all of the powders mentioned in Table 1 were made from glass with a density of 2485kg/m^3.

After calibration a set of experiments were conducted each at a different superficial velocity, u_0. After several measurement at different superficial velocities have been made for one powder the tower was emptied and the same procedure was followed with one of the other powders under consideration. Starting with calibration and ending with the actual measurements. All of the measurements made in the present study were taken over a 60s period. According to Makkawi and Wright this is an acceptable experimental span and will produce stable and reliable results [3].

1.3 The reconstruction program

The commercial code *MATLAB* was employed in developing the reconstruction program. For each experiment where 60s of data were recorded a series of 6000 images were produced. These images were organised and sorted and then it was read in, one by one, into the reconstruction program. For each image (which can be viewed as a 'slice' through the reactor) bubbles were recognised, labeled and saved. Information about the bubble diameter, location and shape were also stored into a growing data structure. Then the next image was read in and the same was done except now it was compared to the previous image. By doing this the program could connect these separate images to produce a three dimensional image of bubbles and other phenomena in a fluidized bed. Quantities like the bubble volume and height could also be determined that would otherwise have been impossible by just using separate images.

The two planes of the ECT system also make it possible to correlate velocities. This could not be done in the present study because of the dramatic changes the fluidized bubbles go through from one measuring plane to the other. Bubble coalesce and split before they reach the second measuring plane. This makes tracing a single bubble no trivial task. If the 6000 recorded images were to be placed one after the other a time stacked image will be produced. This is because it is only time that separates one image being recorded from the next. To convert the time stacked images into a three dimensional image in space only the bubble velocity was needed. Equations predicting the bubble velocity were used to get a semi-empirical estimation of the bubble velocity as described by Kunii and Levenspiel [7]. These equations required the bubble diameter (b_d) as independent variable and the values that were obtained from the reconstruction program was used for this parameter. To estimate the bubble diameter the mean diameter of all the 'slices' of a single bubble was calculated by the program and then the mean of all the bubbles mean diameters were calculated to produce the representative average bubble diameter for a particular experiment.

To determine the bubble velocity three expressions were used that give estimations of the bubble rise velocities in bubbling beds. The bubble velocity for a single bubble is given as

$$u_{br} = 0.711\,(gb_d)^{1/2}\,, \tag{2}$$

according to the Davidson and Harrison model [8]. For bubbles in a bubbling bed the Davidson and Harrison model states

$$u_b = u_o - u_{mf} + u_{br}, \tag{3}$$

with u_o the superficial velocity and u_{mf} the minimum fluidization velocity. The second equation covers all the particles size distributions, Geldart **A** to **D**, and takes the reactor's diameter into account. This equation was proposed by Werther [7] and is expressed as

$$u_b = \psi\,(u_o - u_{mf}) + \alpha u_{br}, \tag{4}$$

where ψ is the fraction of visible bubbles and α compensates for the differences between a single rising bubble and bubbles rising in a bubbling bed. Empirical estimations for these quantities are described by Kunii and Levenspiel [7]. Kunii and Levenspiel proposed the third correlation and is based on experimental data of Geldart **B** particles in a tower with a diameter less than one meter. This correlation is expressed as

$$u_b = 1.6\left((u_o - u_{mf}) + 1.13b_d^{1/2}\right)d_t^{1.35} + u_{br}, \tag{5}$$

with d_t the tower diameter [7]. According to Kunii and Levenspiel equations (3) to (5) must be calculated and the larger value must be used as the bubble rise velocity, u_b. This procedure was followed in the present study.

2 Results and discussion

To present the result in a clear way a velocity coefficient, v_c, was introduced. This dimensionless quantity made it possible to plot most of the obtained data of a particular parameter in one figure. The coefficient used is defined as

$$v_c = \frac{u_b - (u_o - u_{mf})}{\sqrt{g d_t}}, \tag{6}$$

where u_b is the bubble rise velocity, u_o is the superficial velocity, u_{mf} is the minimum fluidization velocity and d_t is the experimental tower diameter.

In Figure 3(a) the bubble diameter ratio as a function of the velocity coefficient is given and all the data obtained, from both plane one and plane two, followed the same trend. A quadratic fit was also made and presented along with the data. The data and thus the quadratic fit seem to be independent of the height above the distributor. This might seem in contradiction with equation (7) but this apparent independence of z is merely an effect of the velocity coefficient.

Numerous experimental correlations have been developed over the years to estimate the bubble size in a fluidized bed and mainly for small experimental towers with Geldart **B** particles [7]. One such a correlation is given by Werther for Geldart **B** particles that is fluidized through a porous plate distributor and is expressed as [7]

$$b_d = 0.853 \left[1 + 0.272 \left(u_0 - u_{mf}\right)\right]^{1/3} \left(1 + 0.0684 z\right)^{1.21} \; [cm], \tag{7}$$

where z is the height above the distributor and b_d the bubble diameter. Equation (7) along with data obtained from the 100–200μm powder are given in Figure 3(b).

From Figure 3(b) the correlation between the experimental data and equation (7) is inadequate. One trivial explanation for this inadequate correlation might be because of the definition of the average bubble diameter in the present study (refer to Section 1.3). Usually the bubble diameter, b_d, is defined as the diameter of an equivalent sphere having the same volume as the actual bubble [8]. With the ECT system and the reconstruction program this definition was impractical for the present study. The operating condition of equation (7) is given as follows

$$\begin{aligned} d_t &> 20cm, & 1 \leq u_{mf} &\leq 8cm/s, \\ 100 \leq d_p &\leq 350\mu m, & 5 \leq u_o - u_{mf} &\leq 30cm/s. \end{aligned}$$

These operating conditions thus does not cover all the parameters of the 100–200μm powder experiments' operating condition and thus discrepancies could be expected.

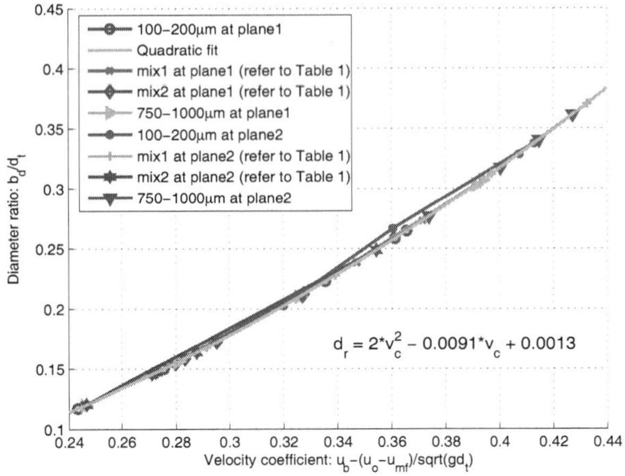

(a)

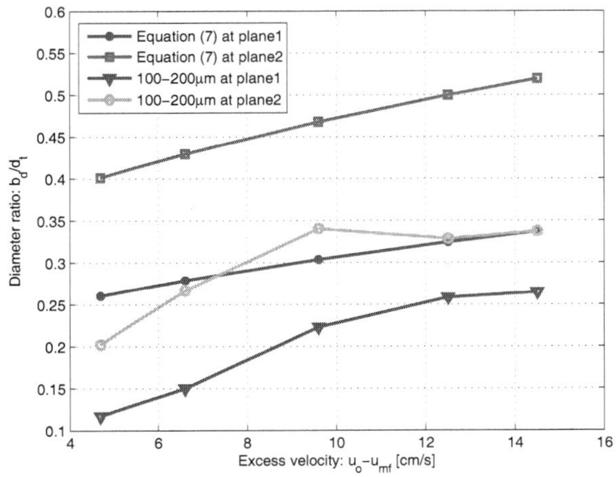

(b)

Figure 3: (a) Dimensionless velocity coefficient against the dimensionless diameter ratio, d_r, for both plane one and plane two. A quadratic fit to the data is also given along with its equation. (b) Theoretical equation (7) against the 100–200μm particles data.

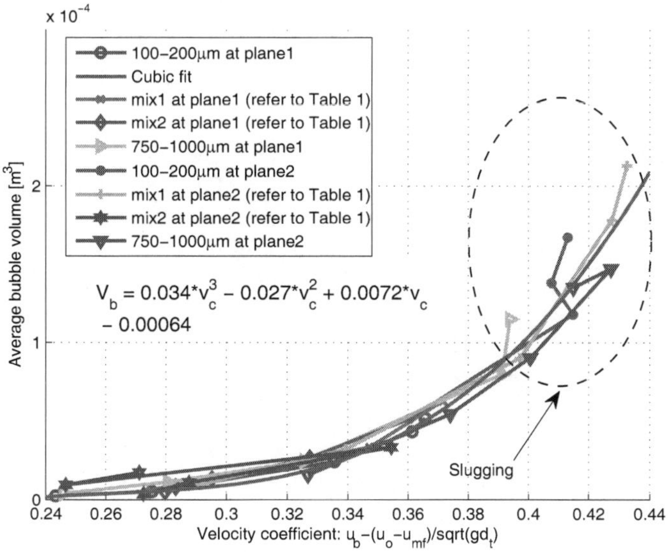

Figure 4: Average bubble volume, V_b, as a function of the velocity coefficient for both plane one and plane two.

From Figure 3 it is not trivial to observe the influence of fines on a powder. Nevertheless the equation produced by the quadratic fit might be a useful tool in determination of the bubble diameter or of the velocity coefficient, depending on which experimental data are available. Overall the fines caused a smaller average bubble diameter which in turn caused a lower bubble rise velocity and thus low velocity coefficient values (refer to Figure 3(a)).

In Figure 4 the average bubble volume is given as a function of the velocity coefficient.

Again very similar behaviour is observed from all the recorded data at plane one and plane two. A cubic fit was found to give a good estimation to all the recorded data points at velocity coefficient values less than 0.38. At velocity coefficient values greater that 0.38 some of the average bubble volume data deviate from the cubic trend. These deviations occurred for all of the data recorded at plane two and for the 750–1000μm powder at plane one as well. These deviation is suspected to be an indication of the onset of the slugging regime. Slugs will cause a lower bubble rise velocity and high average bubble volumes. This is clear in Figure 4 from the higher values of the average bubble volume that was observe for the 100–200μm powder and the *mix 1* powder at plane two and the 750–1000μm powder at plane one. The *mix 2* powder's data at plane two reached a maximum average bubble volume at a velocity coefficient value of about 0.35 and then decreased. This effect is attributed to segregation effects in the bed due to the large particle size distribution that is present. As the superficial velocity is increased the bed mixes better and thus the bed behaviour can change as is observed with the *mix 2* powder. The average bubble volume for the *mix 2* powder thus stayed low compared to the case where just the 750–1000μm powder was present. It is difficult to observe but the same evidence is present in Figure 3(a). When just the 750–1000μm powder was fluidized; evidence of slugging behaviour occurred even at plane one. The presence of smaller particles or fines in a powder results in an overall smaller average bubble. The last data point of the 750–1000μm powder had a lower velocity coefficient and a lower average bubble volume. This is attributed to a slug that collapsed because of instabilities in the forces that keep the slug's shape. This topic is part of on-going research and is not part of the present study.

3 Conclusion

The influence of fines in powders used in a fluidized bed has been investigated. A range of different powders and powder mixtures have been used and these powders were mainly from the Geldart group **B**. The effect of the fines in powders have been studied by means of the bubble behaviour in the fluidized bed. Using an ECT (Electrical Capacitance Tomography) system data were obtained from a fluidized bed in operation and a reconstruction program developed for the present study was used to calculate characteristics of the bubbles in the bubbling bed. These characteristics were the bubble diameter and volume.

From the result it can be concluded that the insertion of fines into a powder has the effect of smaller average bubble sizes in the fluidized bed. This was observed in both the data obtained for the bubble diameter and volume. It was also evident that fine cause a powder that is prone to slugging to require much larger superficial velocities before signs of slugging set in. This was observed in Figures 3(a) and 4. Fines is thus desirable in Geldart type **D** particles as it suppresses the slugging behaviour that these particles would usually exhibit and instead causes smaller bubbles that would lead to better mixing of the bed content.

Finally, two expressions were correlated using the experimental data. For the diameter ratio the expression is given as

$$d_r = 2v_c^2 - 0.0091v_c + 0.0013, \tag{8}$$

where d_r is the dimensionless diameter ratio (b_d/d_t) and v_c is the velocity coefficient (equation (6)). The expression for the average bubble volume is

$$V_b = 0.034v_c^3 - 0.027v_c^2 + 0.0072v_c - 0.00064, \tag{9}$$

where V_b is the average bubble volume. Deviation from these correlations seem to appear with the on set of the slugging regime. As long as the bed is in the bubbling regime these correlation should given an accurate estimation of the fluidized bed behaviour. Whether these equations are generally adequate is not yet clear and further research is needed to validate these empirical correlations.

References

[1] Qiu, C., Hoyle, B.S. & Podd, F.J.W., Engineering and application of dual-modality process tomography system. *Flow Measurement and Instrumentation*, **18**, pp. 247–254, 2007.

[2] Process tomography Ltd., 86 Water Lane, Wilmslow, Cheshire. SK9 5BB, UK, *PTL300-TP-G ECT system, Operation manual*, 2003.

[3] Makkawi, Y.T. & Wright, P.C., Electrical capacitance tomography for conventional fluidized bed measurements-remarks on the measuring technique. *Powder Technology*, **148**, pp. 142–157, 2004.

[4] Makkawi, Y.T. & Wright, P.C., Fluidization regimes in a conventional fluidized bed characterized by means of electrical capacitance tomography. *Chemical Engineering Science*, **57**, pp. 2411–2437, 2002.

[5] Jayarathna, S.A., *Recommendation of a model for simulating and analysis of the influence of particle size distribution on the simulations of bubbling fluidized beds*. Master's thesis, Telemark University College, 2008.

[6] Rautenbach, C., *Modelling of flow through porous packing elements of a* CO_2 *absorption tower*. Master's thesis, Stellenbosch University, 2009.

[7] Kunii, D. & Levenspiel, O., *Fluidization Engineering second edition*. Butterworth-Heinemann series in chemical engineering: Oxford, U.K., 1991.

[8] Davidson, J.F. & Harrison, D., *Fluidized particles*. Cambridge University Press: Cambrige, U.K., 1963.

Experimental study of multiphase flow in a model gearbox

V. Chernoray[1] & M. Jahanmiri[1,2]
[1]*Division of Fluid Dynamics, Department of Applied Mechanics, Chalmers University of Technology, Göteborg, Sweden*
[2]*Department of Mechanical and Aerospace Engineering, Shiraz University of Technology, Iran*

Abstract

This study concerns dynamics of a two-phase flow around a rotating solid body. Under consideration is a model of a gear wheel in a gearbox which rotates and is partially submerged in oil. The flow of interest is complex and involves effects of free surface dynamics, rotation, and formation of bubbles and drops. Occurring flow regimes include laminar, transitional and turbulent. The major focus of the investigation is on details of the developed flow, and the purpose is validation of numerical methods developed for design and optimization of such components. Current experiments are performed in a test rig which is modelling a generic simplified gearbox with a single isolated rotating wheel. The flow measurements are carried out by using particle image velocimetry (PIV) and the test rig is specially designed for this purpose with the optical access maximized. The flow similarity with respect to a real gearbox is fully maintained and the working fluid is a transparent mineral oil. The PIV measurements are performed at four different rotation speeds for two different wheel configurations in order to cover a spectrum of operational conditions needed for numerical modelling. The emphasis is on the result of experiments on a smooth wheel. The measurements are providing velocity distribution around the wheel and details on bubble and drop distribution.
Keywords: bubble and drop dynamics, rotating wheel, splash, surface tension, gear box, PIV measurements.

WIT Transactions on Engineering Sciences, Vol 70, © 2011 WIT Press
www.witpress.com, ISSN 1743-3533 (on-line)
doi:10.2495/MPF110131

1 Introduction

Multiphase flows are present in numerous engineering applications. For better development of applied methods used in prediction and optimization of engineering flows a deep understanding of underlying flow phenomena is crucial. Nevertheless it is known that obtaining knowledge on complex multiphase flows occurring in industrial processes [1] is often a challenging task where both numerical simulations and experimental investigations can be very demanding.

Gear lubrication is a significant concern in a wide range of industries which use a power transmission. SAAB Powertrain Sweden is developing computational fluid dynamics (CFD) models for predicting and optimizing the oil flow around rotating components in gearboxes. The main objective is to optimise the losses in a gearbox and thereby decrease the fuel consumption and exhaust emissions of vehicles. Furthermore, an effective lubrication reduces the operating cost of a gearbox not only by improving the efficiency of transmissions but also by prolonging the component lifetime. Increasing the efficiency of gearboxes has a great market potential not only in the vehicle industry but also in many other industries where the power transmission is used such as the pulp industry, mining industry, wind and hydropower industry.

The gearbox optimization is rather complex problem since in a gearbox the oil is used not only to reduce the friction between the gearwheels but also acting as a cooling fluid. This means that both the oil amount and the oil flow have to be optimised. The numerical prediction of the oil flow in a gearbox is not a trivial task and reliable measurements for validation of CFD models are required.

There are surprisingly few studies which are focussing on the actual details of the gearbox flow. The oil and airflow related losses are contributing to the so-called load-independent spin power losses which include the air windage, oil churning, inertial power loss from an impinging oil jet (for the case of jet lubrication), fluid trapping and squeezing between meshing gear teeth, rotating seals, and bearings [2]. Further discussion on importance of these losses can be found in several recent works devoted to the gearbox efficiency [3–5]. A particular insight into the fluid mechanics of gearboxes is attempted only in a fewer works. In papers [6–8] the studies were performed by using CFD and in papers [9, 10] the CFD was used in combination with experiments. However the experimental data available in these works are very limited.

Current experimental study was initiated in order to complement the above research. The investigation is performed by using PIV and the flow similarity with respect to a real gearbox is maintained.

2 Experimental set-up

The gearbox model with single rotating wheel is shown in Fig. 1. The test rig is representing a simplified model of a gearbox. The major idea is to isolate the splash effects from the wall effects and for this purpose the box size is enlarged as compared to the real gearbox. The rig is specially designed for using PIV which means that the optical access to the flow is maximized. The dimensions of the box are $0.82\times0.52\times0.32m^3$.The gear wheel is made interchangeable and the wheel rotation speed can be varied to investigate different operational conditions. The wheel models are produced by stereo lithography method and used wheels have diameter of 0.22 m and thickness of 28 mm. Two wheel configurations are used, a flat wheel, which can be seen in figure, and a spur gear wheel with 67 teeth.

Figure 1: Model gearbox and gear wheel. showing the experimental configuration and position of the CCD cameras.

Two-dimensional, two-component PIV measurements (2D-2C) are reported in this paper. The laser plane was positioned at wheel symmetry plane (midpoint of the wheel thickness) and perpendicular to the axis of rotation. The camera fields of view are depicted schematically in Fig. 2. Each of two cameras was adjusted for own field of view for easiness of obtaining repeatable experiments.

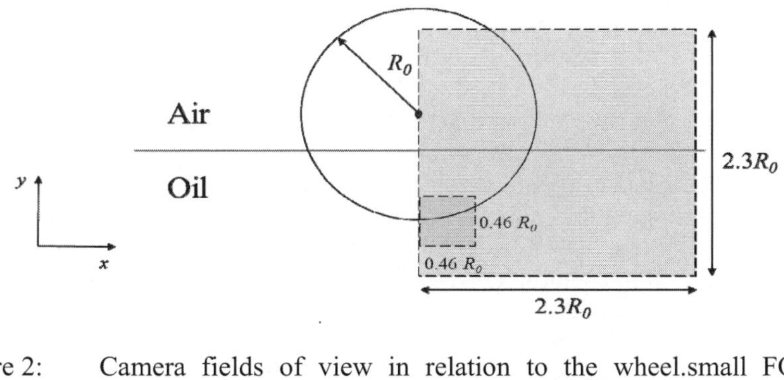

Figure 2: Camera fields of view in relation to the wheel.small FOV is $50\times50\ mm^2$ and large FOV is $250\times250\ mm^2$.

The optical axis of camera is normal to the plane of the light sheet. The camera fields of view are $50\times50\ mm^2$and $250\times250\ mm^2$. To obtain the flow information around the other half of the wheel the wheel rotation direction was changed. Due to the flow symmetry this was equivalent to moving the field of view to other side of the wheel.

The CCD cameras have resolution of 2048×2048 pixel (LaVision Imager pro X 4M) with 7.4×7.4 μm2 pixel size. The laser is pulsed dual-cavity Nd-YAG laser with 400 mJ/pulse at 532 nm (Spectra-Physics). The thickness of the light sheet was approximately 5 mm. The flow was seeded with fluorescent particles (PMMA Rhodamine B) with diameter of1-20 μm and density of 1.5 kg/m^3. The fluorescent light from particles is separated by a low-pass filter.

The working fluid is a transparent synthetic oil which has same viscosity at room temperature as gearbox oil at design temperature. In this way the Reynolds number similarity with a real gearbox is maintained.

3 Results and discussion

3.1 Oil splash visualizations

The complexity of the flow in the model gearbox can be understood from visualizations presented in Fig. 3. The flow is visualized for a single smooth wheel rotated at rates from 50 to 700 rpm, which corresponds to Reynolds number range from 3000 to 40000.This operating range is representative for a city car driving cycle.

Figure 3 shows that the flow is divided into two parts, below and above the oil surface. In the upper part, the flow is governed by the interplay between the rotational inertia, gravity, surface tension, and viscous shear forces. In the below part the same forces except the surface tension are of importance. At low rotation rates (50 rpm) the flow is laminar in both of these parts and the two fluid phases, oil and air, are distinctly separated. The flow problem in this case can be compared to a laminar rotating drum problem [11]. As rotational rate increases, the volume flow of the oil elevated by the wheel increases, and the streams of the oil flowing down from the wheel are also becoming stronger. The oil jet impinging into the oil pool results in air entrainment into the oil in the form of a void. With increased jet momentum, the void grows in size until it becomes unstable and collapses, and the mixing of the two fluids occurs with the formation of bubbles in the oil. The physical mechanism of air entrainment by a liquid jet is described in detail by Soh *et al.* [12].

The jet entrainment can either be laminar or turbulent, as our PIV data show, and the particular case depends on the flow Reynolds number. The laminar jet is undergoing the laminar-turbulent transition after the entrainment into the oil pool. In either case the entraining jet is creating the turbulence in the oil pool. As also shown by PIV measurements at 100 rpm (Re=6000) the laminar-turbulent transition of the boundary layer on the submerged wheel surface occurs.

Due to further increase of the oil stream elevation with increased rotation rate (150 rpm), the stream length increases and the pinch-off of the oil jet occurs, which caused by the Rayleigh-Plateau instability. At rotation rate of 150 rpm the inertia and shear forces become large enough to transport the oil beyond the topmost wheel point which leads to the formation of a secondary oil stream on

Figure 3: Smoothwheel at rotation rates of50, 100, 150, 200, 300, 700 rpm.Rotation direction is counter-clockwise, *Re*=3000-40000.

the left side of the wheel. This stream has larger velocity as compared to the primary stream since the wheel tangential velocity vector on this side is directed downwards, so the inertial and gravity forces acting in same direction downwards and accelerating the stream. At 150 rpm rotation the secondary stream is having a larger momentum as compared to the primary stream and causes a deeper stream entrainment into the pool which in turn leads to the formation of a larger amount of air bubbles to the left of the wheel. At rotation rate of 200 rpm the oil film leaves the wheel surface on the topmost point of the wheel. At this rotation rate the oil film thickness increases and the inertial force overcomes the viscous forces. At 300 rpm the oil splashes are becoming strong enough to reach the top and side walls of the surrounding box. At 700 rpm the oil splashing flow is so strong that the oil from the top wall is running in all directions and the oil streams can be observed on the front window of the box. These streams are responsible for a significant optical distortion seen on picture.

3.2 Air bubble visualizations

The air bubbles in the oil pool are primarily generated by the entraining oil jets. The process of bubble creation was discussed in previous section. In current section the bubble distributions are considered.

In Fig. 4 raw PIV images are presented which are showing the bubbles around the wheel at different rotation rates. Each of visualizations is composed of two images which were taken at different experimental runs, i.e. not simultaneously.

The bubbles are uniformly lighted up by the fluorescent light emitted from the seeding particles and due to this appear as contours. This method of bubble visualization is commonly called as planar fluorescence for bubble imaging (PFBI) and the main advantage of PFBI that it can be applied in flows with large void fractions of bubbles, up to 5-6% [13].

As seen in Fig. 4the air bubbles of various sizes are generated in the oil. The bubble generation starts form rotation rates of 100 rpm, and the concentration of bubbles increases with increased rotation rate. In Fig. 4 one can observe a variety of bubble shapes from nearly spherical to elongated, aerofoil-like shapes. The elongated bubbles are visible near the vicinity of the wheel where the counter action of the inertial/shear forces and the buoyancy acting on bubbles leads to their stretching.

The bubble occupation regions are also varying with the rotation rate. At 200 rpm the splash and bubble creation occur primarily near the wheel centre plane and the bubble layer is localized around this plane. The bubble occupation zone is limited to the region near the wheel. At higher rotation rates the splash streams are becoming more spread so the bubble generation zone increases and the bubble occupation zones also extend significantly. The oil motion leads to further distribution of the bubbles so they are occupying the entire pool. After a longer time interval the large bubble void fraction in the oil pool created far too large obstruction of the light path so the image quality decreased drastically. To overcome this effect, for rotation speeds of 400 and 1200 rpm the measurements were performed only during a short time interval after the transient start of the wheel.

At 400 rpm one can observe big bubbles which are rising up on the sides of the wheel. As one can expect these bubbles are not following the wheel rotation but inducing the vertical velocity and turbulence in the fluid [14] and acting as obstacles for the sidewall boundary layers on the wheel. The lifespan of the biggest bubbles is of the order of seconds and the smallest bubbles are remaining in the pool for significantly longer time, of order of 1 hour. As one can expect the smallest bubbles are following the flow streamlines very well.

For each rotation rate there is an own equilibrium state between the rate of bubble generation and the rate of bubble escape from the bath. Increased rate of the bubble generation leads to a larger bubble void fraction in the pool which in turn increases the rate of bubble merging and escaping. Of great interest is to study this process in detail which is planned in our future work.

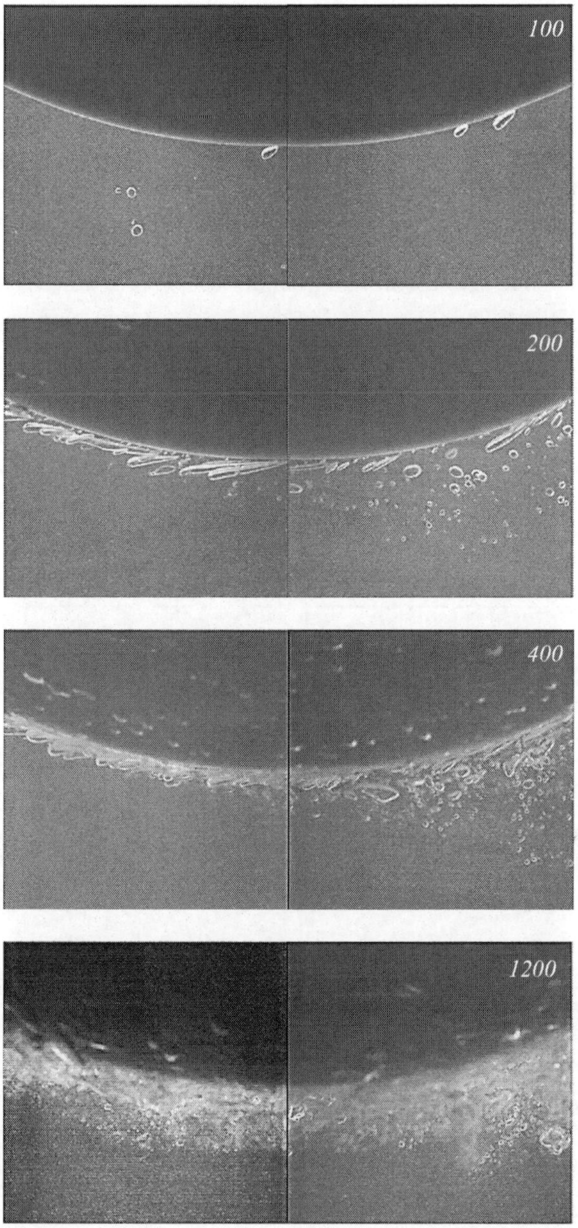

Figure 4: Air bubbles around the flat wheel at rotation rates 100, 200, 400, 1200 rpm. Wheel rotation direction is counter-clockwise. Note the elongated bubble shapes close to the wheel, and increase of bubble concentration at increased rotation rates.

3.3 Velocity fields

The mean velocity fields around the wheel obtained at different rotation rates are presented in Fig. 5. One should notice that in this figure the velocity of the flow in the upper part (above the oil surface) has to be considered only as a qualitative result. Due to the fact that in the region above the oil surface the PIV technique is applied to the oil drops instead of the seeding particles, and presence of strong non-uniformities of illumination intensity and multiple light scattering, the reliability of the data is poor. From other hand the top part of the flow is of minor interest since the motion of the oil droplets is relatively simple and described by ballistic parabolic trajectories.

From Fig. 5 it is clearly seen that as the rotation rate increases the viscosity-dominated flow is changed to the flow with dominated inertial forces. As a consequence, the characteristic scales of the flow change. For the bottom part the characteristic scale is the thickness of the boundary layer. The thickness of the boundary layer decreases with increased inertial forces. For the top part the characteristic scale is the flying distance of the oil droplets. The flying distance increases with increased inertial forces since the initial velocity of the oil drops leaving the surface of the wheel increases.

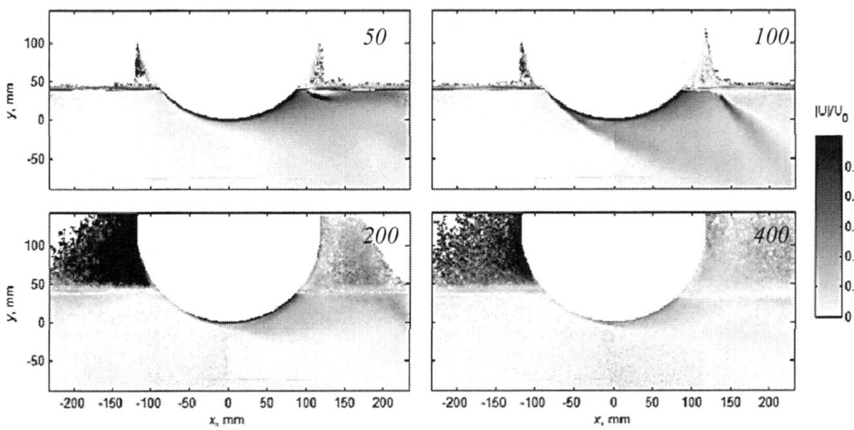

Figure 5: Contours of average velocity magnitude at rotation rates of 50, 100, 200, and 400 rpm.

Analysis of the boundary layer profiles on the wheel surface has revealed that the boundary layer is laminar at 50 rpm, transitional at 100 rpm and turbulent at rotation rates above 200 rpm.

The breakdown of the oil jet which is emitted from the surface of the wheel and enters the pool is illustrated in Fig. 6 for case of 100 rpm rotation rate. One can observe that the initially laminar jet undergoes the instability phase and become turbulent. At higher rotation rates the oil jets entering the pool become turbulent nearly immediately after the entrance.

Figure 6: Instantaneous velocity vector field showing the oil jet entrainment and breakdown. 100 rpm case.

In fact for successful description of the multiphase flow the phases involved have to be discriminated. Ideally for each phase the velocity and the void fraction have to be determined. Among the methods which are commonly used for the discrimination of the bubble phase, see [13–19], none is offering a good enough reliability of the bubble recognition in the case of significantly irregular bubble shapes and high degree of mutual bubble overlap. This means that in our case the bubble discrimination appeared to be a non-trivial task. From other hand the fluid phase discrimination was performed successfully by using common PIV processing methods. The processing steps involve use of very strict settings for the peak ratio criteria (peak ratio 2) and for the median filter criteria (1 r.m.s. of the neighbours). As a result the most of the slow bubbles and bubble shadowed zones were removed from the velocity fields. The result was further improved by applying a3σ threshold to the time-series of images and removing the vectors having far too small or excessive amplitudes as compared to other vectors in the same time-series. A typical result after the processing is illustrated in Fig. 7. The figure shows that the largest and slowest bubbles as well as the bubble shadows are successfully removed from the velocity field. The smallest bubbles follow the oil flow very well and are preserved.

Figure 7: PIV vector field combined with a raw image demonstrating that the slow-moving bubbles and the bubble shadows are not contributing to the velocity field. Every second vector is shown; case of 200 rpm rotation rate.

3.4 Comparison with spur gear wheel

Flow around a spur gear wheel has its own peculiarities. The air is trapped between the teeth of the wheel, see Fig. 8. The volume of the trapped air increases with increased rotation rate. Also, the splash strength increases significantly since larger amount of oil is dragged in the teeth pockets. From other hand, for the submerged part of the wheel the fluid in the pockets between the teeth is moving with same velocity as the wheel itself. Due to teeth the boundary layer developed on the wheel surface is undergoing transition to turbulence quicker. The scaling of the boundary layer thickness is shown in Fig. 9 in comparison to smooth wheel.

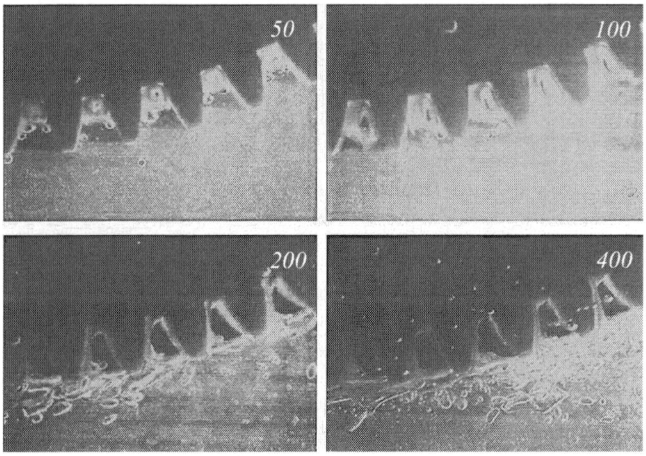

Figure 8: Air bubbles betweenthe teeth and in the vicinity of a spur gearwheel at rotation rates 50, 100, 200, and 400 rpm.

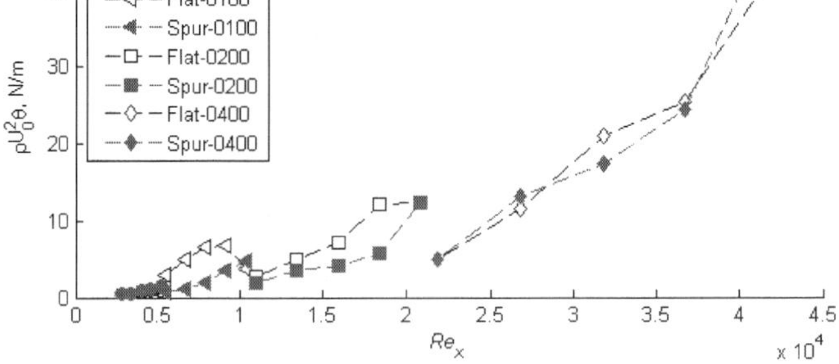

Figure 9: Boundary layer momentum thickness for flat and spur gear wheelsshown in normalized coordinates.

4 Conclusions

Flow around a single rotating wheel within a modelled gearbox is investigated by means of visualizations and PIV. Two-component PIV velocity measurements were successfully performed for several different rotation rates of the wheel. The results of study demonstrate that the flow around the wheel is very complex. The rotation of the wheel results in elevation of the oil by the wheel and creation of the streams of the oil ejecting from the wheel, i.e. the oil splash. The streams can break-up and lead to formation of oil drops. With increased rotation rates the splashed oil streams start to agitate the surface of the oil pool and collide with the gearbox walls. The oil streams entering the oil pool lead to generation of air bubbles in the oil. The penetration of the oil jets into the oil bath and motion of the air bubbles are creating turbulence in the oil.

The splash increases with increased rotation rate due to increased role of inertial forces. From other hand, for the flow inside the pool the prevalence of the inertial forces over the viscous forces leads to thinner boundary layers and more and more confined regions of induced flow very close to the wheel.

Acknowledgements

Authors would like to thank SAAB Powertrain, Sweden for providing the experimental rig and financial support for this project. Many thanks to Ehsan Sistani for taking part in the experiments.

References

[1] Honkanen M., Eloranta H., and Saarenrinne P. Digital imaging measurement of dense multiphase flows in industrial processes. *Flow Meas. Instrum.* **21**, pp. 25–32, 2009.

[2] Petry-Johnson T.T., Kahraman A., Anderson N.E., Chase D.R. An experimental investigation of spur gear efficiency, *ASME J. Mech. Des.* **130**, 062601, 2008.

[3] Changenet C., and Velex P., A model for the prediction of churning losses in geared transmissions-preliminary results, *ASME J. Mech. Des.* **129**, pp. 128–133, 2007.

[4] Eastwick C.N., Johnson G. Gear windage: a review, *ASME J. Mech. Des.* **130**, 034001, 2008.

[5] Höhn B.-R., Michaelis K. and Hinterstoißer M. Optimization of gearbox efficiency. *Goriva i maziva* **48** (4) 441–480, 2009.

[6] Al-Shibl K., Simmons K., and Eastwick C.N. Modelling windage power loss from an enclosed spur gear. *Proc. I Mech E,* Vol. **221**, Part A: J. Power and Energy, 2007.

[7] Hill M.J., Kunz R.F., Noack R.W. Long L.N., Morris P.J., Handschuhk R.F. Application and validation of unstructured overset CFD technology for rotorcraft gearbox windage aerodynamics simulation. *In proc. of American Helicopter Society 64th Annual Forum*, Montreal, Canada, 2008.

[8] Lemfeld F., Fraňa K., Unger J., Numerical simulations of unsteady oil flows in the gear-boxes, *J. Applied Science in the Thermodynamics and Fluid Mechanics* **1**, pp. 27–32, 2008.

[9] Moshammer T., Mayr F., Kargl K., and Honeger C. Simulation of oil flow in gear box housing, *SAE Int.*, 2006-01-1574, 2006.

[10] Li L., Versteeg H.K., Hargrave G.K., Potter T., and Halse C. Numerical investigation on fluid flow of gear lubrication, *SAE Int.*, 2008-01-1650, 2008.

[11] Hasan N., Naser J. Determining the thickness of liquid film in laminar condition on a rotating drum surface using CFD. *Chem. Eng. Sci.* **64**, pp. 919–924, 2009.

[12] Soh W.K., Khoo B.C., Yuen W.Y.D. The entrainment of air by water jet impinging on a free surface, *Exp. Fluids* **39**, pp. 496–504, 2005.

[13] Akmetbekov Y.K., Alekseenko S.V., Dulin V.M., Markovich D.M., Pervunin K.S. Planar fluorescence for round bubble imaging and its application for the study of an axisymmertric two-phase jet. *Exp. Fluids* **48**, pp. 615–629, 2010.

[14] Riboux G., Risso F., Legendre D. Experimental characterization of the agitation generated by bubbles rising at high Reynolds number, *J. Fluid Mech.* **643**, pp. 509–539, 2010.

[15] Sridhar G., Ran B., Katz J. Implementation of PIV to multiphase flow. *Cavitation and Multiphase Flow Forum* ASME-FED-Vol. **109**, pp. 205–210, 1991.

[16] Dehaeck S., van Beeck J.P.A.J., Riethmuller M.I. Extended glare points velocimetry and sizing for bubbly flows. *Exp. Fluids* **39**, pp. 407–419, 2005.

[17] Hassan Y.A, Blanchat T.K., Seeley Jr. C.H., Canaan R.E. Simultaneous velocity measurements of both components of a two-phase flow using particle image velocimetry. *Int. J. Multiphase Flow* **18**, pp. 371–395,1992.

[18] Honkanen M. Turbulent multiphase flow measurements with digital particle image velocimetry: Application to bubbly flows. *MSc thesis*, Tampere University of Technology, 2002.

[19] Lindken R. and Merzkirch W. A novel PIV technique for measurement in multiphase flows and its application to bubbly flows. *Exp. Fluids* **33**, pp. 814–825, 2002.

Experimental analysis of two-phase flow in domestic central heating systems: micro bubble characteristics

A. M. Fsadni[1], Y. T. Ge[1] & A. G. Lamers[2]
[1]*Department of Mechanical Engineering, Brunel University, UK*
[2]*Spirotech bv, The Netherlands*

Abstract

Domestic central heating systems result in micro bubble nucleation and detachment at the boiler wall due to the presence of supersaturated conditions in the system. Such conditions originate due to the widespread use of untreated tap water in such systems and due to the air leakage into the system's closed loop circuit during the daily thermal cycling. The presence of air bubbles in such systems results in a number of undesired effects such as corrosion, blockages, noise and the filling of radiators with air due to the deposit of the micro bubbles in the radiators. As a result, a loss in the heat transfer area of the radiator occurs thus resulting in an inefficient use, consequently increasing the carbon footprint of such widely used systems. Therefore, the capture of micro bubbles at the exit point of the boiler should result in improved system efficiencies. A comprehensive knowledge of the expected bubble production rate and bubble size is important for an efficient deaeration system. To date, no studies have been published on the fundamentals of bubble production and size in wet central heating systems, therefore, the importance of this paper for the building services industry. Experimental results on a test rig incorporating a typical 19 kW (rated) domestic gas fired boiler show typical micro bubble production rates ranging between 0.03 to 0.47 litres of nitrogen gas per hour with a shift in the system heating load. An analysis of the expected bubble diameters was done at different system conditions. Bubbles with average diameters within the range of 0.13 and 0.39 mm were observed. A comparison with Winterton's finite angle force balance model resulted in reasonable predictions.
Keywords: micro bubbles, supersaturated solutions, central heating, dissolved gasses, space heating.

www.witpress.com, ISSN 1743-3533 (on-line)
doi:10.2495/MPF110141

1 Introduction

Micro bubble formation is a phenomenon affecting a number of industries, including the food, pharmaceutical, building and chemical industries. Bubble nucleation finds its origins in the presence of superheated or supersaturated solutions. In a domestic central heating system micro bubble formation is the result of water supersaturated with dissolved nitrogen gas, consequently, leading to bubble nucleation on the boiler wall. The use of the term boiler for a standard domestic central heating unit could suggest that some form of boiling takes place in the system's primary heat exchanger, consequently leading to the formation of bubbles through superheat. However, under no operating conditions does the phenomenon of flow boiling or sub cooled flow boiling take place in a domestic central heating unit. This is necessary so as to avoid excess noise originating from the boiler unit and to prevent excessively high exit temperatures thus ensuring the safe operation of the system within the domestic environment.

Micro bubble nucleation is solely attributed to the presence of gas super saturation levels in the water flowing in the system's closed loop circuit. Long term tests have shown that such systems could result in saturation ratios as high as 1.2 at the boiler wall conditions. Such conditions are present during cold start-ups and after the system filling with tap water. Air is mostly absorbed in the system during the cold cycle. At low temperatures, water can absorb the highest quantity of dissolved gasses [1, 2]. In most systems this occurs during night time when the system's boiler shuts off.

Data on micro bubble characteristics in central heating systems is important as a good knowledge of bubble sizes and their distribution is essential for an efficient passive deaeration process. Passive deaerators are more efficient in capturing larger sized bubbles. Deaeration is an important feature of such systems as bubbles accumulate in radiators and result in cold spots, thus reducing the heat transfer area of the radiator and the overall system efficiency. Bubbles are also known to result in unwanted noise, blockages and corrosion. Domestic central heating systems amount to 16% of the carbon dioxide emissions in the UK [3] and consequently an optimised system performance should have significant environmental benefits. The widespread use of domestic central heating systems was also highlighted by Spreitzer *et al.* [4] who state that almost all residential buildings in Germany make use of a wet central heating system which results in circa 40% of the primary energy consumed in Germany. Furthermore, Sauer *et al.* [5] also state that central heating systems are responsible for 70% of all domestic emissions in Germany.

A bubbly flow in the circuit's forward flow line results from the detachment of micro bubbles from the boiler wall into the system. Bubbly two-phase flow is characterised by the presence of bubbles of maximum size much less than the containing vessel or duct. The bubbles are dispersed in a continuous liquid phase [6]. From our test results no micro bubbles were observed on the return line to the boiler, thus suggesting that micro bubbles dissolve, are deaerated or rise to high points in radiators or vessels while flowing through the system. The analogy between this present study and theories developed for the prediction of bubble diameters in two-phase fully developed turbulent bubbly flow in ducts extended this study to the consideration of the two known models for predicting bubble sizes, these being the Hinze [7] approach as adapted by Winterton and Munaweera [6], and the Winterton and Orby [8] approach. However, these studies are based on experimental data obtained through the artificial insertion of micro bubbles in the flow.

In this paper we will investigate the phenomenon of micro bubbles in wet domestic central heating systems. The average bubble diameters, size distributions and bubble production rates at typical primary heat exchanger saturation ratios will be investigated.

2 Theory

2.1 The solubility of gasses in liquids and supersaturated solutions

Fogg [9] states that the solubility of gases in a liquid is a property of a gas dependent on its partial pressure and on the temperature as well as the nature of the liquid phase. In fact, for most gas-liquid systems there tends to be a linear variation of solubility with the partial pressure as this approaches zero. When defining the solubility of gases, Young *et al.* [2] highlight the point that the distinction between vapor-liquid equilibria and the solubility of gases in liquids is arbitrary and often the distinction between the two is not clear. This issue arises from the general inability to rigorously distinguish between a gas, a vapor and a liquid. However, Gerrard [1] defines gases as all the elements and compounds having a boiling point at 1 atm less than 13°C. Gas solubility data at standard atmospheric and volumetric conditions is available through a number of publications [2].

Jones *et al.* [10] defines a supersaturated solution in relation to quantifying the tendency of a system to produce bubbles. They refer to the saturation data as a function of the temperature for the system. A solution could go into the supersaturated state through the increase of its temperature. In fact, point A in fig. 1 represents a saturated solution at a temperature T_A with a saturation mole fraction amounting to X_b. When the temperature of this solution is increased to T_B, the solution would be in its supersaturated state while still retaining the previous mole fraction of the dissolved gas content. The desorption of gas from the water then causes the state of the system to move gradually from point *B* to point *B'* with a new saturation mole fraction equal to X_i. In view of this the resultant saturation ratio is defined through the relationship in eqn. (1);

$$\alpha = \frac{X_b}{X_i} \tag{1}$$

Furthermore, the super saturation ratio is defined as

$$\sigma = \alpha - 1 \tag{2}$$

Our saturation ratio was calculated through the use of eqns. (3) and (4). as done by Lubetkin and Blackwell [11] who calculated the gas concentration in the fluid through the use of the Henry's law as defined by Gerrard [1]. The partial gas pressure was calculated by subtracting the vapour pressure from the total dissolved gas pressure.

$$C_{gas} = P_g X^T \tag{3}$$

The saturation ratio was calculated through the use of eqn. (1). As shown in eqn. (4), the ratio of the actual gas concentrations was used where C_{sat} refers to the maximum gas solubility at the primary heat exchanger wall temperature.

$$\alpha = \frac{C_{gas}}{C_{sat}} \tag{4}$$

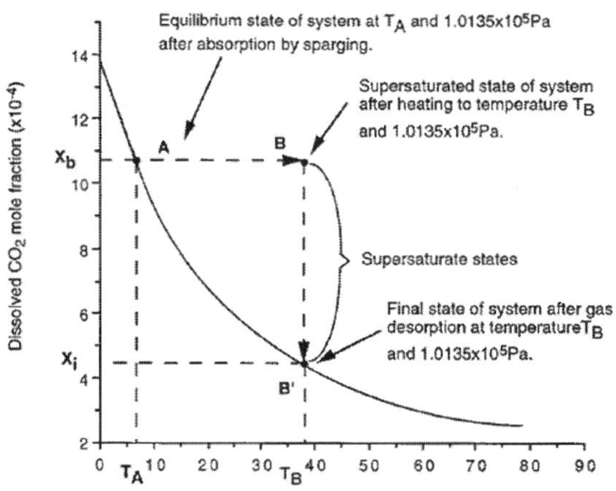

Figure 1: Solubility of carbon dioxide [10, p.30].

2.2 Bubble size prediction

Research in bubble nucleation and growth in supersaturated solutions is not widespread and consequently the number of studies available is limited to very specific and theoretical systems. A model for predicting the detachment radius of bubbles originating in supersaturated duct flow is based on the balance of the drag and surface tension forces acting on the surface of the nucleating bubble on the primary heat exchanger wall, parallel to the wall surface. Winterton's [12] approach for the prediction of bubble detachment diameters is based on this theory. Other models have been developed for predicting the bubble radii such as the approach originally suggested by Scriven [13] and later adapted by Jones *et al.* [10]. These models are based on the symmetric phase growth controlled by heat and mass transfer.

These models were developed for modelling bubble nucleation and require a comprehensive knowledge of the surface conditions and the bubble nucleating time. The lack of knowledge with respect to the bubble nucleating time in the helical coils of the boiler's primary heat exchanger made this model unusable in the present study. The model developed by Winterton [12] is based on the knowledge that bubbles break away from the surface into the flow when the drag force equals the surface tension force. For a finite contact angle Winterton [12] derived eqn. (5) for the experimental conditions of the present study.

$$\frac{r}{R_h} = 1.4\left[\frac{F(\theta)}{We}\right]\mathrm{Re}^{0.24} \tag{5}$$

where $F(\theta) = \sin\theta_0(\cos\theta_r - \cos\theta_a)$ and $We = \frac{(\rho u^2 D_h)}{\gamma}$.

A number of recent studies have adapted the use of the Winterton model with good results. Amongst these are studies done by Al-Hayes and Winterton [14, 15] and Hepworth *et al.* [16]. Al Hayes and Winterton adapted the original Winterton model to include the effect of liquid motion on the bubble growth rate due to the change in the mass transfer coefficient for the gas entering the bubble from the bulk liquid. Their final approach is similar to the original Scriven [13] and Jones *et al.* [10] models and thus requires a comprehensive knowledge of the bubble nucleating time.

3 Experimental facility

A schematic diagram of the experimental set up is shown in fig. 2. The test rig consists of a Valliant eco TEC pro 24 condensing boiler that is connected to standard 22mm copper tubing which supplies a radiator and a buffer vessel. A condensing boiler is used as this is mandatory equipment for new buildings in most European Union member states [3]. Stainless steel sheathed K type thermocouples are used to measure the fluid temperatures along the circuit whereas pressure transducers monitor the system pressure. A pressure transducer monitors the dissolved gas partial pressure in combination with a semi permeable silicone membrane. The system fluid flow rate is monitored through an Electromag 500 Series electromagnetic flow meter. A National Instruments cDAQ-9172 chassis and data modules receive all the signals from the transducers, thermocouples and the electromagnetic flow meter. The three system parameters are controlled as follows:

i The system flow rate or velocity is varied through the use of a ball valve on the supply line. The velocity in the boiler tubes was stepped between a minimum 0.3 m/s and a maximum of 0.8 m/s. This is equal to a system flow rate ranging from 4.5 to 12.5 litres per minute. The system pressure was set at 2.75 bars with a saturation ratio of 1.1 at the boiler wall conditions.

ii The system heating load was varied between a minimum of 7.5 kW and a maximum of 21.5 kW through the step increase in the boiler flame settings. The heating load is equal to a boiler heat flux ranging between 17 and 50 kW/m^2. The inlet temperature was maintained at a constant 55°C to ensure sufficient latent heat recovery through the boiler in-built condenser. The exit temperature ranged between 65 and 84°C while a system pressure of 2.75 bars (abs) was maintained. An average saturation ratio of 1.1 at the boiler wall conditions was maintained.

iii Super saturation ratios, ranging between 1 and 1.2 at boiler wall conditions were achieved through the filling of the top part of the radiator with nitrogen gas. The system pressure was set at 2.75 bars and a system flow rate of 12.5 litres per minute was maintained.

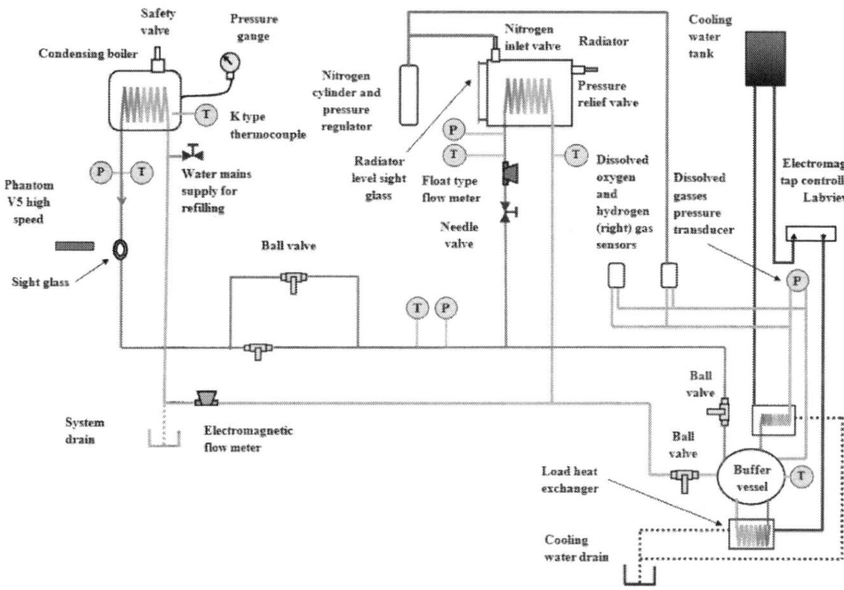

Figure 2: Schematic diagram of experimental set up.

Standard central heating systems make use of untreated steel radiators and copper piping. The untreated radiators result in a limited amount of oxidation due to the dissolved oxygen present in the tap water. The oxidation process releases iron oxide and some hydrogen gas. The analysis of dissolved gases through the use of Orbisphere 3655 oxygen and Orbisphere 3654 hydrogen sensors resulted in very low concentrations of oxygen and hydrogen present in their dissolved form. In fact both gases were present in concentrations of circa 9 PPB. Therefore, nitrogen is evidently the dominant gas and its properties were used for the dissolved gas properties in this study. A square sight glass with internal dimensions of 20x20 mm was used for filming micro bubbles at the exit line of the boiler. A square section was designed to reduce the distortion as a result of viewing bubbles through a curved surface. As discussed by Prodanovic *et al.* [17], such distortions are due to light refraction. A Vision research Phantom V5 high speed camera connected to a PC was used to film and store the video clips. A monozoom (Navitar) microscope lens was used to develop the desired magnification and a shutter speed of 30 μs and a frame speed of 100 frames per second were used. Lighting was provided by two high intensity Everest.VIT ELSV 60 W light sources attached to semi rigid fibre optic light guides.

4 Image analysis

The video films were converted to image frames saved as 'tag image file format' or tiff files using the Phantom Version 606 camera software. The analysis of the bubble diameters was done manually using the Phantom Version 606 Image Analysis software using a representative sample of 100 bubbles for each experimental run [17]. This was done so as to ensure an accurate result due to the general limitations of the automated image processing software.

The bubble production rates were calculated through the use of the image analysis software, Image-Pro Plus developed by Media Cybernetics. An average of 1500 frames were analysed for each experimental run. A macro was written enabling a series of images to be analysed for in focus bubble counts and diameters. The macro included the use of a Sobel filter to enable the distinction between in and out of focus bubbles. The Sobel filter plots the gradient of intensity change between objects and their background through the extraction and enhancement of edges and contours. This is done by expressing intensity differences or gradients between neighbouring pixels as an intensity value. Therefore, objects that are in focus have sharp edges with a high gradient change and consequently result in a high intensity values, whereas out of focus objects do not display such a characteristic. The Sobel filter was used as it is less sensitive to image noise when compared to other filtering techniques [18]. A typical analysed image is illustrated in fig. 3, where in focus bubbles are circled. Experimental uncertainties were calculated based on the method given by Coleman and Steel [19] and estimated as a mean absolute value of 3% for the manual bubble diameter measurements and 12% for the bubble production rates.

5 Results and discussion

The measured bubble diameters are illustrated in fig. 4. These measurements have shown an increase in the bubble diameter with an increase in heat flux or

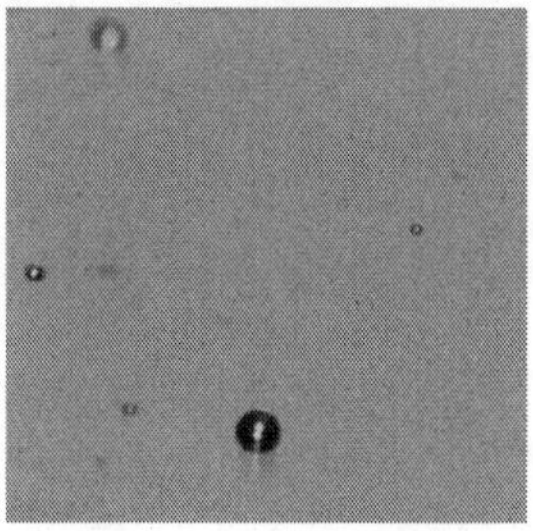

Figure 3: Camera image (post processing – in focus bubbles circled).

the corresponding space heating load and a decrease in velocity. Changes in the super saturation ratios did not result in notable bubble diameter differences. Bubbles were mostly spherical in shape and therefore, no elongation is attributed to the surrounding fluid flow. This is in agreement with recent studies done in bubbly flows [6]. The present study did not permit a real time measurement of the bubble diameters during their detachment from the heat exchanger wall. Therefore, the positioning of the sight glass at the exit end of the boiler with sufficient insulation to ensure adiabatic bulk fluid flow conditions, allowed the authors of the present study to assume that the observed bubble diameters are reasonably equal to the detachment diameters. Furthermore, the effects of bubble coalescence and dissolution during their short time in the bulk fluid flow were assumed to be at a minimum.

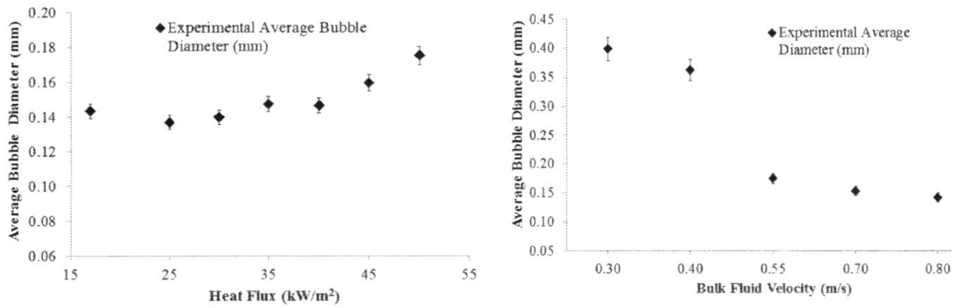

Figure 4: Average bubble diameters with heat flux and velocity in heating tubes.

Winterton's finite contact angle model predicts the bubble diameters with changing system parameters with a mean absolute error of 20%.As illustrated in fig. 5, all data points fall between ± 30% of the model prediction. With due consideration to the limitations of studies in two-phase flow as discussed by Winterton and Munaweera [6] and the general limitations of the present study, the adapted model predicts the expected bubble diameters reasonably well. The Winterton model accurately predicts the decrease in bubble diameter with an increase in bulk fluid velocity for velocities ranging from 0.4 to 0.8 m/s. However, at a lower velocity the error amounted to 45%. The inclusion of the Weber and Reynolds number enables the Winterton model to predict the bubble detachment diameters with changing bulk fluid flow. Similar experimental results were observed by Al-Hayes and Winterton [14] and in studies done in sub cooled flow boiling by Prodanovic *et al.* [17]. Other more recent studies done by Hepworth *et al.* [16] suggest that the bulk fluid velocity could increase the bubble growth rate and consequently lead to higher nucleation rates due to the liquid motion affecting the mass transfer coefficient for the gas entering the bubble on the heat exchanger wall. As illustrated in fig. 4, the increase in heat flux resulted in an increase in the bubble diameter by circa 20%. This is attributed to the higher boiler wall temperatures resulting in an increase in the diffusivity of nitrogen moving into the bubbles growing on the heat exchanger wall. This trend is not well represented in the Winterton finite contact angle force balance model. However, a minimal increase in the predicted diameter is represented through the change in the fluid properties, such as the density, kinematic viscosity and surface tension. Jones *et al.* [10] predict a reduced nucleation time with higher temperature due to the higher diffusivities present at the heat exchanger wall. However, this is not reflected on the predicted bubble detachment diameter. Similar trends were reported by Abdelmessih *et al.* [20] in flow boiling.

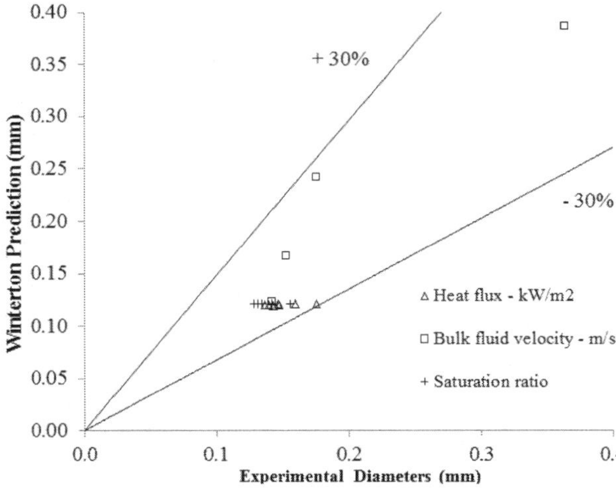

Figure 5: Accuracy plot.

Changes in the saturation ratio did not result in notable differences in the bubble diameters. A higher gas concentration gradient at the boiler's primary heat exchanger wall is expected to result in an increase in the bubble growth rate. However, the force balance theory seems to dominate and thus a minimal effect is seen on the resultant bubble diameter. Hepworth *et al.* [16] reported an increase in the bubble diameter with an increase in their dissolved gas concentration. However, higher saturation ratios were present and consequently, our observed constant bubble diameter with changes in the saturation ratio could be attributed to the narrow super saturation range present in our system.

The bubble distributions illustrated in fig. 6 illustrate the expected trends as seen in the experimental mean diameters whereby smaller bubbles make up a greater percentage of the total bubble count for the lower heat flux levels. Higher heat flux or heating loads could result in increased saturation ratios present at the boiler wall and therefore, a higher bubble production rate. However, the higher percentage of larger bubbles present at high system heating loads would facilitate the deaeration process through a passive deaerator and therefore at higher loads systems are expected to deaerate more efficiently. In fact, passive deaeration systems make use of deaeration chambers where bubbles float up a column where they are subsequently expelled through a float valve.

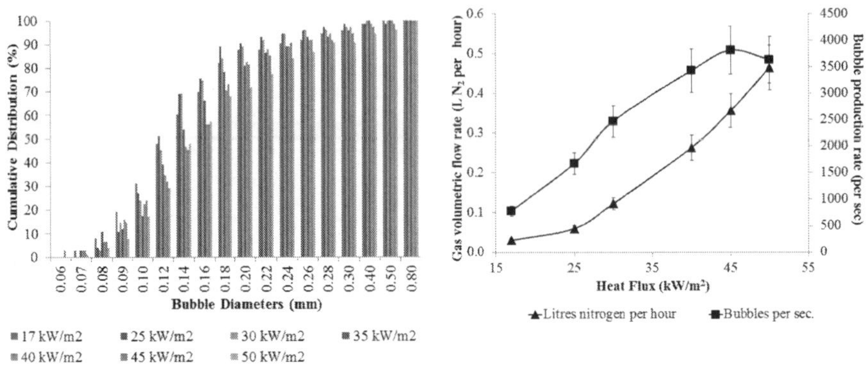

Figure 6: Bubble distribution, volumetric flow and production rate with heat flux.

Fig. 6 also illustrates the bubble production rate and the volume of air exiting the heat exchanger. The system bubble production rates range from 800 to 3800 bubbles per second. This corresponds to a gas volume production ranging from 0.03 to 0.46 litres per hour of operation at a constant system flow rate of 12.5 litres per minute. Tests have shown that the bubbles produced do not return to the boiler. Therefore, such bubbles are expected to dissolve back to the water in colder regions in the system or rise to high spots such as radiators located on the higher floors. Continuous system operation would therefore lead to the formation of a gas filled void at the top part of the radiators, consequently reducing the heat transfer efficiency, prompting the system user to increase the system operation time and exit temperature. This results in an increased carbon footprint. The system could then require the process of bleeding followed by a system re-pressurisation. No literature sources discussing bubble gas volumes in similar systems are known to exist. Therefore the possibility of data comparison here is very limited. However, when considering the total bubble production together with the boiler area that is under super saturation conditions, the average nucleation rate amounts to circa 1.3 bubbles/cm^2 s. This compares well with similar studies, most notably that done by Verschaeren on smooth metal heated surfaces [21].

6 Conclusions

This study has adapted mostly theoretical two-phase studies to more practical grounds that could lead to a better understanding of the two-phase characteristics in wet domestic central heating systems. An analysis of the expected micro bubble diameters has resulted in bubbles with average diameters within the range of 0.13 and 0.39 mm with changing system parameters. Good predictions were achieved through the use of Winterton's finite contact angle force balance prediction for bubble detachment radii in supersaturated solutions. Tests have shown that the average bubble diameter decreases with an increase in the bulk fluid velocity and a reduced heat flux and exit temperature while no change was evident with changes in the saturation ratio. An analysis of the typical bubble size distribution over a range of heating loads and at a typical saturation ratio of 1.1 at the primary heat exchanger wall has shown a larger concentration of smaller sized bubbles at lower heat flux levels. The system bubble production rates ranged from 800 to 3800 bubbles per second while the corresponding gas volume production ranged from 0.03 to 0.46 litres of nitrogen gas per hour of operation at the minimum and maximum system heating loads respectively.

Acknowledgement

The authors would like to thank the Engineering and Physical Science Research Council, Spirotech b v., The Netherlands for supporting this research work.

Notation

C_{gas}	Gas concentration in system, (standard cm^3/Litre Water)
C_{sat}	Maximum gas concentration at boiler wall temperature, (standard cm^3/Litre Water)
D_h	Helical tube hydraulic diameter, (m)
P_g	Partial pressure of dissolved gas and vapour pressure, (Pa)
r	Bubble detachment radius, (m)
R_h	Helical tube hydraulic radius, (m)
Re	Reynolds number, (-)
u	Bulk fluid velocity, (m/s)
X_b	Bulk gas concentration, (mol/m^3)
X_i	Saturation gas concentration, (mol/m^3)
X_T	Gas solubility factor, (standard cm3/Litre Water/bar)

Greek letters

α	Saturation ratio, (-)
γ	Surface tension, (N/m)
θ_a	Advancing contact angle, (degrees)
θ_o	Static contact angle, (degrees)
θ_r	Receding contact angle, (degrees)
ν	Kinematic Viscosity, (m^2/s)
ρ	Water density, (kg/m^3)
σ	Super saturation ratio, (-)

References

[1] Gerrard W., *Solubility of gases and liquids*, Plenum Press, New York, 1976.

[2] Young C.L., Battino R., Clever H.L., *The solubility of gases and liquids – introductory information, Nitrogen and Air, solubility data series – Volume 10*, Pergamon Press, Oxford, 1982.

[3] *The Building Regulations, Conservation of Fuel and Power – Part L1A*, Office of the Deputy Prime Minister, 2006.

[4] Spreitzer K., Rückbrodt D., Straky H., Observer-based estimation of the water mass flow through a central heating boiler, *Proceedings of the American Control Conference Anchorage AK*, **37,** pp. 5054-5059, 2002.

[5] Sauer J., Schmeink C., Wastell D.G., Feedback Quality and Environmentally Friendly Use of Domestic Central Heating Systems, *Journal of Ergonomics*, **50.(6)**, pp. 795-813, 2007.

[6] Winterton R.H.S., Munaweera J.S., *Bubble size in two-phase gas-liquid bubbly flow in ducts, Chemical Engineering and Processing*, **40,** pp. 437-447, 2001.

[7] J.O. Hinze, *Fundamentals of the hydrodynamic mechanism of splitting in dispersion process*, AIChE Journal, **1,** pp. 289-295, 1955.

[8] Winterton P., Orby P., Bubble size in bubbly flow in ducts and bubble columns, *Proceedings of the 10th International Heat Transfer Conference*, **6,** pp. 295-299, 1994.

[9] Fogg P.T.G., Some Aspects of the Solubility of Gases in Liquids, *Monatshefte für Chemie*, **134,** pp. 619 – 631, 2003.

[10] Jones S.F., Evans G.M., Galvin K.P., The cycle of bubble production from a gas cavity in a supersaturated solution, *Advances in Colloid and Interface Science*, **80,** pp. 27-50, 1999a.

[11] Lubetkin. S., Blackwell M., The nucleation of bubbles in supersaturated solutions, *Journal of Colloid and Interface Science*, **26,** pp. 610-615, 1988.

[12] Winterton R.H.S., Sizes of bubbles produced by dissolved gas coming out of solution on the walls of pipes in flowing Systems, *Chemical Engineering Science*, **27,** pp. 1223-1230, 1972a.

[13] Scriven L. E., On the dynamics of phase growth, *Chemical Engineering Sciences*, **10,** pp. 1-13, 1958.

[14] Al-Hayes R.A.M., Winterton R.H.S., Bubble growth in flowing liquids, *International Journal of Heat and Mass Transfer*, **24,** pp. 213-221, 1981a.

[15] Al-Hayes R.A.M., Winterton R.H.S., Bubble diameter on detachment in flowing liquids, *International Journal of Heat and Mass Transfer*, **24,** pp. 223-229, 1981b.

[16] Hepworth N.J., Boyd J.W.R., Hammod J.R.M., Varley J., Modelling the effect of liquid notion on bubble nucleation during beer dispense, *Chemical Engineering Science*, **58,** pp. 4071-4084, 2003.

[17] Prodanovic V., Fraser D., Salcudean M., Bubble behaviour in sub cooled flow boiling of water at low pressures and low flow rates, *International Journal of Multiphase Flow*, **28,** pp. 1-19, 2001.

[18] Image-Pro Plus, *Start-Up Guide*, Media Cybernetics, 2010.

[19] Coleman, H.W., Steele, W.G., *Experimentation and uncertainty analysis for engineers*, 2nd ed., John Wiley& Sons Inc., New York, 1999.

[20] Abdelmessih A.H., Hooper F.C., Nangia S., Flow effects on bubble growth and collapse in surface boiling, *International Journal for Heat and Mass Transfer*, **15,** pp. 115-125, 1972.

[21] Verschaeren R., *Degassing nucleation sites in water flow under practical heating conditions*, Master in Mechanical Engineering, Technical University Eindhoven, 2010.

Experimental and computational study of pressure drop and void fraction in a bubbling fluidized bed.

C. K. Jayarathna & B. M. Halvorsen
Institute of Process, Energy and Environmental Technology, Telemark University College, Norway

Abstract

The aim of this work is to study flow behaviour in a fluidized bed with different mixtures of particles. Mathematical simulation is an alternative way to study the fluidization behaviour. Experiments are performed in a cylindrical bed with a uniform air distribution. Spherical glass particles with different mixtures of particles are used in the experiments. The pressure and void fraction variations are observed for two different powders and mixtures of the powders. The commercial CFD code Fluent 6.3 is used for the corresponding simulations. Eulerian-Eulerian model are used to simulate a multiphase bubbling fluidized bed. The influence of particle size distribution on the bubble size, pressure and void fraction variation along the bed has been investigated computationally. The computational results are compared to the experimental data and the discrepancies are discussed.
Keywords: fluidized bed, particle size distribution, pressure drop, CFD, fluent.

1 Introduction

Fluidized beds are widely used in industry because of good mixing and large contact area between phases. It enhances the chemical reactions, heat transfer and mass transfer. Liquid like behaviour of the bed particles are giving smooth operation conditions and well mixed beds are obtaining isothermal properties, hence the operation can be easily controlled.

Particle mixing in a fluidized bed is very much depending on bubble size distribution and the particle characteristics [1]. The behaviour of particles in

WIT Transactions on Engineering Sciences, Vol 70, © 2011 WIT Press
www.witpress.com, ISSN 1743-3533 (on-line)
doi:10.2495/MPF110151

fluidized beds depends on a combination of the particle size and density. Geldart fluidization diagram [2] is used to identify characteristics associated with fluidization of powders. Geldart Group A particles are easily fluidized and give a high bed expansion which means high void fraction before bubbles appear. This type of powders is mostly used as catalyst in fluidization system. Geldart group B particles give low bed expansion, and bubbles will appear as soon as the gas velocity reaches the minimum fluidization velocity [3].

Earlier studies have shown that powders with a range of particle sizes cannot be characterized only based on mean diameter [4–7]. The fluidization properties are highly influenced of the bulk density and changes in bulk density with the particle distribution as well. The bubble behaviour and the bubble size depend on pressure and the void fraction variations along the bed. Pressure readings conveys the information regarding the hydrodynamics of the fluidized bed, detecting the regime transitions, to investigate the chaotic behaviour of the fluidized bed and monitoring the quality of fluidization [1].

In this work the pressure and void fraction variations for different mixtures of particles are studied.

2 Computational model

Computational studies have been performed on a two dimensional fluidized bed. The simulations are performed by using the commercial CFD code Fluent 6.3. The model is based on an Eulerian description of the gas and the particle phases. The combinations of models used in this work are presented in Table 1. Jayarathna et al. [7] made a computational study of the influence of particle size distribution on flow behaviour in fluidized beds and by studying different combinations of models they concluded that the combination presented in Table 1 gives the most realistic flow behaviour.

Table 1: Recommended combination of models [10].

Property	Model
Drag	Syamlal and O'Brien
Granular viscosity	Syamlal and O'Brien
Granular bulk viscosity	Constant
Frictional viscosity	Schaeffer
Frictional pressure	Based-ktgf
Solid pressure	Ma-ahmadi
Radial distribution function	Ma-ahmadi

The Syamlal and O'Brien drag model [13] is used to express the solid-gas interaction. The model is expressed in eqn. (1):

$$\Phi_{sg} = C_D \frac{3\varepsilon_s \varepsilon_g \rho_g \left|\vec{U}_g - \vec{U}_s\right|}{4 v_r^2 d_s} \tag{1}$$

where ε_g and ε_s are the gas and solid fractions, ρ_g is the gas density, U_g and U_s are the gas and solid velocities and d_s is the particle diameter. The terminal velocity correlation for the solid phase, v_r, is a function of void fraction and Reynolds number [10]. The drag factor developed by Dalla Valle [11] is presented in eqn. (2):

$$C_D = \left(0.63 + \frac{4.8}{\sqrt{\mathrm{Re}_s / v_r}}\right)^2 \tag{2}$$

The granular viscosity includes a collisional and a kinetic viscosity term. The kinetic term is given in eqn. (3):

$$\mu_{s,kin} = \frac{\varepsilon_s d_s \rho_s \sqrt{\Theta_s \pi}}{6(3e_{ss} - 1)} \left[1 + \frac{2}{5}(1 - e_{ss})(3e_{ss} - 1)\varepsilon_s g_{0,ss}\right] \tag{3}$$

and the collisional term is presented in eqn. (4):

$$\mu_{col} = \frac{4}{5} \varepsilon_s d_s \rho_s g_{0,ss} (1 + e_{ss}) \sqrt{\frac{\Theta_s}{\pi}} \tag{4}$$

where d_s e_s and Θ_s are the particle diameter, elasticity coefficient and the granular temperature of solid phase s respectively. The radial distribution function is presented by $g_{0,ss}$. The radial distribution function included in the Syamlal and O'Brien [13] symmetric equation is expressed by Ma and Ahmadi [12].

Two particle phases are included in the simulations of mixtures. Syamlal-O'Brien symmetric is used to express the particle-particle momentum exchange [13].

3 Experimental and computational set-up

Experiments and corresponding simulations are performed. The set up is presented in this chapter.

3.1 Experimental set-up

A lab-scale fluidized bed with a uniform air distribution is constructed. The bed is cylindrical and is made of Lexan glass. The diameter and the height of the bed are 0.072 and 1.4 m respectively. The gas flow rate is controlled by a pressure reduction valve, and measured by a digital flow meter. The pressure can be measured at eight positions in the bed.

Glass particles with two different mean particle sizes are used in this study. The particles have the size range 100–200 μm (small particles) and 400–600 μm (large particles). The particle density is 2485 kg/m^3. The bed and particle parameters are presented in Table 2.

Experiments have been performed with 100% small particles, 100% large particles and mixtures of small particles with 20, 40, 50, 60 and 80% of large particles. The aim is to study how the different fractions of particle sizes influence on the minimum fluidization velocity and the bed expansion. Before the experiments with thc mixtures started, the powders were well mixed and 2 litres of a compact mixture were weighted and filled into the bed. The void fractions at start and at minimum fluidization were calculated based on the weight and the volume.

Table 2: Experimental data.

Bed design			
Height Diameter		1.4 m 0.072 m	
Particles (Spherical glass particles) Density: 2485 kg/m^3			
Particle range	100–200 μm (small)	400–600 μm (medium)	Mixture
Mean particle size	154 μm	488 μm	
% large particles			20, 40, 50, 60, 80

3.2 Computational set-up

The simulations are performed with particles with diameters equal to the mean diameters of the glass powders used in the experiments. The simulations are performed with mixtures of 0%, 20%, 40%, 50%, 60%, 80% and 100% large particles. Two particle sizes are used to simulate the mixtures of two powders with different mean particle size. The simulations are run with the same velocities and initial bed heights as used in the experiments. Two-dimensional Cartesian co-ordinate system is used to describe the geometry. The width and the height of the computational bed are 0.072 m and 1.4 m respectively. A grid resolution test is performed to find a suitable grid size. The grid is uniform and the size of a cell is 3x3 mm. The simulations have been run for 7 seconds. The simulations have been run with different gas flow rates.

4 Results

This chapter presents the experimental and computational pressure fluctuations and average void fractions variations of the different particles mixtures with different superficial air velocities.

4.1 Void fraction variations

All the particle samples are having an identical maximum compact volume. The volume of the particles has expanded when it is filled smoothly in to the fluidized bed. Figure 1 shows the void fractions and the compositions of large and small particles in the mixtures. It is possible to observe that the samples with only small and only large particles have higher void fractions than the particle mixtures with both large and small particles. This may be due to the repulsive forces between the small particles [2]; which means that they are having some barriers to reduce the voids as shown in Figure 2(a). The large particle cannot reduce the voids between the particles, because of their geometry as shown in Figure 2(b). The repulsive forces are not significant for large particles.

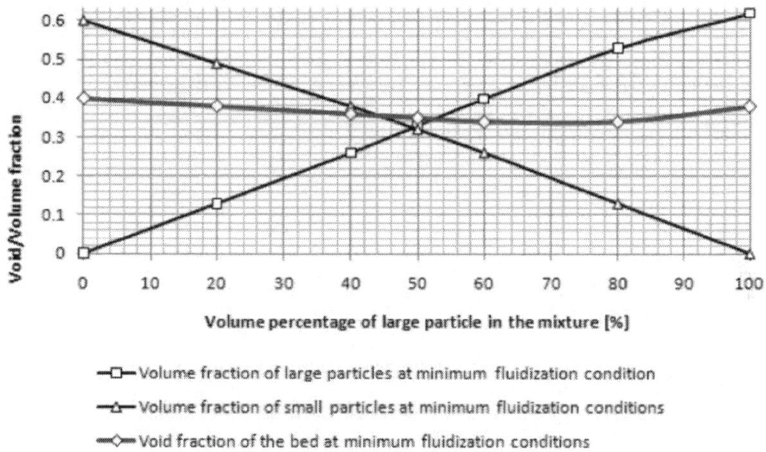

Figure 1: Void fractions and compotions of particles.

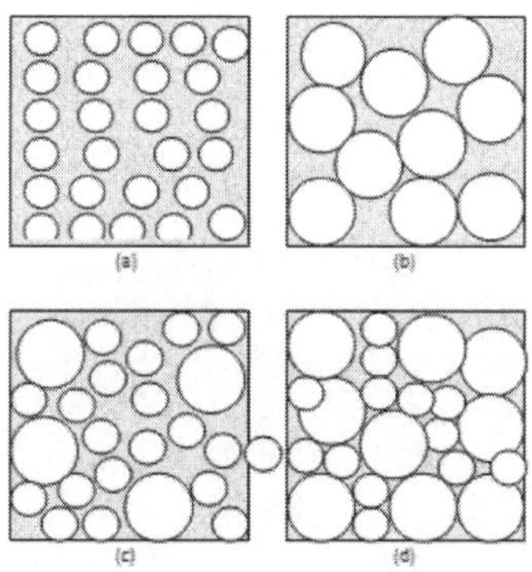

Figure 2: Possible particles location with two different sizes of particles.

In the mixtures of both small and large particles, lower void fractions are observed. The lowest void fractions are observed for the mixtures with 50% to 90% large particles. The minimum void fraction is found for the mixture of 70% large and 30% small particles. That means the mixtures of both large and small particles are more packed than the mono sized particle samples. There may be less space between particles in the mixtures with more large particles than in the mixtures with more small particles as shown in Figure 2(c) and (d).

4.2 Pressure variation

The pressure (gauge) is measured and calculated at height 3.5 cm, 33.5 cm and 63.5 cm in the fixed bed. Figure 3 shows the experimental and computational variations in the pressure versus percentage of large particles in the sample at different heights. The simulations give slightly lower pressure than the experiments. This may be because the particle samples used for experiments consist of a wide range of particle sizes while the simulations consider only the mean particle size to represent each mixture. The pressure along the bed is decreasing from the bottom to the top due to the weight of the particles above the measuring point [3].

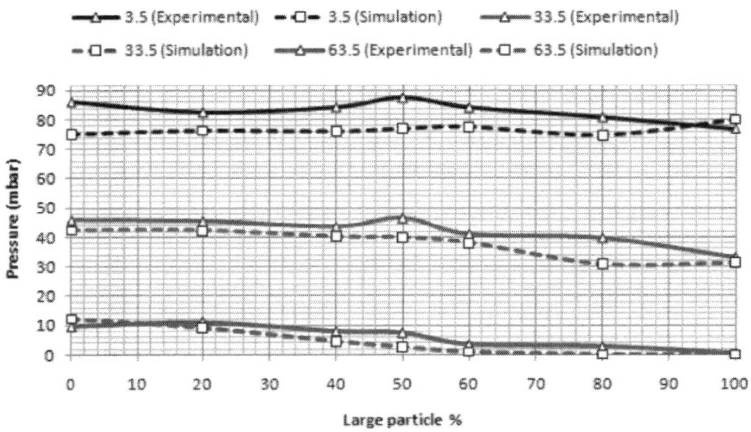

Figure 3: Pressure variations of the observations and the computations at three different bed heights. Fixed bed.

Figure 4 shows the total pressure at bed height 23.5 cm as a function of superficial air velocity for several selected particle mixtures. The figure includes both experimental and computational results. The curves from the experiments have steep gradients until they reach very high velocities whereas the gradients of the curves from the simulations are less steep. In both the experiments and simulations the total pressure at height 23.5 cm is increasing with increasing superficial air velocity. This is due to the bed expansion and the change in bubble distribution in the bed when the velocity is increased. The main contribution to the variation in pressure is the weight of the particles above the measuring point. The distribution of voids and bubbles depends on the particle mixture and the superficial velocity. Bubble distribution for different flow conditions are further discussed in Chapter 4.3. The 40% experimental curve is rather flat at higher superficial air velocities which indicates that the average void fraction in the lower part of the bed is rather constant when the velocity exceeds about 0.2 m/s.

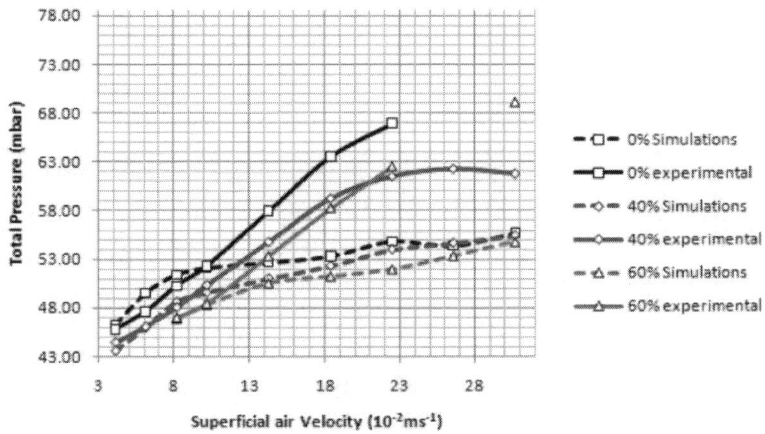

Figure 4: Pressure variations of observations and computations with respect to superficial air velocity at 23.5cm bed height.

4.3 Analysis of fluidized bed behaviour based on simulations

Fluidized bed behaviour is analyzed based on computer simulations for different particle mixtures. The simulations are performed for different superficial velocities with the lowest velocity as the minimum fluidization velocity observed in the experiments. Figure 5 shows the simulated fluidized bed of 100% small particles for three different gas velocities. No bed expansion is observed at velocity 0.029 m/s which is the experimental minimum fluidization velocity for the corresponding particle mixture. The computational minimum fluidization velocity is observed at superficial gas velocity 0.05 m/s. The bed expands to almost the double of the initial height when the velocity is increased to 0.307 m/s. It is possible to observe large air bubbles and very high bed expansions for the samples with 0%, 20% and 40% large particle mixtures as shown in Figure 5 and 6. These particle mixtures give slugging and turbulent beds at high superficial gas velocities.

Figure 7(a) shows the bed behaviour for a mixture of 50% small and 50% large particles after 7 seconds. The initial bed height is 0.511 m and the superficial velocities are varied from 0.039 m/s to 307 m/s. These simulations are performed for a mixture with a total volume fraction of large and small particle phases of 0.32 and 0.35 respectively. The initial and boundary conditions are the same as used in the experiments with the same particle mixture. The experimental minimum fluidization value for this particle mixture is 0.039 m/s whereas the minimum fluidization velocity in the simulation is observed at velocity 0.055 m/s. The deviation can be due to the several different particle sizes in the real mixture. In the simulation with air velocity of 0.039 m/s the gas percolates through the void spaces between stationary particles. This means the bed is at fixed bed conditions. At the next simulation the air flow of 0.045m/s is used, particles moved apart and they moved in restricted regions. This time the

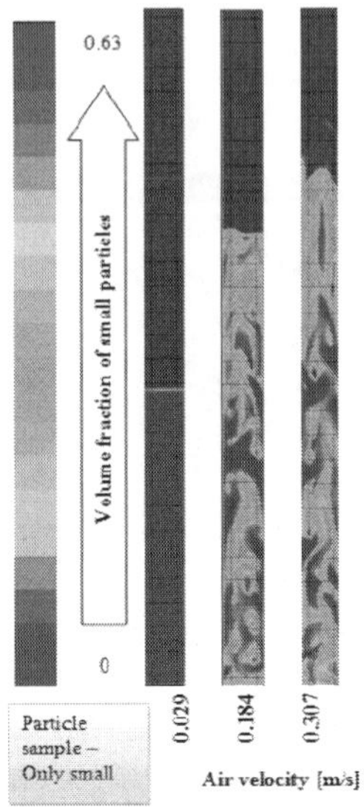

Figure 5: Simulated fluidized beds for several superficial air velocities for small particles.

bed has converted to an expanded bed and it is possible to observe a small increment of the bed height. The bed starts to fluidized when the air flow is increased to 0.055m/s. At higher velocities (0.102m/s and 0.143m/s) slugging started and the bed height is increased further, gas bubbles coalesce and grow as they rise. Figure 7(b) shows the flow behaviour for a mixture of 60% large particles. This mixture gives about the same bed expansion and bubble behaviour as the mixtures with a lower content of large particles. It is also observed that mixtures of 0%, 20%, 40%, 50% and 60% large particles have about the same minimum fluidization velocities [14, 15]. This indicates that the smallest particles in the mixture influence significantly on the flow behaviour even when the fraction of small particles are lower then the fraction of large particles.

Figure 8(a) and (b) show the flow behaviour for mixtures of 80% and 100% large particles respectively. These powders give a low bed expansion and smaller and more isolated bubbles. By studying Figure 8(a), it can be seen that still with only 20% small particles the flow behaviour are very much influenced by the small particles. The mixture with 80% large particles has started to fluidize at velocity 0.080 m/s whereas the 100% large particles start to fluidize at a velocity above 0.184 m/s.

At slugging conditions bubbles are moving upward by using a zigzag path as shown in Figure 9. This can be clearly visualized from the computer animations. All the generated air bubbles at the bottom of the bed are moving into low pressure zones. Bubbles are moving upward because of the generated highest pressure at bottom of the bed. When the bed is extremely narrow, bubbles are having a Zigzag path to escape from the bed as it is having obstacles to move directly upward.

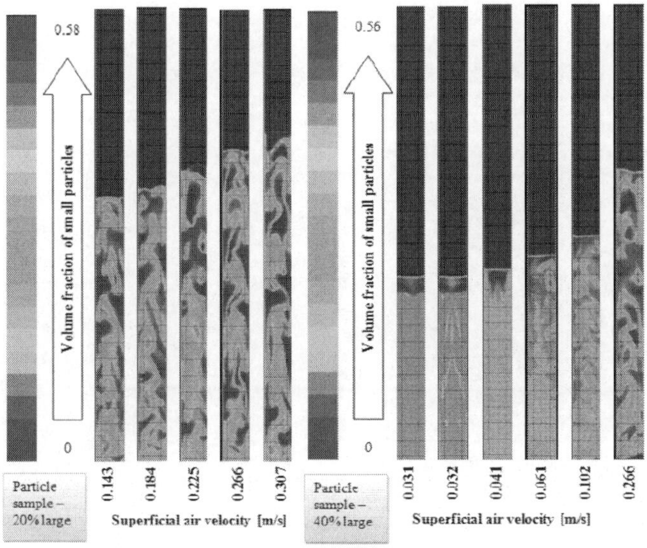

(a) (b)

Figure 6: Simulated fluidized beds for several superficial air velocities for 20% (a) and 40% (b) large particle mixture.

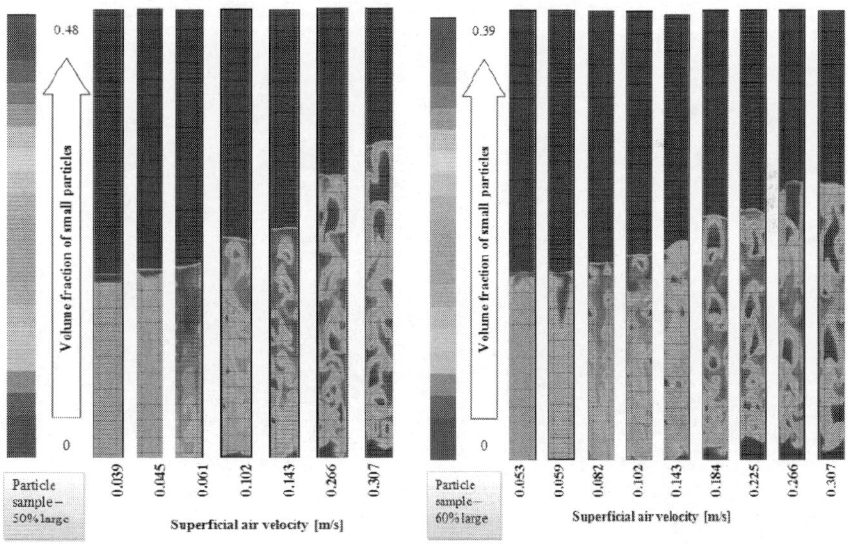

(a) (b)

Figure 7: Simulated fluidized beds for several superficial air velocities for 50% (a) and 60% (b) large particle mixture.

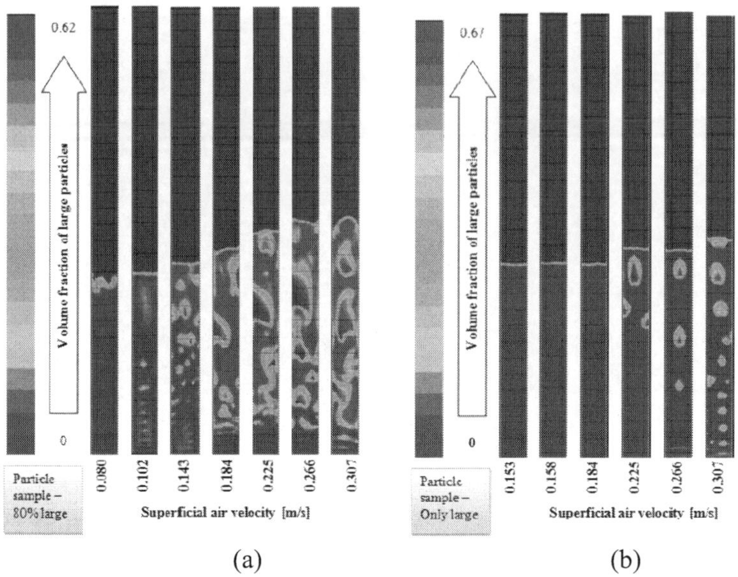

Figure 8: Simulated fluidized beds for several superficial air velocities for 80% (a) and 100% (b) large particle mixture.

Figure 9: Zig zag bubble leaving pattern.

5 Conclusion

A series of experiments and simulations are carried out. Experiments are performed in a cylindrical bed with a uniform air distribution. Different mixtures of spherical glass particles with mean diameter of 154µm (small particles) and 488µm (large particles) are used in the experiments. Corresponding simulations are performed by using the commercial CFD code Fluent 6.3. In addition to the

www.witpress.com, ISSN 1743-3533 (on-line)

pressure variations along the bed, and void fraction is calculated at minimum fluidization velocity. The void fractions for only small and only large particles are higher than for the mixtures of both particles sizes. Lowest void fractions are observed for the mixtures with 50%-90% large particles.

Fluidized bed behaviour is analyzed based on computer simulations for different particle mixtures and different superficial velocities. The highest bed expansion is observed for only small particles and the lowest for only large particle samples. Big air bubbles and high bed expansion are observed for the 0%, 20%, 40% and 60% large particle mixtures. More isolated and relatively small bubbles are formed in the fluidized bed with only larger particles. The fluidization and bubble behaviour are influenced significantly more by the small particles than then of the large particles in a mixture. At the slugging conditions bubbles are moving upward by using a zigzag path. According to the observation, the computational fluidized bed behaviour agrees well with the experimental data.

References

[1] Bai, B., Ommen, J.R.V., Nijenhuis, J., Coppens, M.O., Characterization of the void size distribution in fluidized beds using statistics of pressure fluctuations, *Powder Technology*, **160**, pp 81-92, 2005.

[2] Geldart, D., Gas *Fluidization Technology*, John Wiley & Sons Ltd., 1986.

[3] Kunii, D., Levenspiel, O., *Fluidization Engineering*, Second Edition, Butterworth-Heinemann, Newton, US, 1991.

[4] Halvorsen, B.M., Lundberg J., Mathiesen, V. (2008), Computational study of fluidized bed, *Conference on Heat Transfer, Fluid Mechanics and Thermodynamics*, HEFAT 2008, June 30–July 2, Pretoria, South Africa.

[5] Ariyarathna, D.G.A.S.U, Wo, W.J., Halvorsen, B.M. (2008), Verification of the importance of introducing particle size distributions to bubbling fluidized bed simulations, *SIMS 2008*, October 7th-8th, Oslo, Norway.

[6] Wu, W.J., Ariyarathna, D. G. A. S. U., Halvorsen, B. M.(2008), Experimental study of effects of particle size distribution on bubble behaviour for validation of CFD modelling of bubbling fluidized bed, *SIMS 2008*, October 7th-8th, Oslo, Norway.

[7] Jayarathna, S.A., Jayarathna, C.K., Wu, W.J., Halvorsen, B.M., (2008), Influence of particle size distributions on CFD simulations and experiments of bubbling fluidized beds. *AIChE Annual Meeting 2008*, November 16-21, Philadelphia, US.

[8] Ariyarathna, D.G.A.S.U., Recommendation of a Model for Simulating & Analysis of the Influence of Particle Size Distribution on the Simulations of Bubbling Fluidized Beds. In Thesis for the degree of MSc. Ing, pp. 22-46, 2008.

[9] Syamlal, M., The Particle-Particle Drag Term in a Multiparticle Model of Fluidization., National Technical Information Service, Springfield, VA, 1987.

[10] Garside, J. & Al Dibuouni, M.R., Velocity-Voidage Relationships for Fluidization and Sedimentation, I&EC Process Des. Dev., **16**, pp. 206-214, 1977.
[11] Dalla Valle, J.M., Micromeritics, Pitman, London, 1948.
[12] Ma, D., Ahmadi, G., Thermodynamical Formulation for Dispersed Multiphase Turbulent Flows, *Int. J. Multiphase Flow*, **16**, pp. 323-351, 1990.
[13] Syamlal, M., O'Brien, T.J., A Generalized Drag Correlation for Multiparticle Systems, Morgantown Energy Technology Center, 1987.
[14] Jayarathna, A.C.K., Experimental and computational study of the effect of particle size distributions on the flow behaviour in fluidized beds, In Thesis for the degree of MSc. Ing., 2009.
[15] Jayarathna, A.C.K., Halvorsen, B.M., Experimental and computational study of minimum fluidization velocity and bed expansion in a bubbling fluidized bed, SIMS 2009, October 7th-8th, Fredericia, Denmark.

Extension of a model for the distribution of voidage around bubbles in a fluidized bed

C. Rautenbach, C. Jayarathna, M. C. Melaaen & B. M. Halvorsen
Institute for Process, Energy and Environmental Technology, Telemark University College, Norway

Abstract

Up to date most models predicting the voidage around bubbles in fluidized beds assume a simple two-phase flow model. A classic example of a model using this basic assumption is the Davidson and Harrison bubble model. In more recent experimental studies it has been shown that there exists a shell of lower voidage around a bubble in a fluidized bed. Better understanding of these phenomena might play a significant role in understanding other very important effects associated with bubbles in fluidized beds. An example of this might be better understanding the wake of a bubble rising in a fluidized bed and also to better predict mixing in a fluidized bed. The amount of contact between the gas and the solids in a gas-solid fluidized bed is paramount for designing an effective fluidized bed reactor.

In the present study a model describing the voidage distribution in front and behind a bubble in a fluidized bed was reexamined and extended. The model proposed by Buyevich *et al.* used some basic assumptions and previously derived models to compose the model. In the present study the model is extended to two dimensions to give a non-uniform particle distribution around the bubble. Thus voidage distributions all around the bubble could be determined. This new distribution of the voidage could be used to determine a new form of the bubble surface and thus a new boundary condition for the fluidized bubble.

The results from the model are compared with simulations carried out at the TUC in Porsgrunn, Norway. The results are discussed and evaluated.
Keywords: solid fraction, voidage distribution, bubble model, fluidization.

WIT Transactions on Engineering Sciences, Vol 70,
www.witpress.com, ISSN 1743-3533 (on-line)
doi:10.2495/MPF110161

1 Introduction

1.1 The momentum conservation equation and some basic assumptions

In some special cases the dispersed phase in a gas-solid fluidized bed may be modeled as a dense gas [3]. The model extended in the present study is based on the closure of the conservation equation for the dispersed phase in a fluidized bed based on this dense gas assumption. Conservation equations for the mass, momentum and fluctuation energy can be postulated in this manner but for the present study only the momentum balance is of interest. The conservation of momentum in the dispersed phase in a fluidized bed can thus be expressed as

$$\rho_1 \left(\phi \partial / \partial t + \langle \phi \mathbf{w} \rangle \cdot \nabla \right) \mathbf{w} = \nabla \mathbf{P} + n \langle \mathbf{f} \rangle + \phi \rho_1 \mathbf{g}, \quad (1)$$

with ρ_1 the particle phase density, ϕ the solid fraction, $\mathbf{w}$ the particle velocity, $\mathbf{P}$ the tensor of the particulate stresses and $\mathbf{f}$ the interphase interaction force per particle [3]. The number concentration of particles are represented by n and $\mathbf{g}$ is gravitational acceleration. The averaged values in equation (1), namely the total volume flux, $\langle \phi \mathbf{w} \rangle$, and the random forces experienced by the particles, $\langle \mathbf{f} \rangle$, can be approximated by $\phi \mathbf{w}$ and $\mathbf{f}$ respectfully, given that the fluctuations of these quantities are relatively small [3]. For simplicity of the model these random forces and the total volume flux was assumed as non-fluctuating quantities. Another simplification proposed by Buyevich [3] is to neglect the quasi-viscous stresses. As a result the gradient of particle stresses can be written as $-\nabla p_1$, where p_1 represents the pressure in the dispersed phase. This approximation resembles the ideal-fluid approximation given by Euler [3]. The total conservation of momentum can thus be expressed as

$$\rho_1 \left(\phi \partial / \partial t + \phi \mathbf{w} \cdot \nabla \right) \mathbf{w} = -\nabla p_1 + n \mathbf{f} + \phi \rho_1 \mathbf{g}. \quad (2)$$

The random forces experienced by the particles consists of three entities namely the *drag force*, *buoyancy* and *inertial* effects which may originate from several different origins.

The *drag force* per particle can generally be expressed as

$$\mathbf{f}_d = m \left[F_1(\phi) + F_2(\phi) u \right] \mathbf{u}, \quad (3)$$

with $\mathbf{u} = \mathbf{v} - \mathbf{w}$ and where $\mathbf{v}$ is the interstitial velocity. Thus $\mathbf{u}$ is the fluid slip velocity [3]. In equation (3) m represents the particle mass. Several correlations exist for the prediction of F_1 and F_2 as a function of the solid fraction, ϕ, and other physical parameters. Any adequate drag model can be used and in the present study the model suggested in the work by Buyevich *et al.* [2] will be used. The functional value of F_1 will be more significant in the Darcy regime while F_2 will be more dominant in the Forchheimer regime. In the transition from one regime to the other both terms in the drag correlation will be of importance. For the sake of simplicity low particle Reynolds number flow was assumed in the present model. Thus only the Darcy regime will be of importance and thus F_2 will be assumed negligible.

www.witpress.com, ISSN 1743-3533 (on-line)

According to Buyevich this correlation for F_1 can be expressed in terms of a self-consistent model for finally dispersed concentrated suspensions and is given by

$$F_1 = \frac{9}{2}\frac{\rho_0\mu}{\rho_1 d_p^2}\frac{1}{(1-\phi)^n}, \tag{4}$$

where μ is the fluid viscosity, d_p is the effective particle diameter and ρ_0 is the gas density. For this application a value of $\frac{5}{2}$ is assumed for the value of n in equation (4) which corresponds to a value of 4.5 for the exponent in the Richardson and Zaki equation [2].

The contribution of *buoyancy* to the total force experienced by a particle can be expressed as

$$\mathbf{f}_b = -\frac{4\pi}{3}d_p^3\rho(\phi)\,\mathbf{g}, \tag{5}$$

where $\rho(\phi) = (1-\phi)\rho_0 + \phi\rho_1$ and is representative of the mean density of the mixture.

The last force to be considered is fluid *inertia*. This force will be ignored in the total interphase interaction force and is an acceptable assumption when working with gas as the ambient fluid [3]. Moreover this force will be much less significant compared to the influence of the *drag* and *buoyancy* forces.

Thus the total random interphase interaction force per unit volume of the mixture without fluctuations can be expressed as

$$n\mathbf{f} = n(\mathbf{f}_d + \mathbf{f}_b) = \phi\rho_1\left[F_1(\phi) + F_2(\phi)u\right]\mathbf{u} - \phi\rho(\phi)\mathbf{g}, \tag{6}$$

with n representing the number of identical spherical particles per unit volume of the mixture [3].

1.2 First approximations and closure

Following the procedure outlined in the work by Buyevich *et al.* [2] the Davidson and Harrison model [1] will be used as a first approximation for the particle velocity, $\mathbf{w}$, as well as for the relative interstitial gas velocity, $\mathbf{u}$. According to a simple filtration model for a homogeneous porous body containing a spherical void in two dimensions it follows that

$$\mathbf{u} = u_0(1 + \frac{2R^3}{r^3})cos\theta\,\overrightarrow{r} - u_0(1 + \frac{R^3}{r^3})sin\theta\,\overrightarrow{\theta}, \tag{7}$$

where R is the radius of a sphere with the same volume as the bubble at a particular height in the fluidized bed. Equation (7) is written in spherical coordinates where the radial and angular direction is defined in Figure 1. In the unperturbed dense phase the interstitial velocity will be given by

$$u_0 = (1-\phi_0)^{n+1}u_t, \tag{8}$$

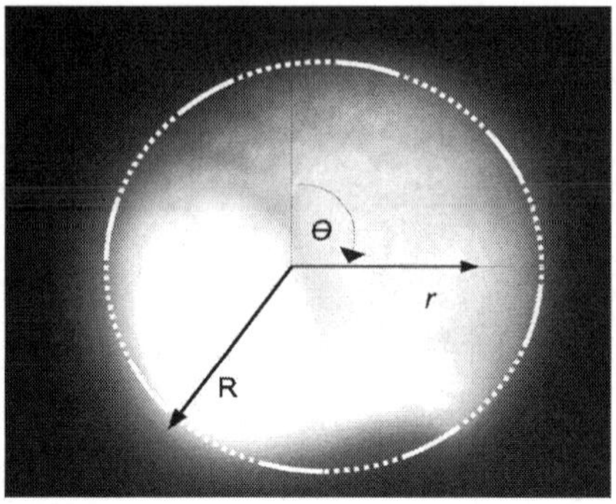

Figure 1: Schematic illustration of the proposed model alongside an image of a real fluidized bubble.

where u_t is the terminal fall velocity of a single particle in a stagnant gas and can be expressed as

$$\frac{\rho_1}{\rho_0}\frac{2d_p^2}{9\mu}g. \tag{9}$$

This particular model will be assumed to be adequate in the present study but other correlations might produce better results. Further work on this particular topic is planed in future work. The particle velocity, **w**, is given by the ideal flow model around a sphere and is expressed as

$$\mathbf{w} = U_b\left\{\left(\frac{R^3}{r^3}-1\right)cos\theta\overrightarrow{r} + \left(sin\theta+\frac{R^3}{2r^3}sin\theta\right)\overrightarrow{\theta}\right\}, \tag{10}$$

in two dimensions with $U_b = 2/3\sqrt{gR}$ from the Davidson and Harrison model [1].

The only term in equation (2) that still needs closure is the particulate pressure term. Different approximations also exist for the particulate pressure. In work done by Buyevich [4] models are suggested for the closure of the particular pressure. In the present study the particulate pressure will be assumed to be given by

$$p_1 = G(\phi)(\phi/\upsilon)T \tag{11}$$

as described by Buyevich [3]. In equation (11) ϕ/υ represents the number concentration of particles with υ the volume of a sphere. The function $G(\phi)$ describes the increase of the pressure in a dense gas consisting of hard spheres as compared to that of a dilute gas [2]. This function is based on a statistical mechanical model of Carnahan and Starlling [2]. Thus it follows that

$$G(\phi) = \frac{1+\phi+\phi^2-\phi^3}{(1-\phi)^3}. \tag{12}$$

In equation (11) the "temperature" of the pseudo-gas is represented by T. Again different models may be used to model this analogy to a molecular system and in the present study it will be assumed to be given by

$$(\phi/v)T = 6.83 \times 10^{-3}\phi M^2\langle\phi^2\rangle\rho_1 u^2, \tag{13}$$

with

$$M = \frac{n+1}{1-\phi} + \frac{g}{F_1(\phi)u}; \langle\phi^2\rangle = \phi^2\left(1 + 2\phi\frac{4-\phi}{(1-\phi)^4}\right)^{-1}. \tag{14}$$

In this model for the particular pressure, pulsation energy transfer and work done in expanding the particle pseudo-gas is ignored. Gradients in time and space of average variables are also ignored [2]. From literature it is clear that this particular model is not the best model for predicting the particulate pressure but it will suffice for now as a good starting point for expanding the existing model described by Buyevich *et al.* [2].

2 The Buyevich model

In the original work done by Buyevich *et al.* [2] only the voidage distribution in one dimension was described. Thus only the voidage directly in front of and at the back of the bubble could be predicted. The present study aimed to use the same basic approach to create a more general model that would be capable of predicting the voidage all around the bubble, in other words, for all values of r and θ (refer to Figure 1). As a starting point of the extended model only two dimensions will be discussed here. The two dimensional solution was found by solving equation (2) along several different radial lines from the center of the bubble. As an initial control the results were tested with the results published by Buyevich *et al.* for the voidage along the vertical axis of the bubble [2]. This will be the case with $\theta = 0$. In Figure 2 these results are shown for the voidage distribution in front of and at

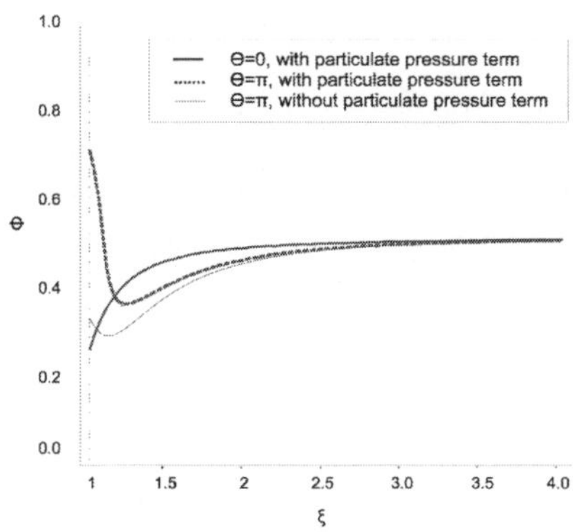

Figure 2: The prediction of the Buyevich bubble model with a free bed solid fraction, ϕ_0, of 0.5 and a bubble radius, R, equal to one. The prediction is given in front of- and at the back of the bubble.

the back of the bubble. In Figure 2 the ξ represents the dimensionless distance and is defined as $\xi = r/R$, with R as the bubble radius.

The radius of the bubble will be taken as a sphere with approximately the same volume as that of the real bubble. In practice it is not always trivial to determine the radius of this equivalent sphere but the concept will be used in the present study. Figure 2 is just a control to check that the new two dimensional equations simplify back to the results obtained by Buyevich *et al.* [2]. Exactly the same result is retrieved with a solid fraction equal to 0.5 and a bubble radius equal to one. Some differences were observed in the prediction at the rear of the bubble. The results published by Buyevich did not include the particulate pressure term in the total conservation of momentum equation. In some cases it is acceptable to assume that this term is negligible but it is clear that this term has a significant effect on the prediction of the model. In front of the bubble the results of the model seems fairly physical and is a monotonously increasing function of the solid fraction. The only effect of neglecting the particulate pressure term here would be a higher solid fraction at the bubble boundary. At the back of the bubble the particulate terms causes a very high solid fraction at the bubble interface. Whether this is physical or not is not clear as a solid fraction of 0.7 seems very high if the free bed solid fraction is a mere 0.5. Further research on the physicality of this phenomena is needed.

3 Simulation correlations with the two dimensional model

In an attempt to quantify the results obtained with the model around the bubble, CFD simulations were used to produce data. The simulation data should give some indication of the validity of the model in question. The commercial code *Fluent Ansys 12.1* was used to perform the simulations. The Eularian approach was employed and the Syamlal *et al.* drag model was used to simulate the drag [5]. A uniform particle size equal to $483\mu m$ was assumed and the material was chosen as glass. This set-up created a fluidized bed with a free bed solid fraction, ϕ_0, of 0.63. The fluidization medium was modeled as air. All of the simulations carried out was in a two dimensional rectangular column with a diameter of 23cm.

In Figure 3(a) a bubble is shown that was retrieved using simulations. To produce a single bubble the bed was modeled to be at minimum fluidization conditions and then an extra supply of gas was fed into the bed via a jet that was located in the middle of the bed.

The data was extracted as several points along several radial lines drawn from the estimated center of the bubble. The radial lines that were investigated are given in Figure 3(b) and the particular point used are illustrated in Figure 3(c). At each point the solid fraction was extracted. The data from the extracted points were compared with the model along each radial line in term of the dimensionless distance, ξ. The results are given in Figure 4. The radial lines that were investigated are located at $\theta = 0, \pi/6, \pi/3$ and $\pi/2$ respectively (refer to Figure 3(b)). The equivalent spherical bubble radius was taken as 7cm. This is an empirical estimation. A definition for the bubble boundary should also be defined.

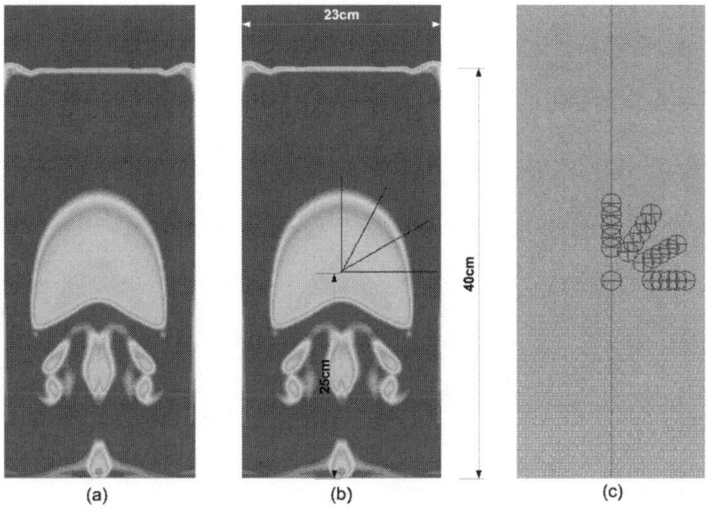

Figure 3: (a) A bubble created in a fluidized bed at minimum fluidization conditions, (b) radial lines investigated to quantify the accuracy of the model and (c) the point at which the void fraction data were extracted.

From Figure 4(a)–(b) it is clear to see that the correlation without the particulate pressure term provides a much more accurate result. This may be a misleading result as the value of ξ is strongly connected to what is defined as the bubble boundary. If the equivalent spherical bubble's radius was taken as less that 7cm the correlation with the particulate term might have been a better fit. Simply from this result it is clear that further research is needed in clearly defining the bubble boundary.

There were a few unexpected and interesting results from the simulations. First of all the solid fraction inside of the bubble was much higher than expected. In most models commonly used, a simple two phase model is assumed. In other words, there is only a uniform dense phase outside of the bubble and the gas phase inside of the bubble. All gas in excess of the minimum fluidization velocity is assumed to pass through the bed as bubbles [6]. This assumption seems to be very crude in light of the simulations. These phenomena also make it more difficult to define the bubble boundary. Even at what was assumed to be the center point of the bubble a solid fraction of 0.12 was observed. From Figure 4 it is clear to see that the first two data points on each radial line had a some what different gradient than the other data points. This change in gradient might be an indication of the bubble boundary, in which case, the new two dimensional model does relatively well in prediction the boundary. Contrary to the previous published results it seems that the bubble boundary cannot always be assumed to start at a solid fraction of zero [2]. As the angle, θ, was tending toward $\pi/2$ the model correlation became worse (refer to Figure 4(c) and (d)). Actually the model predictions depicted in Figure

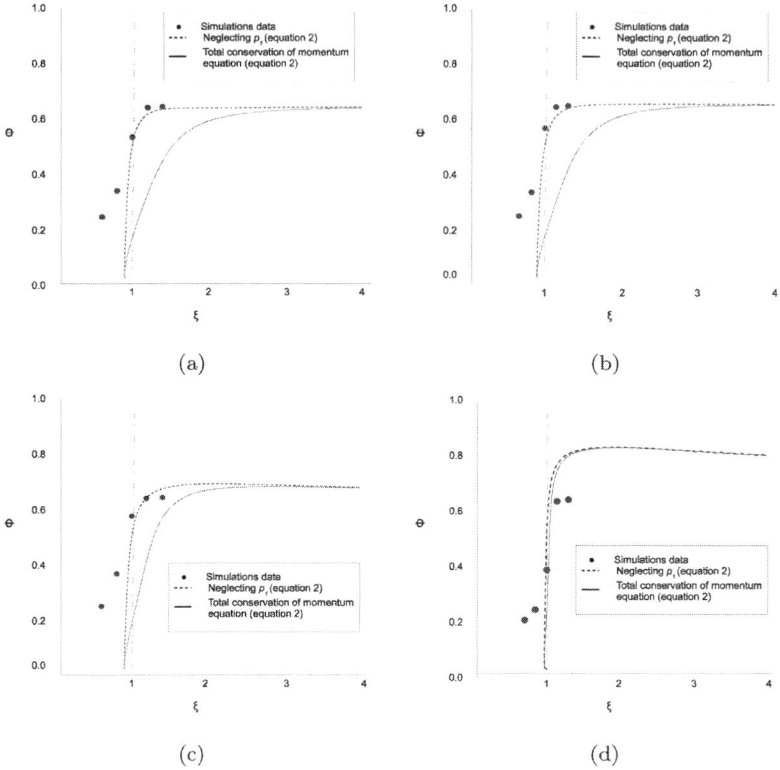

Figure 4: Solid fraction distribution along the radial lines (a) $\theta = 0$, (b) $\theta = \pi/6$, (c) $\theta = \pi/3$ and (d) $\theta = \pi/2$.

4(d) were calculated using the value $\pi/2.1$ as the model produced a solid fraction distribution equal to zero at $\pi/2$. This result is unphysical and is a consequence of the trivial model used to predict the relative interstitial velocity, $\mathbf{u}$. From equation (7) it is clear to see that the velocity will be zero at $\pi/2$ in the radial direction and when the velocity is modeled as zero it follows that the drag and consequently the solid fraction will also be modeled as zero. This problem at $\pi/2$ raises doubts about the physicality of the model's perditions as $\theta \to \pi/2$. This phenomena might also be the explanation for the relatively bad predictions as $\theta \to \pi/2$ (refer to Figure 4(c)).

In Figure 5(a) and (b) the same result are show as in Figure 4 but on a bigger scale.

In Figure 5 is a section of higher solid fraction just adjacent to the bubble boundary. Further research on the physicality of this result is in progress. If it can be proven that this region of higher solid fraction does indeed exist, it might help the scientific community understand bubble interaction and coalescence in a more clear way.

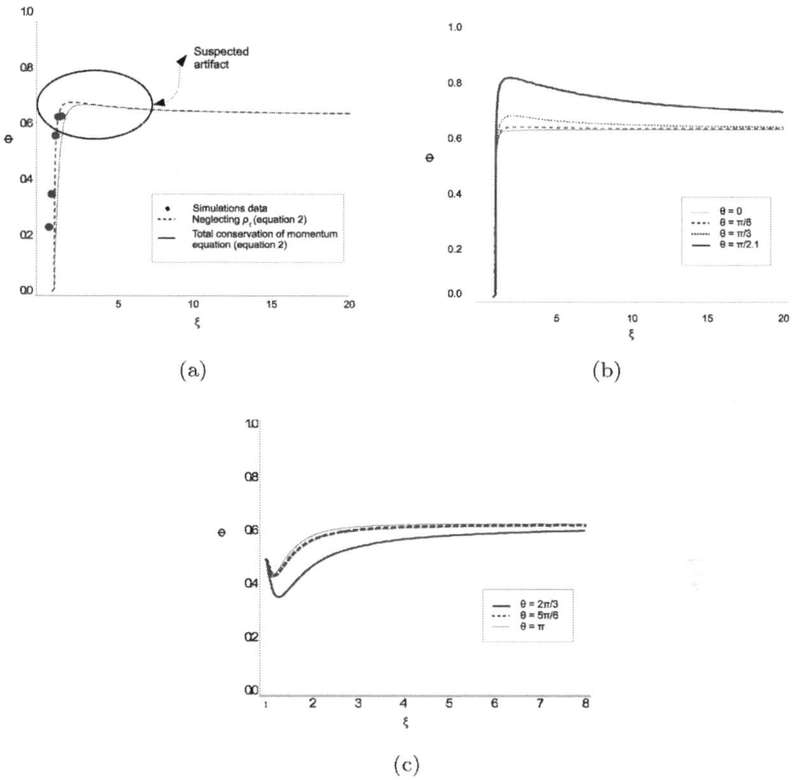

Figure 5: (a) The solid fraction distribution along the radial line at $\theta = \pi/3$, (b) the solid fraction distributions along several radial lines and (c) gives the solid fraction distribution in the rear quadrant of the bubble.

In Figure 5(c) the model predictions for $\theta = 2\pi/3, 5\pi/6$ and π are given. In this quadrant behind the bubble the opposite effect is observed. Adjacent to the hypothetical bubble boundary a band of lower solid fractions exists. This effect is again more pronounced for values of θ close to $\pi/2$. Whether this is just an artifact or a real wake effect is not yet clear as no data was acquired in this region.

4 Conclusion

Bubbles in fluidized beds are one of the major phenomena through which mixing takes place. Better understanding of physical phenomena in and around bubbles in a fluidized bed is of paramount importance. Knowledge in this regard can help engineers and scientists to design and develop more effective fluidized bed reactors.

In the present study a model describing the voidage distribution in front and behind a bubble in a fluidized bed was reexamined and extended. The model suggested is not complete or satisfactory. This work is only the initialization of a research project to model the flow behavior in and around a bubble in a fluidized bed. This model, originally suggested by Buyevich *et al.* [2], might be developed into a very useful and convenient predictive tool.

Time the model gives very good results in the area close to the front of the bubble. This was established by using data that was acquired using CFD simulations. Suspected limitations of the models used to predict the relative interstitial velocities, equation (7), might of been the cause of the inaccuracy of the model at $\theta = \pi/2$. Never the less, these problems have the possibility of being addressed in the future and thus render a very accurate and useful model for predicting the solid fraction distribution all around a bubble in a fluidized bed.

References

[1] Davidson, J.F. & Harrison, D., *Fluidized particles*. Cambridge University Press: Cambrige, U.K., 1963.
[2] Buyevich, Y.A., Yates, J.G., Cheesman, D.J. & Wu, K.T., A model for the distribution of voidage around bubbles in a fluidized bed. *Chemical Engineering Science*, **50(19)**, pp. 3155–3162, 1995.
[3] Buyevich, Y.A., Fluid dynamics of coarse dispersions. *Chemical Engineering Science*, **49(8)**, pp. 1217–1228, 1994.
[4] Buyevich, Y.A., Particulate pressure in monodisperse fluidized beds. *Chemical Engineering Science*, **52(1)**, pp. 123–140, 1997.
[5] Syamlal, M., Rogers, W. & O'Brien, T.J., Technical note. *MFIX Documentation theory guide*, U.S. Department of Energy, Office of Fossil Energy, Morgantown Energy Technology Center, Morgantown, West Virginia, 1993.
[6] Kunii, D. & Levenspiel, O., *Fluidization Engineering second edition*. Butterworth-Heinemann series in chemical engineering: Oxford, U.K., 1991.

Section 4
Heat transfer

Spray quenching of specimen for ring heat treatment

P. Stark, S. Schuettenberg & U. Fritsching
Research Centre "Distortion Engineering",
Foundation Institute for Materials Science (IWT), Bremen, Germany

Abstract

The quenching process of metallic workpieces within manufacturing and heat treatment can be optimized by applying locally and temporally adapted quenching cooling conditions. Locally variable heat transfer conditions at the workpiece surface are realizable by the regulation of adjustable and flexible flow fields on the basis of impinging spray or jet flows. For the analysis of heat treatment, the heat transfer rates with respect to the applied flow parameters and their influence on the cooling conditions are described. The simulation is assisted by experimental analysis of impinging multiphase jets and sprays.
Keywords: spray cooling, quenching, heat treatment, heat transfer coefficient.

1 Introduction

A main step in the manufacturing process of e.g. machine and gear components is the heat transfer process. Large metallic rings are used in many technical applications. Due to their function as a main component of bearings high demands on material quality and precision are made during the production process. Large distortion and/or shape deviation occur during the process steps of heating, rolling and quenching that commonly require material allowance to enable an additional reworking process.

Thus, a new approach to minimize the occurring distortion has been developed in which the heat necessary for the forming process shall be used to compensate the distortion during quenching within the hardening process. The minimization and compensation of workpiece distortion can be realized by impressing asymmetric cooling conditions by the use of flexible flow fields based on liquid jets or sprays. Especially the cooling by two phase sprays of gas

WIT Transactions on Engineering Sciences, Vol 70,
www.witpress.com, ISSN 1743-3533 (on-line)
doi:10.2495/MPF110171

and liquid, as schematically showed in figure 1, enables to generate specific local heat transfer conditions on workpiece surfaces. The quenching process in flexible jet nozzle fields has originally been developed for gaseous flows [1]. But experimental investigations on relevant parts result in insufficient experiences because of the limited achievable heat transfer by this quenching method, which is too low for successful distortion compensation. However, by controlled quenching in liquid media (such as water or hardening oil) and by means of jet or spray quenching, the heat transfer process can be heavily intensified and to generate a much higher distortion compensation potential [2].

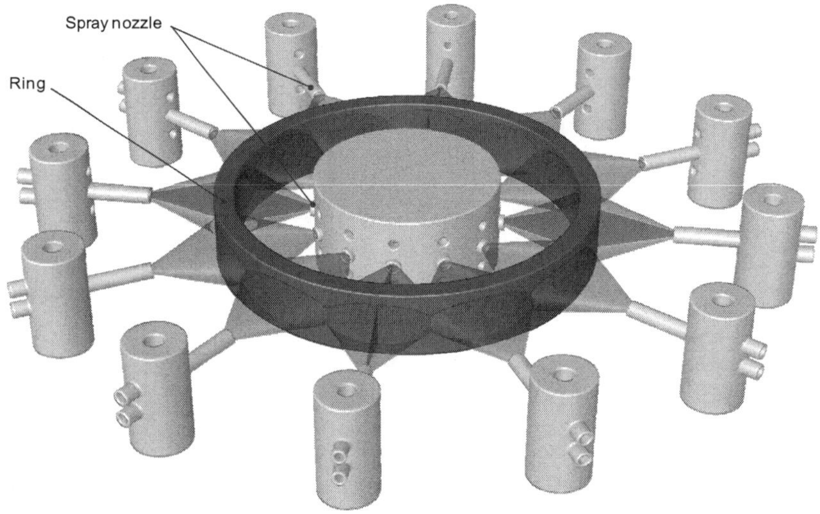

Figure 1: Spray quenching of heated rings.

In liquid jet quenching processes, the generated flow velocity on the impingement area outside a liquid jet should be strong enough to suppress any boiling phenomena (vapour layer, nucleate boiling) in the impingement region on the heated surface. Here, pure convective/single phase heat transfer from the workpiece surface to the incident flow will generate extremely high cooling rates and thus high heat transfer rates. For the numerical simulation these processes, Computational Fluid Dynamics (CFD) simulations are done. To achieve accurate results within the simulation of the heat transfer on the ring segment's surface, the complete flow field of the jet array could be rebuilt in a 3D-CFD-simulation. The chart of figure 2 shows the results of heat transfer coefficient (HTC) allocations along a vertical scanline through the centres of impinging jets on a heated ring segment surface. It can clearly be seen that the impinging liquid jet causes significant higher HTC-profiles ($\alpha_M \approx 25$ kW/(m²K)) in comparison to the results of the gas quenching process ($\alpha_M \approx 1$ kW/(m²K)). The high level of the local HTC shows the high distortion compensation potential of the liquid jet quenching process. However, quenching experiments showed that the impressed local HTC are too high for a useful control of asymmetric quenching conditions.

In consequence, two-phase atomizers instead of single-phase liquid jets are utilized to allow the adjustment of impressed heat transfer conditions with the choice of spray parameters (liquid mass flow and gaseous pressure). This quenching technique enables to set up local heat transfer conditions to generate HTC results in a useful range between the achievable results for pure gas respectively liquid quenching processes, as shown in figure 2.

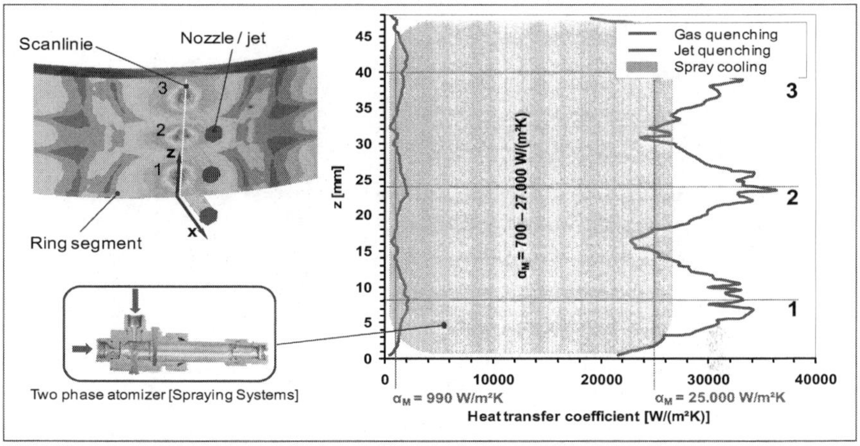

Figure 2: Heat transfer coefficients for different jet and spray cooling techniques.

Through the use of multiphase atomizers for impressing intensive but controlled local heat transfer on surfaces, it is also possible to avoid vapour layer formation on heated workpiece surfaces [3, 4]. A spray of fine liquid droplets and an overlaid gaseous flow is impinged on the heated surface. Within this process, the efficiency of a complete evaporative cooling should be ideally reached, because the liquid fraction of the spray is restricted so that no closed liquid film with a vapour layer underneath is built.

For experimental investigations on hot rings, a flexible nozzle field shall be utilized to create local gradients in heat transfer, so that the occurring internal stresses lead to a compensation of the distortion while the surface hardness is not dropping below given requirements. To detect and compare the occurring distortion of the quenched parts, the shape deviations of the specimens with respect to the applied quenching conditions are measured through a mechanical measuring system before and after the cooling process.

2 Methods

2.1 Experimental setup

To analyse the spray cooling process, a twin fluid atomizer with water and air is used. This nozzle (type: CasterJet, manufacturer: Spraying Systems Co.) is typically applied for cooling in steel casting processes and is assigned by very high mass fluxes (up to 1600 kg/h) for the liquid phase. The spray pattern of this atomizer type is a flat-spray cone, which is the optimal geometry for a homogeneous cooling over the circumference of the given ring segment. A spray pattern at typical operation parameters (water mass flow 100 kg/h, air pressure p = 0.3 MPa) and position related to a workpiece surface is shown in figure 3. The given operation parameters (air pressure and water mass flow) and especially the operation boundaries are examined and defined in experimental investigations.

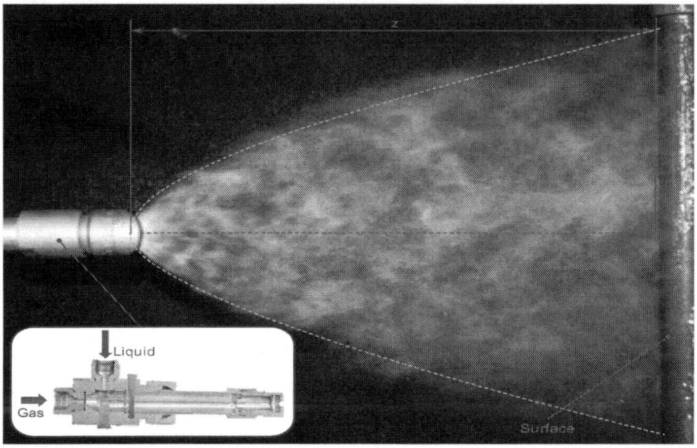

Figure 3: Two phase atomizer arrangement for spray cooling.

For efficient estimation of the spray process, a spray characterization in combination with evaluation and cooling curve based measurements of heat transfer coefficients are performed. The spray characterization consists of drop diameter measurements by using Laser Diffraction Techniques and droplet velocity examinations done by Particle Image Velocimetry (PIV). The measurement of liquid mass flux distributions is done by using patternators.

The cooling behaviour of 1/16 segments of a large ring (figure 4) with respect to the applied spray quenching conditions is investigated experimentally. Both inside and outside of the ring two spray nozzles are arranged perpendicularly to the surface, so that the flat spray profile is only widening in the circumferential direction. The area of impingement is located at half of the ring's height. A constant nozzle distant of z = 140 mm provides an overlapping region of about 30% of the width of one single spray cone on both sides of the ring segment. The ring segment is heated for 90 minutes at T_{oven} = 900°C under ambient atmosphere. The spray nozzles were activated at t = 0 s as soon when the specimen is reaching the desired position in the nozzle field. During quenching, the temperature at different positions inside of the specimen have been measured through thermocouples (NiCr-Ni, Type K), both in the core (diameter = 1 mm) and in the near-wall region (diameter = 0.5 mm). The distances from the surface of the near-wall thermocouples are in the range of 1 - 1.5 mm. The measurement frequency is 15 Hz.

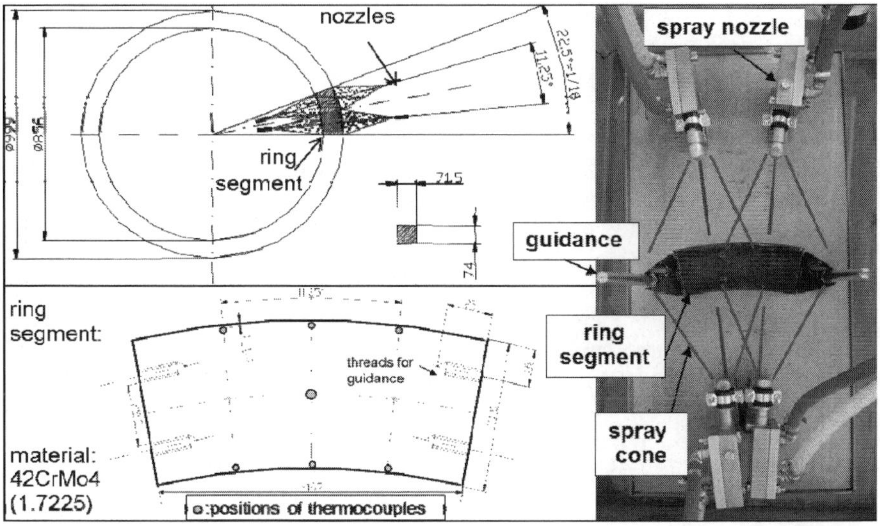

Figure 4: Experimental setup for quenching of ring segment with sprays.

2.2 Determination of heat transfer coefficients

The basis for high heat transfer rates in spray cooling processes is a high evaporation efficiency. It increases for decreasing drop diameters and is often used in dependency to the Weber number, whereby the drop behaviour during workpiece surface impact can be characterized and evaluated by approaches of Bolle and Moureau [5] and Berg [6]:

$$We = \frac{u^2 \cdot \rho \cdot d}{\sigma} \tag{1}$$

Here, u is the drop velocity [m/s], ρ is the density of the liquid phase [kg/m^3], d is the drop diameter [m] and σ the surface tension [N/m]. For Weber numbers above 80, a liquid film is built on the heated surface. The film next collapses in a number of small drops which results in a high heat transfer. For setting up high evaporation efficiency, it is necessary to adjust the spray for getting Weber numbers above 80.

For evaluation of heat transfer coefficients, the empiric approach of Puschmann [3] was the chosen:

$$\alpha_C = \dot{m}_S \cdot 16{,}8 \cdot u^{0.12} \cdot d^{-0.29} \tag{2}$$

Here, α_c is the convective heat transfer coefficient [W/(m^{2}K)] and m_s is the impingement density [kg/(m^{2}s)]. Puschmann could show that the results for calculated heat transfer coefficients in spray cooling processes are directly proportional to the water impingement density for heated surfaces above the Leidenfrost temperature. The drop diameter and velocity only have a secondary role in comparison to the water impingement density.

Based on measured temperature curves at the near-wall thermocouples, the underlying time-dependent heat transfer coefficient (HTC) on the surface can be determined by applying a numerical approach. It is based on the observation that the temperature in the near-wall region is decreasing rapidly, while the reaction of the temperature in the core is heavily delayed due to the thickness and the relatively low thermal conductivity of the solid (Biot number $Bi > 1$). By also neglecting cross flows of energy in the axial direction of the ring, each of the measurement positions is treated individually as a one-dimensional stripe of the solid with the length $(r_{outer} - r_{inner})/2$ involving the core and the very tip of the thermocouples.

In this approach, energy can only be removed through that boundary representing the outer surface of the specimen. All other walls are set to adiabatic. Starting from a homogenous temperature distribution of $T = T_{oven}$, the heat transfer coefficient on the surface is iteratively calculated for each time step until the simulated temperature at the position of the thermocouple corresponds to that one from the experiment within a $\pm 0.1°C$ interval. Within this loop, the current temperature distribution at time t is used as the initial condition to determine the heat transfer coefficient at time $t + \Delta t$.

This approach enables to calculate the time-dependent heat transfer coefficient at the wall for all cooling curves at the beginning of the quenching process. For a two-sided flat plate with the thickness d cooled from both sides, it is valid for [7]:

$$Fo = \frac{a \cdot t}{(d/2)^2} \leq 0.04 \tag{3}$$

Here, the Fourier number Fo is calculated with the thermal diffusivity a [m^2/s] of the steel, the time t and the characteristic length d. Setting $d = (r_{outer} - r_{inner})$ leads to a validity period of $t \leq 28$ s for the ring segment.

2.3 Numerical simulation of the temperature distributions

The determined time-dependent heat transfer coefficients can be used as a boundary condition for the simulation of the cooling of the specimen. Due to the limited number of thermocouple measurements, these values are linearly interpolated in circumferential direction between these known locations at each time step. In axial direction, no gradient in heat transfer is applied. At the other surfaces (upper, lower and end faces) which are not directly exposed to an incoming flow, low values for the heat transfer coefficients of 2000 W / (m^2K) are assumed and kept constant for the entire calculation. Starting from a homogenous temperature distribution of 900°C, the cooling until equilibrium ($T_{ambient}$ = 25°C) is simulated. Temperature dependent material properties of the material (42CrMo4) are applied during the calculation of the unsteady cooling of the specimen.

3 Results

3.1 Heat transfer coefficients based on cooling curves

The cooling behaviour at different position in the ring segment has been measured for a symmetric and an asymmetric nozzle setup at different positions inside of the specimen (see figure 5):

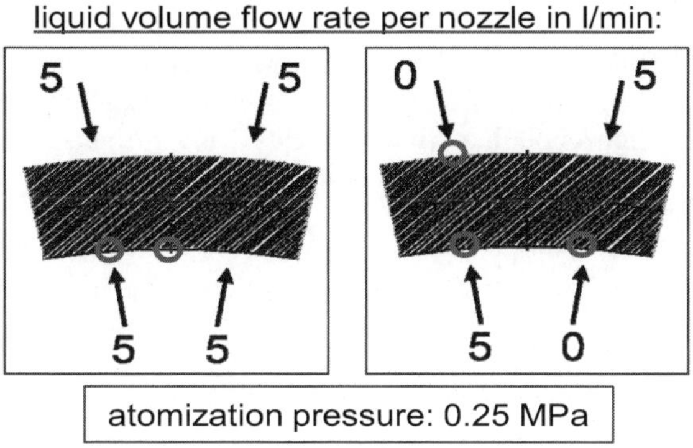

Figure 5: Symmetric (l.) and asymmetric spray nozzle setup, positions of the thermocouple measurements.

The cooling curves measured in the centre of a spray cone show the same result regardless whether the symmetric or asymmetric respectively the convex or concave surface is analysed. Thus, they are used as the reference curve in figure 6, in which the measured cooling curves are compared.

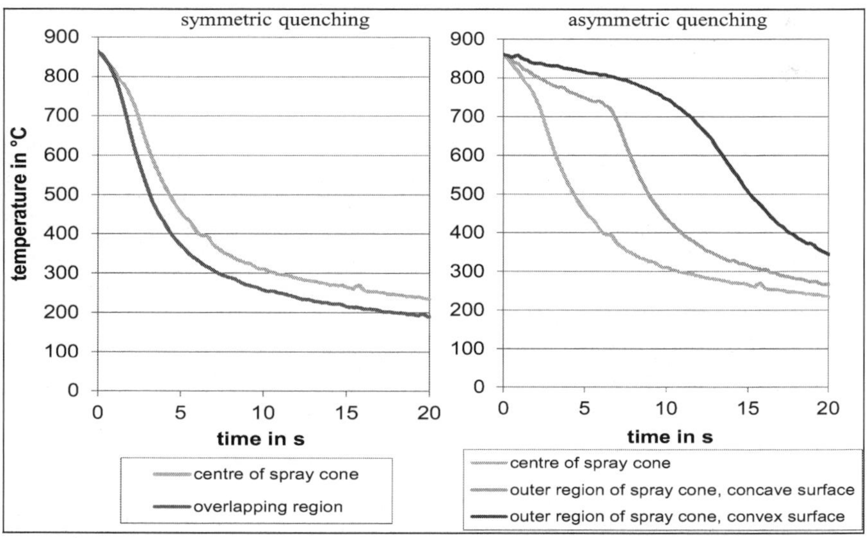

Figure 6: Measured cooling curves for spray quenching.

At symmetric quenching conditions, the cooling behaviour is qualitatively the same at both measuring position, but a slightly faster temperature decrease can be observed in the overlapping region.

In case of asymmetric quenching conditions, the cooling behaviour in the outer regions of the spray cones follows a differing trend. On both sides of the ring segment, a period of only slow temperature decrease may be observed within the first few seconds of the quenching process followed by a more abrupt decrease of the temperature, which is typical for occurring boiling phenomena.

Due to the geometric shape of the surfaces this trend is more emphasized on the outer side of the specimen. There, the convex surface shape leads to a more acute angle of impingement of the spray, so that the stagnation pressure of the flow may not be high enough to transport away any formed vapour film. After the collapse of the vapour film, the cooling trend is very similar at all measurement positions.

The corresponding cooling trends in the core regions down to a temperature of 400°C are not differing significantly. It takes 82 s in case of symmetric quenching conditions, whereas 95 s are necessary when the asymmetric quenching conditions are applied.

The heat transfer coefficient has been determined by applying the numerical approach described in chapter 2.2. The results are summarized in figure 7.

For almost all curves, the heat transfer coefficient increases rapidly at the beginning of the quenching process increases, until it stays at a constant level. For the symmetric quenching conditions, this constant region is reached after

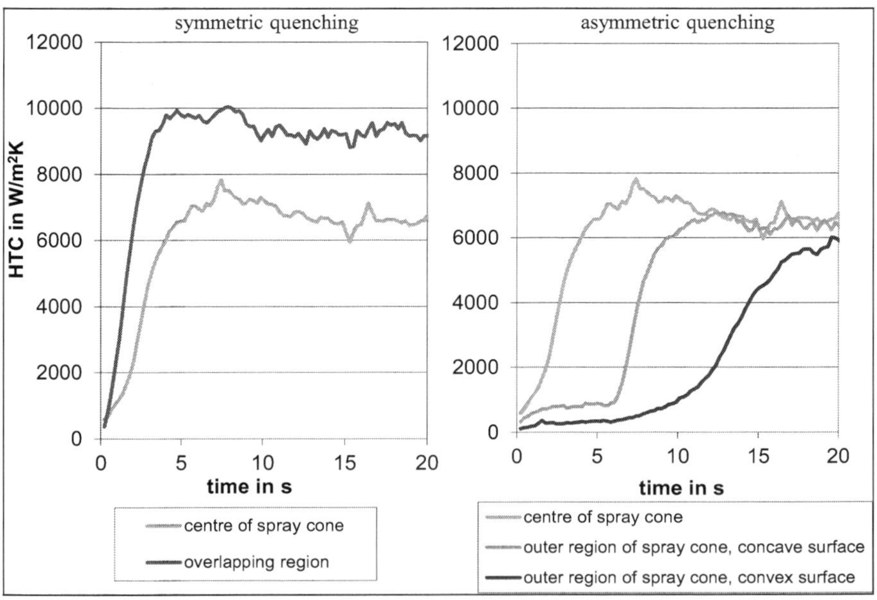

Figure 7: Determined time-dependent heat transfer coefficient for spray quenching.

about the same period of time (≈ 4 s) both in the core region of the spray cone as well as an in the overlapping region. In the overlapping region, this constant HTC has a value of about 9100 W / (m²K) compared to 6700 W / (m²K) in the centres of the single spray cones.

For the asymmetric quenching conditions, the reached constant value of 6700 W / (m²K) is almost equal at all measured location, but the described effect of the boiling phenomena on the hot surface in the outer regions of the spray cones leads to a time shift of about 5 s between the three measurement positions, concerning the point in time when the low HTC ≤ 2000 W / (m²K) are quickly increases to the higher values in the constant regime of the HTC.

3.2 Heat transfer coefficients based on calculation approach validation

The CasterJet nozzle was used for measurements by different water mass flows and air pressures. The result of the drop diameter measurements on the spray axis for a distance of z = 140 mm is shown in the left charts of figure 8 in dependence of the mean volumetric diameter d_{30} with the water mass flow and air pressure. It can be seen that the d_{30} increases with increasing water mass flow, but it shrinks in proportion to increased air pressures.

The local drop velocities inside the spray were measured by the use of PIV method. The right charts of figure 8 shows exemplarily results of velocity measurements for typical water mass flows by air pressure 0.15 MPa. The trend of the averaged velocities for an increased distance to the spray axis shows a maximum on position r = 0 and a light decrease of the velocity for outer ranges.

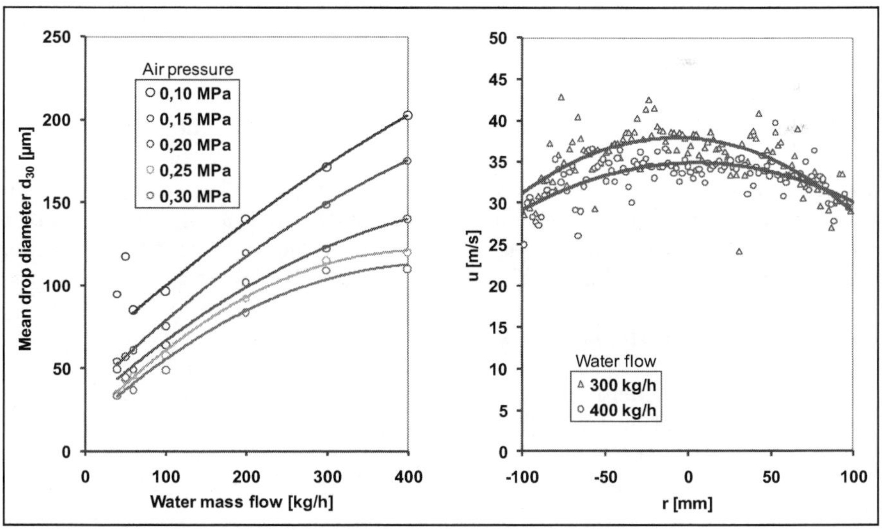

Figure 8: Measured mean drop diameter (l.) and drop velocity dependent to spray parameters.

With knowledge of the drop diameter and velocities on different points inside the spray it is possible to calculate local Weber numbers by eqn (1). The calculated Weber numbers should validate the applicability of the used approach for calculating local heat transfer coefficients by eqn (2). The calculations result in Weber numbers higher than 85 for every spray parameter combination (water mass flow and air pressure). Based on this, eqn (2) is valid.

By [3] the calculated heat transfer coefficients are direct proportional to the water impingement density. The impingement density was measured on different positions inside the spray for the distance of z = 140 mm. Figure 9 shows results on the dependence of the water mass flow and used air pressures.

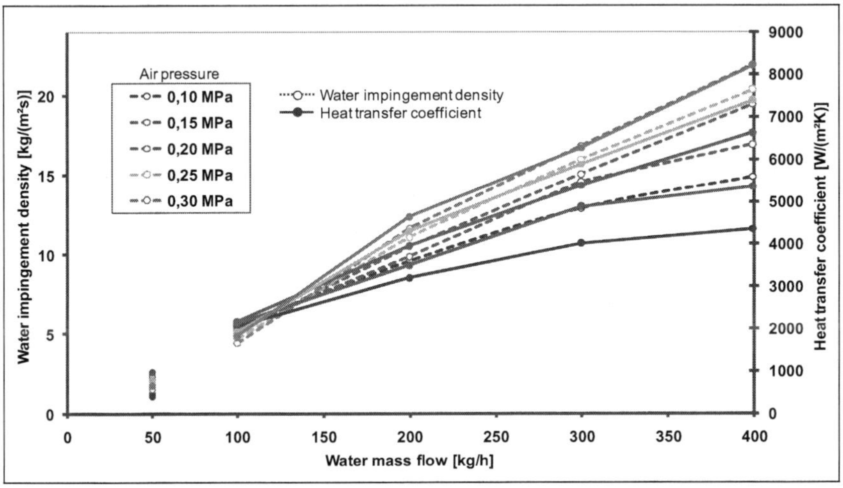

Figure 9: Heat transfer coefficients and water impingement density.

The determined results from the spray characterization for drop diameter, velocities and impingement densities are used to calculate heat transfer coefficients by eqn (2) for the position r = 0. The results of these calculations are also included in figure 9 for direct comparison. Based on this, a direct proportionality of the water impingement density to the heat transfer coefficients can be validated under consideration of the adjusted air pressures.

3.3 Temperature distributions in ring during quenching

The temperature distributions inside of the ring segments are calculated by using the determined local heat transfer coefficients as the boundary condition in the numerical simulation (see figure 10).

For symmetric and asymmetric quenching conditions, the temperature distributions are shown at a certain time step, so that the maximum temperature in the core region is 400°C. The corresponding cooling process takes 75 s in case of symmetric quenching with four activated spray nozzles, resp. 90 s for the asymmetric quenching conditions with only two used nozzles (compare to chapter 3.1). A response of the temperature distribution to the applied quenching intensities can still be observed both in the core and in the near-wall regions of the specimen. An additional comparison of the experimentally and numerically determined cooling curves in the core regions shows good agreements within a ±30°C interval.

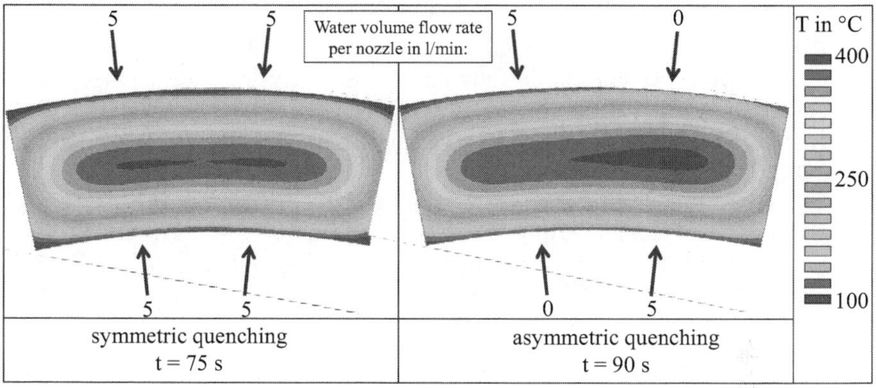

Figure 10: Simulated temperature distributions at cutting planes through the centres of the ring segments (atomization pressure p = 0.25 MPa).

4 Conclusion

By creating asymmetric quenching intensities on the specimen's surface, local gradients in the internal cooling behaviour inside of the body can be caused which may lead to increased distortion behaviour of the quenched specimens.

The general merits of the spray quenching technique and the influence of the spray parameters on the local cooling conditions were presented. The decisive precondition for this approach is the knowledge of the local spray conditions and their influence on the temporal and spatial distribution of the heat transfer rates. Here, two different approaches to determine the heat transfer coefficient were shown.

The effect of the temperature distributions and local cooling rates during quenching of a ring is demonstrated.

Acknowledgement

The present work was executed in the framework of the Collaborative Research Centre (SFB 570) "Distortion Engineering" at the University of Bremen (projects B4 and T5). The authors would like to thank the German Research Foundation (DFG) for the financial support.

References

[1] Schuettenberg, S., Frerichs, F., Hunkel, M., Fritsching, U., Int. J. Materials and Product Technology; 24, 2005, 259-269

[2] Schüttenberg, S., Hunkel, M., Fritsching, U., Zoch, H.-W., Proc. of 5th International and European Conference on Quenching and Control of Distortion, 25 - 27 April 2007, Berlin, S. 257-264

[3] Puschmann, F.: "Experimentelle Untersuchung der Spraykühlung zur Qualitätsverbesserung durch definierte Einstellung des Wärmeübergangs", Dissertation, 2003, Otto-von Guericke-Universität, Magdeburg, Germany

[4] Krause, C., Wulf, E., Nürnberger, F., Bach, F.-W., Forsch. Ingenieurwes. 2008, 72, 163-173

[5] Bolle, L., Moureau, J.C., Multiphase Sci. Technol. 1982, 1, 1-97

[6] Berg, M., "Zum Aufprall, zur Ausbreitung und Zerteilung von Schmelzetropfen aus reinen Metallen", Dissertation, 1999, Universität Bremen, Germany

[7] Verein Deutscher Ingenieure: VDI-Wärmeatlas, 9. Auflage, Springer-Verlag Berlin (u.a.), 2002

Effect of inclination on flow and thermal characteristics of air-water two-phase flow: an experimental investigation

M. J. Vaze & J. Banerjee
S V National Institute of Technology, Surat, India

Abstract

An experimental analysis towards establishing the effect of inclination on flow and thermal characteristics of air-water two-phase flow is presented. The experimental setup consists of a mixing section, a visualization section, a test section and an outlet section. The test section consists of two sets of pressure transducers, heater element and RTD sensors for surface temperature and inlet and outlet bulk fluid temperature measurements. The test setup can be inclined maximum up to 30° of inclination. Various flow patterns like stratified, slug, plug, wavy, annular and other mixed type are observed and captured at 0°, 2°, 5° and 7° for the range of the liquid (Re_{SL}) and gas (Re_{SG}) superficial Reynolds numbers of 500-10000 and 500-30000 respectively. Flow regime map is established based on this flow visualization. For inclined configuration the stratified regime is observed to be replaced by slug and slug/wavy type of flow regime and the transition between slug-plug and slug-slug/wavy takes place at higher Reynolds numbers. Surface and bulk fluid temperatures are recorded for constant wall heat flux condition, by varying the superficial Reynolds numbers of gas and liquid phases to establish the local and average two-phase heat transfer coefficient. It is observed that Re_{SL} has a pronounced effect on heat transfer coefficient in comparison to Re_{SG}. With the change in inclination this effect becomes more evident. A heat transfer correlation is deduced from the present experimental observation which includes the basic parameters like Reynolds number and Prandtl number, pressure drop and inclination effect.

Keywords: two-phase flow, flow pattern, heat transfer, superficial Reynolds number, inclination.

WIT Transactions on Engineering Sciences, Vol 70,
www.witpress.com, ISSN 1743-3533 (on-line)
doi:10.2495/MPF110181

1 Introduction

Alternating flow of gas and liquid (slug flow) is a concern for petroleum industry while crude oil is transport from the well through the pipelines. Although such flows are majorly for horizontal orientation of the pipe, it is evident that the topography of the land produces situations where the orientation of the pipe can be inclined. To properly design such flow lines thus it is necessary to establish the flow and thermal characteristics of two-phase flow through pipes with smaller inclinations.

Introduction of small inclination creates the component of gravitational force which is exactly perpendicular to the flow in horizontal case. The normal component provides segregation of the phases as in the case of horizontal flow while the other component acts parallel and opposite to the flow. Therefore, flow through pipe inclination causes lower velocity than in the corresponding horizontal configuration. These create differences in flow patterns and in situ properties which causes changes in flow and thermal characteristics.

Flow pattern map for horizontal flow was developed by many investigators in the past. Flow pattern map proposed by Taitel and Dukler [1] and the transition boundaries predicted analytically by them are valid for horizontal and near horizontal flow. Beggs and Brill [2] studied inclined gas-liquid flow through pipes and their work has been widely applied in the petroleum industry. This work provides empirical correlations for the prediction of holdup and pressure drop. Hasan and Kabir [3] conducted experiments in 12.5 cm pipe and annular flow channels with inner tube diameters of 4.8, 5.7 and 8.7 cm for deviations up to 32° from vertical. These experiments for very low water flow rates were conducted by feeding air into a stagnant water column. The model developed by Hasan and Kabir [3] performs comparable to the Beggs and Brill model [2]. Experimental conclusions made by Barnea *et al.* [4] revealed the agreement between the experiment and theory proposed by them for inclined flow. Kokal and Stanislave [5] showed that flow pattern transition line is more sensitive to inclination angle and uphill flow regimes were predominantly intermittent while downhill were stratified. A study on inclined two-phase flow was conducted by Tshuva *et al.* [6] for various upward inclinations in 2.4 cm diameter, 3 m long pipes. It was found that the flow distribution can be either symmetric or asymmetric depending on the flow conditions and pipe inclination. The CFD analysis of the effect of inclination was also carried out by Vaze and Banerjee [7] for 0°-7° inclination. They observed that stratified flow for inclined orientations was replaced by slug flow.

In non-boiling systems, the enhancement of convective heat transfer by injecting a gas phase in confined liquids has been observed by Dorresteijn [8] and also by Elamvaluthi and Srinivasa [9]. Mosyak and Hetsroni [10] had shown that the temperature difference between the top and the bottom side of the pipe was quite large for horizontal pipes. It drastically decreased due to an inclination of the pipe. Hetsoroni *et al.* [11] developed relation between the flow parameters and heat transfer using IR thermography for diabatic air–water flow. Large difference of circumferential wall temperature was found in their experiments.

This irregularity was found to be due to the distribution of air and water phase in the pipe. The quantitative information regarding the increase in heat transfer with the increase in inclination angle was not addressed by them.

The objective of the present work is to systematically investigate the flow and thermal characteristics of air-water two-phase flow for various inclinations. Experiments are carried out for 2^0, 5^0 and 7^0 orientation of the pipe. The mass flow rate of gas and liquid phases are varied to visualize the different flow patterns. These visualized images are utilized to generate flow pattern map for each of these inclinations. Due to the distribution of phases in the fluid in various flow regimes, the heat transfer characteristics of each regime differ. Bulk and surface temperatures are measured under constant heat flux condition to establish the local variation of heat transfer for each regime for different inclination of the pipe. A generalized correlation for average heat transfer is then established in terms of fluid and flow properties and inclination.

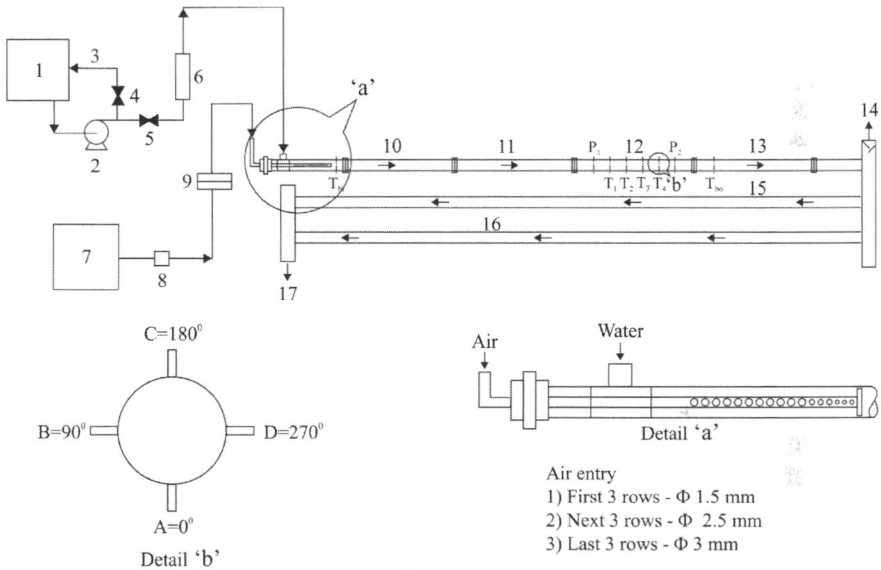

T_1	RTD station 1
T_2	RTD station 2
T_3	RTD station 3
T_4	RTD station 4

1	Storage tank	7	Screw compressor	13	Outlet section
2	Centrifugal pump	8	Pressure regulator	14	Air vent duct
3	By-pass line	9	Orifice meter/Rotameter (air)	15	Return line 1
4	Ball valve	10	Mixing zone/ Calming zone	16	Return line 2
5	Gate valve	11	Visualization zone	17	Line to the storage tank
6	Rotameter (water)	12	Heat transfer test section		

Figure 1: Schematic diagram of experimental setup.

2 Experimental procedures

The established experimental setup (Figure 1) is discussed in an earlier paper by the authors (Vaze and Banerjee [12]). The test setup has the facility to incline the pipe maximum up to 30° by using an in-house electric jack. A 230 V ac supply is provided to the gear motor through reverse-forward switch to facilitate the lifting of the setup. Flow rates, flow patterns, pressure and temperatures at different locations of the setup are recorded after setting the pipe at required angle of inclination.

3 Flow visualization

One of the most important aspects of two-phase flow is the geometric distribution of the phases under flowing conditions. In addition to the flowing conditions, it also depends on the angle of inclination. State of the art literature does not provide sufficient information regarding their functional relationship with the angle of inclination. Unlike the horizontal flow, gravity effects dominate for two-phase distribution in inclined pipes. Due to the higher density difference as in the case of air-water two-phase flow, gravity forces are more important than the other types of forces. The relative contribution of gravity forces compared to inertia forces increases with the increase in inclination. This could result in shifting of the transition line among the various flow regimes. Thus, small change in inclination can cause significant changes in flow and other characteristics of the two-phase flow through the system. Depending on the characteristic form of the flow patterns, seven flow patterns are identified. These are slug, plug, wavy, annular, wavy/annular, slug/wavy and slug/bubbly/annular. In the present investigation images are captured using Nikon D60 camera and

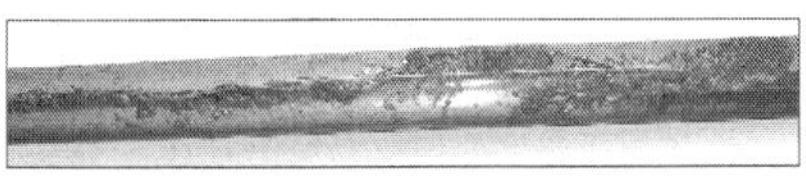

Slug ($Re_{SG} = 832$, $Re_{SL} = 5268$)

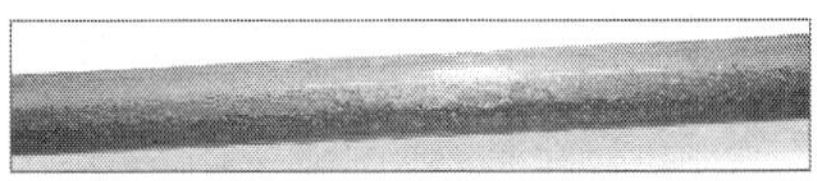

Annular ($Re_{SG} = 29342$, $Re_{SL} = 2950$)

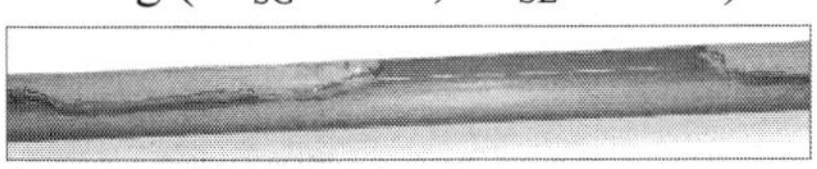

Plug ($Re_{SG} = 832$, $Re_{SL} = 10000$)

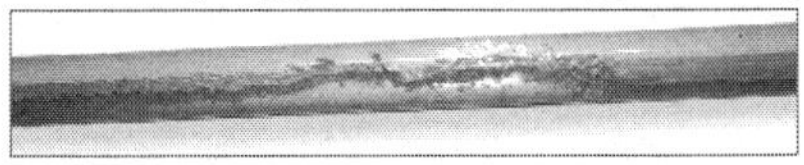

SW ($Re_{SG} = 8602$, $Re_{SL} = 3715$)

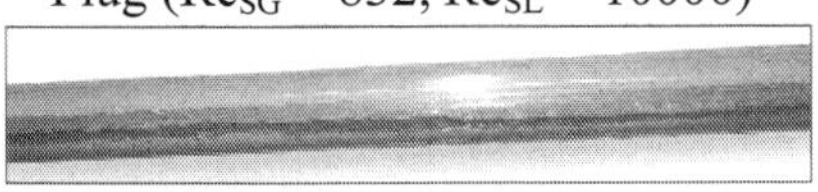

Wavy ($Re_{SG} = 15630$, $Re_{SL} = 705$)

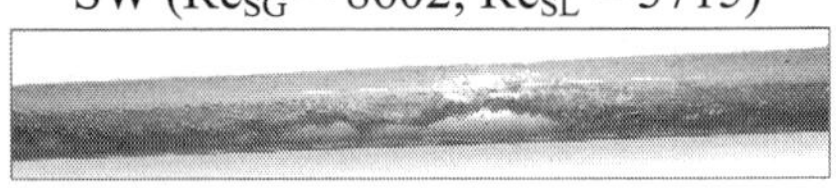

WA ($Re_{SG} = 20538$, $Re_{SL} = 4487$)

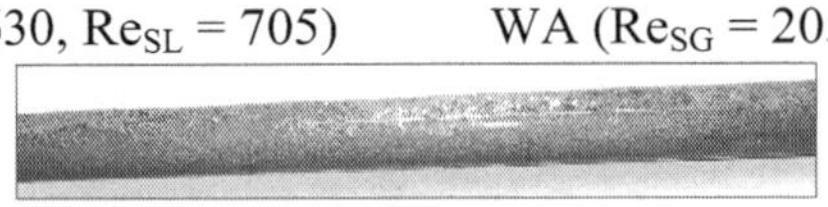

SBA ($Re_{SG} = 16626$, $Re_{SL} = 8474$)

Figure 2: Visualized flow patterns for 2° inclination.

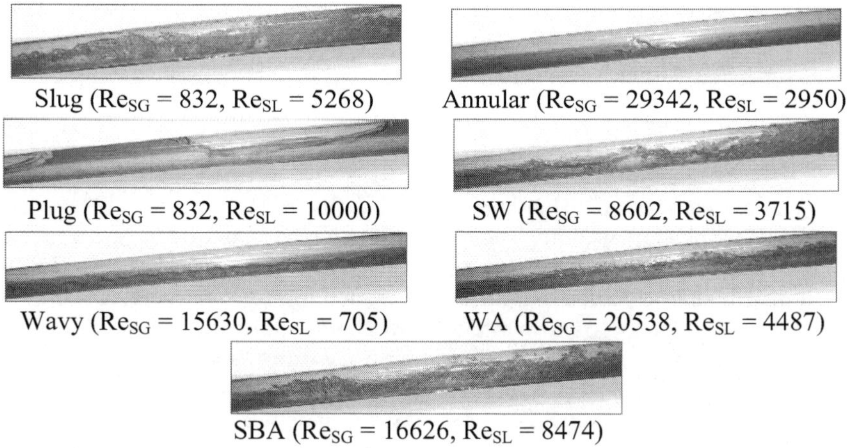

Slug (Re_{SG} = 832, Re_{SL} = 5268) Annular (Re_{SG} = 29342, Re_{SL} = 2950)

Plug (Re_{SG} = 832, Re_{SL} = 10000) SW (Re_{SG} = 8602, Re_{SL} = 3715)

Wavy (Re_{SG} = 15630, Re_{SL} = 705) WA (Re_{SG} = 20538, Re_{SL} = 4487)

SBA (Re_{SG} = 16626, Re_{SL} = 8474)

Figure 3: Visualized flow patterns for 5° inclination.

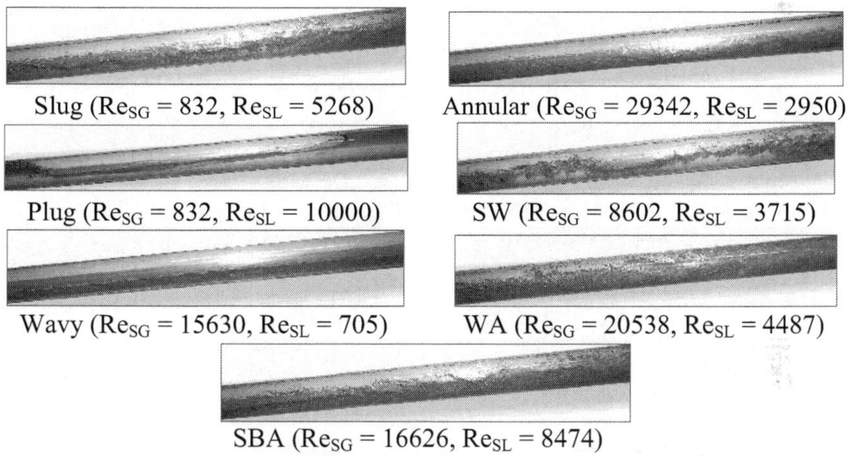

Slug (Re_{SG} = 832, Re_{SL} = 5268) Annular (Re_{SG} = 29342, Re_{SL} = 2950)

Plug (Re_{SG} = 832, Re_{SL} = 10000) SW (Re_{SG} = 8602, Re_{SL} = 3715)

Wavy (Re_{SG} = 15630, Re_{SL} = 705) WA (Re_{SG} = 20538, Re_{SL} = 4487)

SBA (Re_{SG} = 16626, Re_{SL} = 8474)

Figure 4: Visualized flow patterns for 7° inclination.

corresponding flow rates are recorded. Image analysis then establishes the flow patterns. Flow rates for each phase are converted to superficial Reynolds numbers of the respective phases.

Flow patterns that are observed for 2°, 5° and 7° inclinations along with corresponding Re_{SG} and Re_{SL} are described in Figures 2-4. The abbreviations for flow patterns used in the figures and tables are illustrated in the nomenclature. The important difference between horizontal and inclined flow patterns is the chaotic nature of the flow. Except plug and annular type of flow pattern, the bubble and droplet entrainment increases with the increase in inclination. Another important observation is the complete absence of stratified flow pattern for inclined flow. This is due to the effect of gravity on liquid which does not remain constant everywhere inside the fluid domain. This creates slug even at lower superficial Reynolds numbers.

4 Flow regime map

Various flow patterns are visualized by varying the mass flow rates of the phases and representative images are shown in Figures 2-4. These flow patterns are captured and the flow rates are measured. These mass flow rates are converted to non-dimensional Reynolds numbers corresponding to superficial conditions. Around 300 combinations of superficial Reynolds numbers are considered for which various flow patterns are captured. Various symbols are used to distinguish between various flow patterns. A thick transition line is marked on the flow map on visualized flow patterns to distinguish the regimes of flow patterns. An uncertainty analysis carried out following UKAS [13] shows an overall uncertainty of 2% with 95% of confidence for the estimation of superficial Reynolds number for both air and water phase in the experiments. The effect of upward inclination of the pipe on flow pattern transition is carefully examined. The transition lines for 2^0, 5^0 and 7^0 inclinations are overlapped on common flow regime map to understand the effect of inclination on flow regimes. The shift in the transition lines due to inclination of the pipe is shown in Figure 5. Close inspection of Figure 5 reveals following observations.

Upward inclination causes the intermittent flow to take place over a much wider range of flow conditions. Stratified flow is much sensitive to the inclination. Stratified regime is not present for all the inclinations considered in the present case. Portions of stratified regime are replaced by slug and slug/wavy

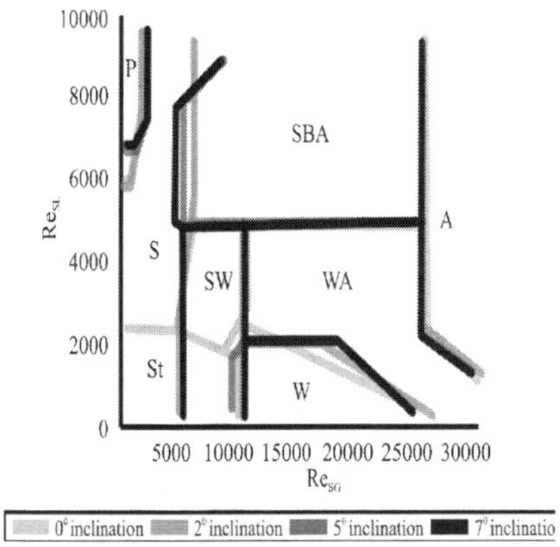

Figure 5: Variation of transition lines with inclination on flow regime map.

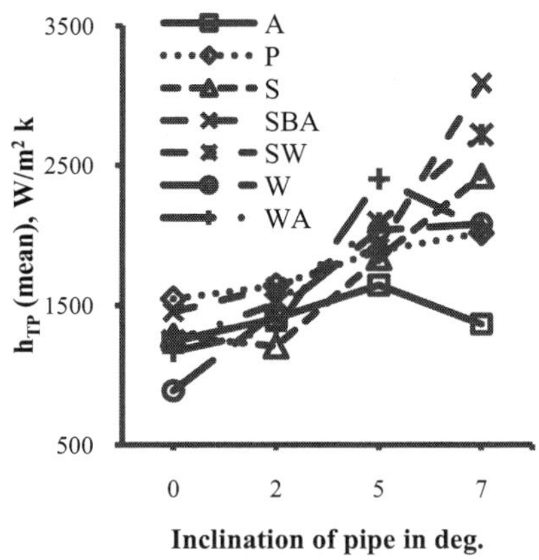

Figure 6: Variation of mean h_{TP} with pipe inclination for various flow patterns.

type of flow regimes. Due to this, the top surface of the pipe is wetted frequently by liquid even at lower Re_{SG} and Re_{SL}. This will ensure lower temperature of the top surface for constant wall heat flux condition. This in turn will eliminate the sudden rise in top surface temperature. Slug-plug transition line shifts towards – higher Re_{SL} at higher inclination. This is caused by accumulation of liquid near the entry zone. For the flow to happen this requires more turbulence which creates chaotic slug. When sufficient inertia is formed by the liquid the chaotic nature reduces and plug is formed. Slug-slug/wavy transition line shifts towards higher Re_{SG} for higher inclination. Other transition lines remain more or less same. The characteristics of such flows are significantly different from the horizontal case due to the effect of gravity Curling of air over the liquid surface is observed during slug flow. Also, there is presence of flow reversal. This produces substantial change in heat transfer.

5 Heat transfer characterization

The surface and bulk fluid temperature measurements are carried out to establish the heat transfer characteristics of air-water two-phase flow with small change in inclination of the pipe (0°-7°). The range of Reynolds numbers considered are 500-10000 for Re_{SL} and 500-30000 for Re_{SG}. Uniform wall heat is applied to the pipe using four resistance type surface heaters each having 2.5 kW capacities at 230 V. Before starting the two-phase heat transfer study, energy balance and in-situ calibration for RTD sensors are carried out for horizontal pipe to establish the amount of heat loss to the surrounding and proper installation of RTD sensors.

Surface temperatures are measured at four different cross sections with four thermocouples inserted at 90° of interval at each section. Bulk fluid temperatures are measured at inlet and outlet of heating zone for different mass flow rates of gas and liquid phase. Various flow patterns are visualized and mapped on flow regime map. According to the flow pattern, heat transfer behavior of the air-water two-phase flow at various inclinations is shown in Figure 6. The non-boiling two-phase heat transfer data are collected systematically by varying air and water flow rates and the pipe inclination. It is observed that that two-phase heat transfer (Nu_{TP}) increases with Re_{SL} for most of the flow patterns. It is also observed that Nu_{TP} is a function of inclination.

For the entire range of superficial Reynolds numbers corresponding to specific flow patterns, the average value of heat transfer h_{TP} is deduced and is represented as a function of angle of inclination in Figure 6. It is observed that the two-phase heat transfer coefficient increases with the angle of inclination. Only for the annular and wavy/annular flow pattern, it increases up to 5° of inclination and then for 7° of inclination the mean h_{TP} reduces. This might be due to the increase in content of air inside the domain which is less conductive to the heat transfer. Summary of heat transfer data for various inclinations are tabulated in Table 1.

Maximum heat transfer rate is obtained for slug/bubbly/annular flow at 7° of inclination while slug/wavy type of flow pattern shows consistently high heat

Table 1: Summary of heat transfer study for all inclination.

Angle of Inclination		0°	2°	5°	7°
St	Re_{SL}	705-1884	-	-	-
	Re_{SG}	832-9956	-	-	-
	$h_{TP,}$	450.23-2613.27	-	-	-
	Mean $h_{TP,}$	731.91	-	-	-
S	Re_{SL}	2044-7660	705-7660	705-7660	705-7660
	Re_{SG}	832-5932	832-3892	832-3892	832-3892
	$h_{TP,}$	701.17-2615.47	691.3-3786.68	886.07-4585.68	495.38-8553.96
	Mean $h_{TP,}$	1279.57	1202.62	1839.36	2424.42
P	Re_{SL}	6057-7660	6057-7660	6057-7660	6375-7660
	Re_{SG}	832-1635	832-2413	832-2413	832-2413
	$h_{TP,}$	1466.79-1846.68	1252.92-2309.27	1066.93-2953.72	1294.28-3723.74
	Mean $h_{TP,}$	1545.86	1640.2	1889.92	2018.01
W	Re_{SL}	705-1894	705-2044	705-2044	705-2044
	Re_{SG}	9956-23473	12297-23473	12297-23473	12297-23473
	$h_{TP,}$	605.71-1190.69	771.05-4865.74	1125.37-4306.01	652-8035.1
	Mean $h_{TP,}$	886.5	1448.45	2037.49	2082.29
A	Re_{SL}	1595-7660	1744-7660	1149-7660	1000-7660
	Re_{SG}	26406-29340	26407-29341	26407-29341	26407-29341
	$h_{TP,}$	811.93-1851.25	1008.25-2844.93	1210.94-4071.17	987.45-1868.52
	Mean $h_{TP,}$	1237.24	1406.76	1645.37	1363.98
SW	Re_{SL}	2044-4693	705-5425	705-5425	705-5111
	Re_{SG}	4595-10940	7182-9956	7182-9956	7182-9956
	$h_{TP,}$	794.54-5913.31	814.22-5317.63	756.43-4568.84	523.05-9661.14
	Mean $h_{TP,}$	1239.01	1515.18	2087.74	2722.75
WA	Re_{SL}	705-4487	705-5425	705-5111	705-5111
	Re_{SG}	12297-29340	12297-29341	12297-29341	14670-29341
	$h_{TP,}$	708.18-2528.83	795.64-3019.44	891.21-3542.52	939.25-8938.14
	Mean $h_{TP,}$	1168.11	1324.73	2402.52	2054.61
SBA	Re_{SL}	4693-7660	5470-7660	5425-7660	5425-7660
	Re_{SG}	7182-23473	7182-23473	7182-23473	7182-23473
	$h_{TP,}$	817.94-3385.44	979.26-3645.94	1414.99-5927.12	1526.94-9991.9
	Mean $h_{TP,}$	1453.43	1630.91	1913.23	3094.3

transfer coefficient. The maximum heat transfer coefficient is in the mixed flow regimes like slug-bubbly-annular regimes. Turbulence plays major role for the higher heat transfer rate in this regime.

The obtained heat transfer results are utilized to develop a correlation for non-dimensional heat transfer (Nu_{TP}) in terms of superficial gas and liquid Reynolds numbers, fluid properties, Lockhart-Martinelli parameter and inclination parameter. All these parameters are introduced in terms of ratio of gas to liquid parameters. The effect of gravity, surface tension and inclination angle is combined together in the inclination parameter. An assumed correlation for two-phase Nusselt number is expressed as

$$Nu_{TP} = C\left(\frac{Re_{SG}}{Re_{SL}}\right)^{m}\left(\frac{Pr_{g}}{Pr_{l}}\right)^{n}\left(\left.\frac{\mu_{b}}{\mu_{w}}\right|_{TP}\right)^{p} X^{q}\left(1+\mathrm{Eo}\,\mathrm{Sin}\,\alpha\right)^{r} \quad (1)$$

Table 2: Values of constant and exponents in the proposed correlation.

Flow	Re_{SL}	Re_{SG}	C	m	n	p	q	r 0°	r 2°	r 5°	r 7°
St	705-1894	832-9956	0.196	0.4915	-2.524	0.0297	0.088	0	-	-	-
W	705-1894	9956-23473	16.88	-0.25	1.0298	0.0342	-0.369	0	0.695	0.685	0.585
S	2044-7660	832-7182	405.4	0.1417	1.3658	0.0627	-0.404	0	-0.671	-0.226	-0.698
P	6057-7660	832-1635	93.9	0.4646	1.293	0.0127	0.746	0	0.4	0.726	0.579
A	1595-7660	26406-29340	18.8	-0.745	-1.26	0.016	-0.119	0	-0.295	-1.449	-0.245
WA	705-4487	12297-29340	0.177	-0.609	-3.403	0.019	-0.143	0	-0.115	0.0244	-0.015
SW	2044-4693	4595-10940	0.261	-0.473	-2.876	0.0194	0.0369	0	0.459	0.574	0.645
SBA	4693-7660	7182-23473	3.89	0.124	-1.376	0.0136	0.1335	0	0.787	0.856	0.996

The constant C and power coefficients m, n, p, q and r are established by curve fitting of our experimental results. The values of constants and exponents obtained by curve fitting are tabulated in Table 2. The comparison between predicted Nu_{TP} and the experimental Nu_{TP} signifies the validity of the predicted correlation. Sample parity plot for 7° inclination is shown in Figure 7. Similar parity plots are also developed for other inclinations. It is observed that 90 percent of the data points fall within ± 30% zone thereby establishing the validity of the developed correlation.

6 Conclusions

The following conclusions are drawn from the present work:

1. With the increase in inclination angle the stratified flow disappears and is replaced by intermittent flow. The transition boundaries shift to higher

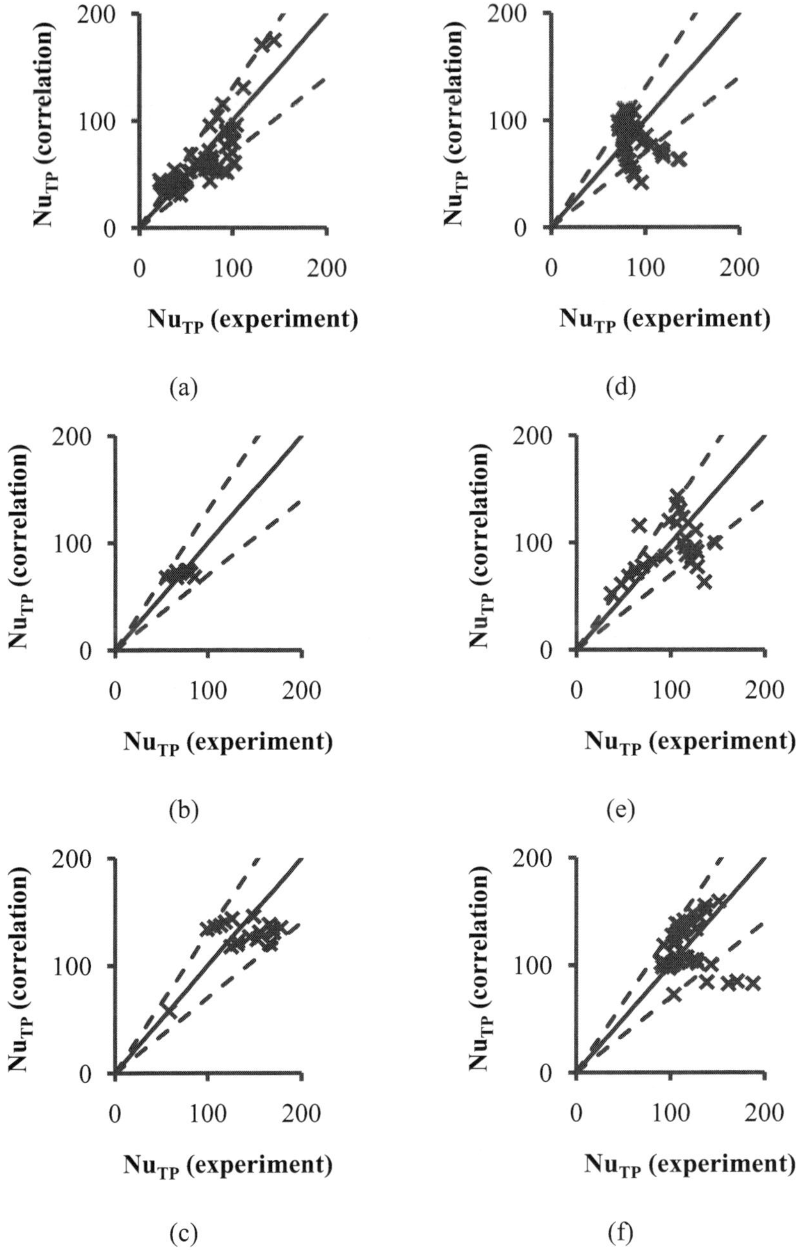

Figure 7: Parity plot between Nu_{TP} (correlation) and Nu_{TP} (experiment) for 7° inclination for various flow patterns (a) Slug (b) Plug (c) Wavy (d) Annular (e) SW (f) WA.

Reynolds numbers. This is due to the fact that increase in angle of inclination causes gravity component to act opposite to flow which in turn generates decelerating effect. It is found that the heat transfer for the inclined pipe is enhanced compared to that in a horizontal pipe.

2. Angle of inclination significantly affects the heat transfer rate except for the annular and wavy/annular flow. Exception is due to the increase in liquid thinning with the inclination and reversing of the flow.
3. A correlation for non-boiling two-phase heat transfer mechanism is proposed based on the experimental observations. This mechanism relates the two-phase Nusselt number in terms of the gas and liquid superficial Reynolds numbers, gas and liquid Prandtl numbers, wall and bulk two-phase viscosities, pressure drop and inclination factors.

Nomenclature

C	constant	**Greek letters**	
Eo	Eotvos number	α	angle of inclination in degree
h	heat transfer coefficient, W/m^2	μ	viscosity, Pa-s
m, n, p, q, r	exponents	**Abbreviations**	
Nu	Nusselt number	A	annular
Pr	Prandtl number	P	plug
Re	Reynolds number	RTD	resistance temperature detector
X	Lockhart-Martinelli parameter	S	slug
Subscripts		SBA	slug/bubbly/annular
b	bulk	St	stratified
l, g	liquid, gas	SW	slug/wavy
SL, SG	superficial liquid, superficial gas	W	wavy
TP	two-phase	WA	wavy/annular
w	wall		

References

[1] Taitel, Y. and Dukler, A. E., A model for predicting flow regime transition in horizontal and near horizontal gas liquid flow. *AIChE J.*, 22, pp. 47-55, 1976.

[2] Beggs, H.D. and Brill, J.P., A study of two-phase flow in inclined pipes. *J. Pet. Technol.,* pp. 607–617, 1973.

[3] Hasan, A.R. and Kabir, C.S., Predicting multiphase flow behavior in a deviated well. *SPE Prod. Engg.,* 3, pp. 474–482, 1988.

[4] Barnea, D., Shoham, O. and Taitel, Y., Flow pattern transition for gas-liquid flow in horizontal and inclined pipes. *Int. J. Multiphase Flow*, 6, pp. 217-225, 1980.

[5] Kokal, S. L. and Stanislave J. F., An experimental study of two-phase flow in slightly inclined pipes – I Flow patterns. *Chem. Engg. Sci.*, 44, pp. 665-679, 1989.

[6] Tshuva, M., Barnea, D. and Taitel, Y., Two-phase flow in inclined parallel pipes. *Int. J. Multiphase flow*, 25, pp. 1491–1503, 1999.

[7] Vaze, M. J. and Banerjee, J., CFD analysis of two-phase flow through inclined pipes, 34th National conference on Fluid Mechanics and Fluid Power, BIT Ranchi, India, December 10-12, Paper No. 147, 2007.

[8] Dorresteijn, W. R., Experimental study of heat transfer in upward and downward two-phase flow of air and oil through 70 mm tubes. Proc. 4th Int. Heat Transfer Conf. Vol. V, B5.9, Elsevier, Amsterdam, 1970.

[9] Elamvaluthi, G. and Srinivasa, N. S., Two-phase heat transfer in two-component vertical flows. *Int. J. Multiphase Flow*, 10, pp. 237-242, 1984.

[10] Mosyak, A. and Hetsroni, G., Analysis of dryout in horizontal and inclined tubes. *Int. J. Multiphase Flow*, 25, pp. 1521–1543, 1999.

[11] Hetsroni, G., Mewes, D., Enke, C., Gurevich, M., Mosyak, A. and Rozenblit, R., Heat transfer to two-phase flow in inclined tubes. *Int. J. Multiphase Flow*, 29, pp. 173-194, 2003.

[12] Vaze, M. J. and Banerjee, J., Experimental visualization of two-phase flow patterns and transition from stratified to slug flow. *Proceedings of the Institution of Mechanical Engineers, Part C, Journal of Mechanical Engineering Science*, 225, pp. 382-389, 2011.

[13] UKAS (United Kingdom Accreditation Service), The expression of uncertainty and confidence in measurement, M3003, 2nd ed., January 2007.

Section 5
Incline flows

Geometrical and kinematic properties of interfacial waves in horizontal heavy oil-water stratified flow

M. S. de Castro, C. C. Pereira, J. N. dos Santos
& O. M. H. Rodriguez
Department of Mechanical Engineering,
Engineering School of São Carlos, University of São Paulo, Brazil

Abstract

The oil-water stratified flow pattern is common in the petroleum industry, especially in offshore directional wells and pipelines. Previous studies have shown that the phenomenon of flow pattern transition in stratified flow can be related to the interfacial wave structure (problem of hydrodynamic instability). The study of the wavy stratified flow pattern requires the characterization of the interfacial wave properties, i.e., its nature (kinematic or dynamic), average shape, celerity and geometric properties as a function of holdup and phases' relative velocity. However, the data available in the literature on wavy stratified flow is scanty, especially when oil is viscous. This paper presents geometric and kinematic interfacial wave properties as a function of holdup and relative velocity in the wavy stratified flow. The experimental work was conducted in a glass test line of 12 m and 0.026 m i.d., oil (density 828 kg / m^3 and viscosity of 0.3 Pa.s at 20°C) and water as the working fluids at horizontal.
Keywords: liquid-liquid flow, oil-water flow, stratified flow pattern, interfacial wave structure.

1 Introduction

Liquid-liquid flows are present in a wide range of industrial processes; however, studies on such flows are not as common as those on gas-liquid flows. The interest in liquid-liquid flows has recently increased mainly due to the petroleum industry, where oil and water are often transported together for long distances.

WIT Transactions on Engineering Sciences, Vol 70, © 2011 WIT Press
www.witpress.com, ISSN 1743-3533 (on-line)
doi:10.2495/MPF110191

Pressure drop, heat transfer, corrosions and structural vibration are a few examples of topics that depend on the geometrical configuration or flow patterns of the immiscible phases. The stratified flow pattern is present in directional oil wells and pipelines and is characterized by the heavier and lighter phases at the bottom and top part of the pipe, respectively, divided by an interface. The interface in stratified flow can be smooth or wavy. The wavy structure was studied in gas-liquid flow by Bontozoglou and Hanratty [1] and Bontozoglou [2]. One of the findings is that the two-phase friction factor of wavy stratified flow can be about fifty times as high as the friction factor of smooth stratified flow. Li *et al.* [3] studied the gas-liquid stratified flow pattern and showed that the interfacial waves have significant influence on heat transfer and pressure drop.

An experimental work on liquid-liquid flow patterns can be seen in Trallero *et al.* [4], where data for horizontal flow were presented including stratified and semi-stratified flow, dispersions and emulsions. Those authors did not differ stratified wavy from stratified with mixing at the interface. Elseth [5] presented a more detailed oil-water horizontal flow pattern classification, dividing Trallero's patterns into several sub-patterns. That author observed the liquid-liquid wavy stratified flow pattern. Alkaya *et al.* [6] continued the work of Trallero *et al.* [4], but now introducing the effects of inclination. A wavy stratified flow pattern was reported. All the quoted authors have dealt with relatively low viscosity oils. On the other hand, Bannwart *et al.* [7] studied very viscous oil-water horizontal flow and reported among others the stratified flow pattern. Interfacial waves were spotted, but details on wave's geometrical properties were not given.

The study of the hydrodynamic stability of liquid-liquid stratified flow and therefore transition to dispersed flow has been carried out by Brauner *et al.* [8], Trallero [9] and Rodriguez *et al.* [10]. Although it seems clear that interfacial wave properties are important for the complete understanding of stratified flow, that information is quite scanty. Some data are presented in the paper of Al-Wahaibi and Angeli [11], but it lacks in details. Therefore, the goal of this work is to investigate geometrical and kinematic properties of interfacial waves in viscous oil-water stratified flow pattern.

In Sections 2 and 3 the experimental setup and procedure are presented. The image processing technique used to measure wave properties is explained in Section 4. In Section 5 data of wavelength, amplitude and celerity of waves are offered. Section 6 presents the mean interfacial wave shape. Finally, the conclusions are drawn in section 7.

2 Experimental setup

The hydrophilic-oilphobic glass test line of 26-mm i.d. and 12-m length of the multiphase flow loop of the Thermal-Fluids Engineering Laboratory of the Engineering School of Sao Carlos at the University of Sao Paulo (LETeF) was used to observe different oil-water flow patterns. A by-pass line allowed the usage of the quick-closing-valves technique to measure *in-situ* volumetric fraction of water and oil. The test section has a transparent box, filled with water, used to film the flow. A schematic view of the flow loop and details related to the holdup measurement technique can be seen in Rodriguez *et al.* [12].

The water used in the experiments has density of 0.988g/cm³ and viscosity of 1cP at temperature 20°C. The oil used has density of 0.828g/cm³ and viscosity of 300cP at 20°C. The oil-water interfacial tension was of 0,034N/m. The oil-water contact angle with the borosilicate glass was 29° (hydrophilic-oilphobic). The viscosity was measured with a rheometer Brokfield™, model LVDV-III+ with rotor SC4-18. The interfacial tension and contact angle were measured with an optical tensiometer of KSV™, model CAM 200.

3 Experimental procedure

The flow was filmed in order to perform the optical measurement of geometrical properties and celerity of the interfacial wave. The film was shot using a high speed video camera (CamRecord 600 from Optronis). The camera was installed on a pedestal attached to the structure. Two xenon lamps were used to illuminate the flow. After reaching steady state the flow was recorded (60 frames/second for 60 seconds) and then a homemade LabView™ based program was used to extract the properties of the wave from the images.

4 Data treatment

The Labview™ based software was used to identify the wave corresponding to the oil-water interface and predict its amplitude (α), wavelength (λ) and celerity (c). Each frame of the film (figure 1) is binarized so that the oil part becomes black and the water part becomes red. Bubbles and other noises are taken off the images. After those steps the interface is captured via a colour contrast technique.

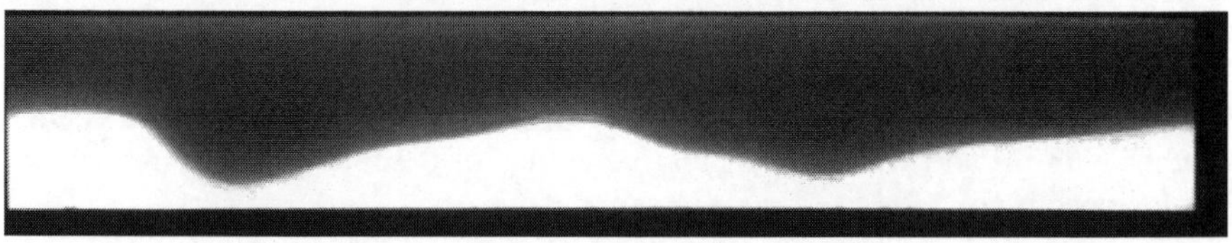

Figure 1: Picture of the observed oil-water wavy stratified flow.

After the identification of the interface the analysis of the wave characteristics depends on the class of wave, related to the internal diameter of the pipeline (D):

Short wave: $\lambda < D$; it is characterized by a strong corrugated interface, its slope indicates imminence of breaking and its behavior is tridimensional.

Intermediate wave: $D < \lambda < 10D$; it is characterized by a clear wavy structure.

Long wave: $\lambda > 10D$; the interface is almost flat, with small amplitude in relation to the water height.

Using the Fourier transform, the spectrum of waves are represented in the frequency domain. Two cut frequencies were used, one for low-frequency waves and one for high-frequency waves. High frequencies were ignored since they were considered noise (ripples). Long waves were not analyzed because they were not entirely visualized in the film section. Hence, short waves and long waves were discarded and only intermediate waves were analyzed.

The wavelength and wave amplitude are calculated using the definition of wave crest as the pixel in the spatial domain where the first derivative is zero, the next pixel has derivative lower than zero and the previous pixel has derivative greater than zero. The wavelength is the length difference between the abscissas (x-coordinate) of two followed crests. Since the interfacial wave is asymmetric, the definition of amplitude is not quite simple. It differs whether it is measured from the crest to the right or left valley (minimum value in the y-coordinate). To avoid this ambiguity, this paper uses the mean between the two calculated amplitudes (from the crest to the right valley and from the crest to the left valley) as the definition of wave amplitude.

The celerity is calculated using cross-correlation between two frames lagged in time (Bendat and Piersol [13]). Figure 2 presents two snap shots used for the measurement of wave celerity.

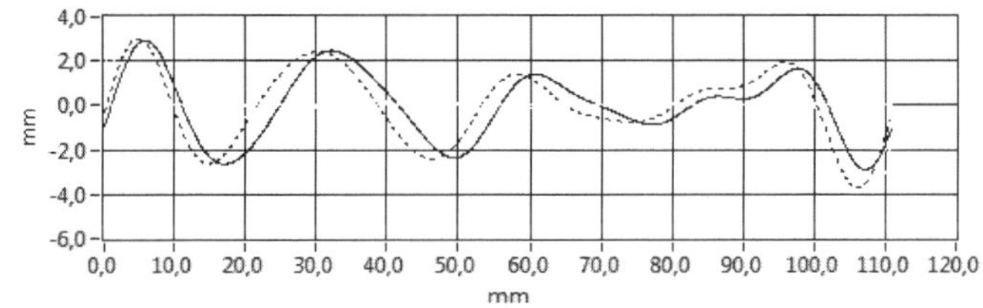

Figure 2: Waves lagged in time.

The PDF (probability density function) of the collected signals was used to analyze the data of amplitude, wavelength and celerity. Around 100 waves were detected by the software in each film.

Two different PDFs of the amplitude data are presented in figure 3, one with oil and water superficial velocities of 0.07m/s and 0.15m/s (figure 3(a)) and another with oil and water superficial velocities of 0.13m/s and 0.15m/s (figure 3(b)), respectively.

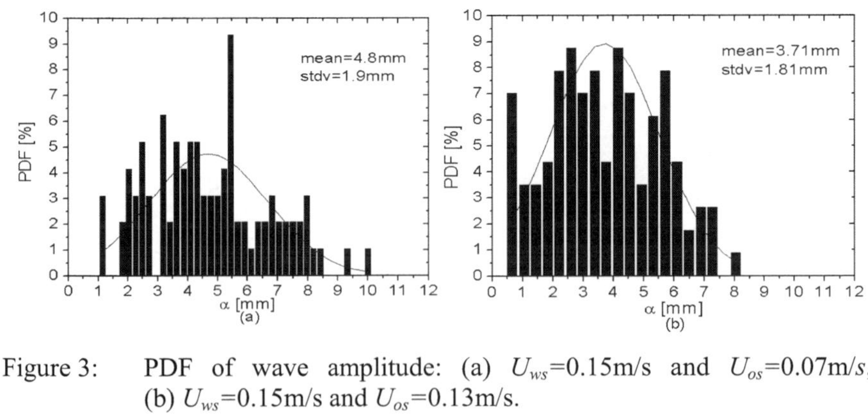

Figure 3: PDF of wave amplitude: (a) U_{ws}=0.15m/s and U_{os}=0.07m/s, (b) U_{ws}=0.15m/s and U_{os}=0.13m/s.

The amplitude distribution differs from a Gaussian in both cases. The mean amplitude value is different, but the data have similar distributions (between 1mm and 10mm).

Two different PDFs of the wavelength data are presented in figure 4, one with oil and water superficial velocities of 0.15m/s (figure 4(a)) and another with oil and water superficial velocities of 0.15m/s (figure 4(b)), respectively.

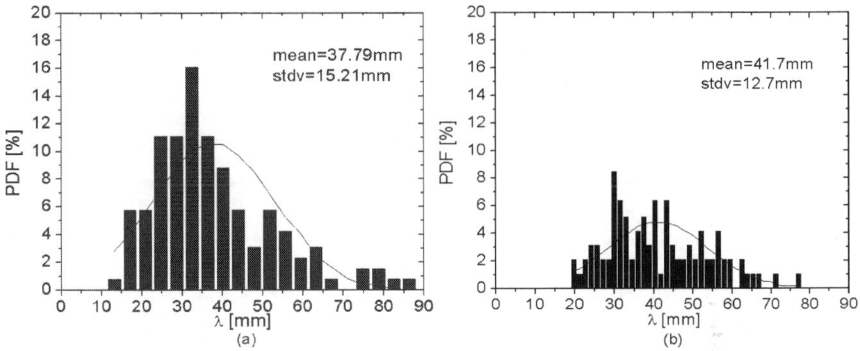

Figure 4: PDF of wave length: (a) U_{ws}=U_{os}=0.15m/s, (b) U_{ws}=0.15m/s and U_{os}=0.07m/s.

The wavelength PDFs also differ from a Gaussian and have different mean values. While figure 4(a) presents distribution between 10mm and 90mm with higher concentration between 10mm and 40mm, figure 4(b) presents distribution between 20mm and 80mm, with a more uniform concentration. It is important to notice that there are long wavelength waves, but they represent a small percentage.

Two different PDFs of the wave celerity data are presented in figure 5, with oil and water superficial velocities of 0.05m/s and 0.15m/s (figure 5(a)) and another with oil and water superficial velocities of 0.13m/s and 0.15m/s (figure 5(b)), respectively.

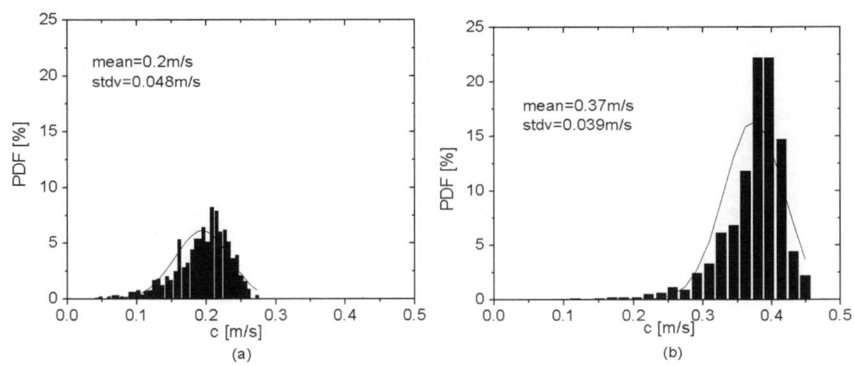

Figure 5: PDF of wave celerity: (a) U_{ws}=0.15m/s and U_{os}=0.03m/s, (b) U_{ws}=0.15m/s and U_{os}=0.13m/s.

One may see that they have different mean values, but similar distribution. Figure 5(a) shows the distribution between 0.05m/s and 0.28m/s, whereas figure 5(b) shows it between 0.1 and 0.45m/s. The latter presents a more concentrated distribution around the mean value. Both figures differ from a Gaussian distribution.

5 Interfacial wave properties

Table 1 organizes the water and oil superficial velocities at which the experiments were taken.

Table 1: Superficial velocities of oil and water tested.

Experiment	**Superficial velocity of oil (U_{os}) [m/s]**	**Superficial velocity of water (U_{ws}) [m/s]**
001	0.03	0.15
002	0.05	0.15
003	0.07	0.15
004	0.10	0.15
005	0.13	0.15
006	0.15	0.15

As the wave properties seemed to have relation with relative velocity and holdup the data are presented as function of these variables. Relative velocity is represented by *Vrel* (eqn. (1)), where U_w and U_o are the *in situ* velocities of oil and water, respectively, and oil holdup is represented by ε_o.

$$Vrel = U_w - U_o \tag{1}$$

5.1 Interfacial wave length (λ)

The normalized mean wavelength (λ/D) is presented as a function of the oil holdup and relative velocity in figure 6(a) and (b), respectively. One may notice that the amplitude decreases with increasing the oil holdup and relative velocity. The relative velocity is supposed to have a destabilizing effect in the stratified flow (Kelvin-Helmholtz stability criterion, Taitel and Dukler [14]). Short wavelength waves have been related to transition from core-annular flow to dispersed flow (Rodriguez and Bannwart [15], and Rodriguez and Oliemans [16]). Therefore, the highest the *Vrel*, the highest the destabilizing effect, consequently the shorter the wavelength. On the other hand, high oil holdup implicates in low water volumetric fractions, therefore the effect of gravity in the denser water phase is lower. The gravity has a stabilizing effect, as can be seen in shallow water stability (Whitham [17], Drazin and Reid [18]). So, as the oil holdup increases, the stabilizing effect of gravity decreases and the wavelength becomes shorter.

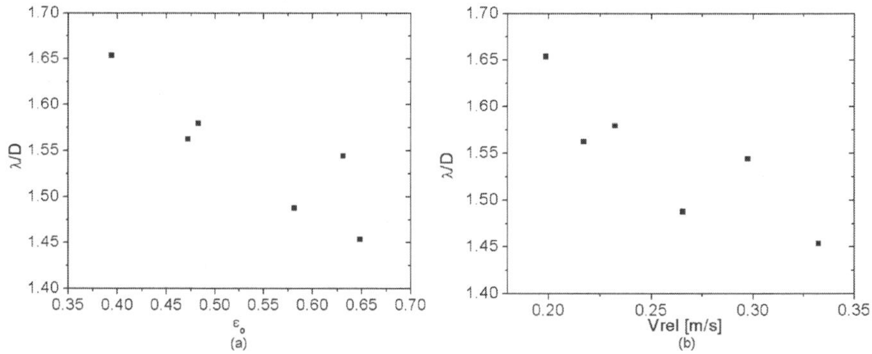

Figure 6: Dimensionless mean wavelength (λ/D) as a function of: (a) oil holdup (ε_o), and (b) relative velocity (*Vrel*).

5.2 Interfacial wave amplitude (α)

The normalized mean wave amplitude (α/D) is presented as a function of the oil holdup and relative velocity in figure 7(a) and (b), respectively. Looking at the graphs of figure 7 one may notice that the amplitude decreases with increasing the oil holdup and relative velocity. It would be expected that for higher oil holdups and relative velocities the flow should become more unstable, which may have been translated into higher wave amplitude. However, it was not seen in the present work. Apparently, shorter waves (refer also to figure 6) have lower amplitudes in the stratified flow investigated in this work.

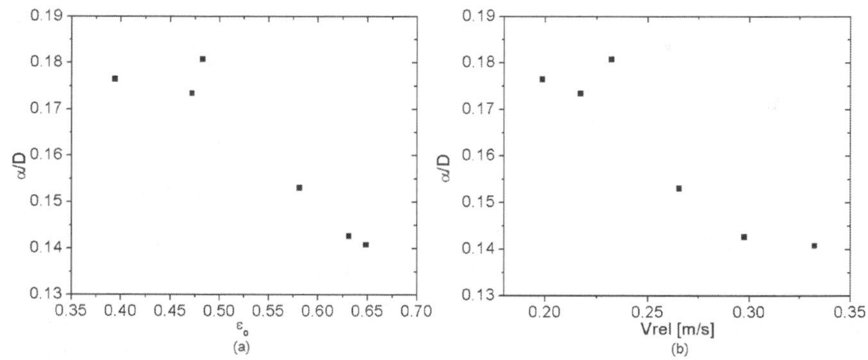

Figure 7: Dimensionless mean wave amplitude (α/D) as a function of: (a) oil holdup (ε_o), and (b) relative velocity (*Vrel*).

5.3 Interfacial wave celerity (c)

The mean interfacial wave celerity is normalized by the mixture velocity ($U_{ws}+U_{os}$) and presented as a function of the oil holdup and relative velocity (figure 8(a) and (b)). The wave celerity increases with increasing both the oil holdup and relative velocity. Higher wave celerity would be related to instability of the interfacial wavy structure, as there is a celerity limit for a stable wave in two-phase flow (Wallis [19]).

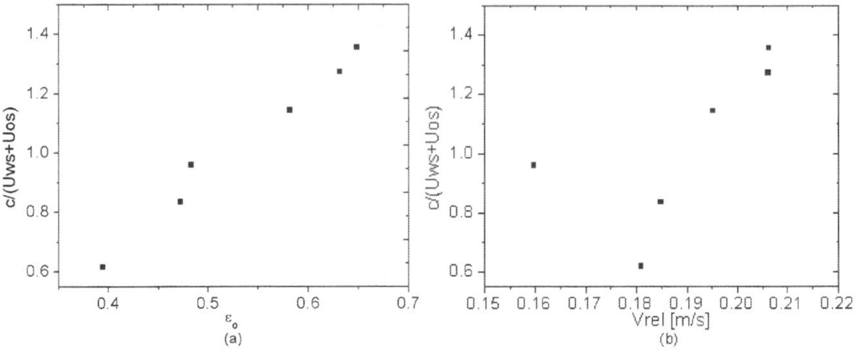

Figure 8: Normalized wave celerity ($c/(U_{ws}+U_{os})$) as a function of: (a) oil holdup (ε_o), and (b) relative velocity (*Vrel*).

5.4 Uncertainties

This section *presents* the uncertainties of the collected dα_ta. The values of uncertainties *are*: 56%, 40% and 40%, respectively, for wave amplitude, wavelength and celerity. The values correspond to 2 times de standard deviation.

The uncertainties of the oil holdup are lower than 3% and for the relative velocity the uncertainty is of 1.41%.

6 Mean interfacial wave shape

The mean interfacial wave shape was found by averaging all the wave shapes captured in the filming for a constant pair of superficial velocities of oil and water. The program divides the waves present in each frame of the film into the previously defined three wave classes, according to the image treatment procedure explained in section 4. Every pair of superficial velocities was related to a number (i) of waves with different amplitudes and wavelengths. Every x-coordinate (j) (abscissa) of the i wave (x_{ij}) is normalized (x^*) by its own wavelength (λ_i), eqn. 2:

$$x^* = \frac{x_{ij}}{\lambda_i} \tag{2}$$

Every y-coordinate (j) (ordinate) of the i wave (y_{ij}) is normalized (y^*) by the pipe internal diameter (D), eqn. 3:

$$y^* = \frac{y_{ij}}{D} \tag{3}$$

The mean dimensionless wave shape is formed by averaging all the y* points of all waves at every x*-coordinate. The mean wave shape is obtained by multiplying the y* and x* points of the mean dimensionless wave shape by the pipe diameter and by the mean wave length given by the PDF for each pair of superficial velocities, respectively.

Mean waves shapes are presented in Fig. 9 for the same water superficial velocity, 0.15m/s, but for different oil superficial velocities, 0.03m/s and 0.15m/s.

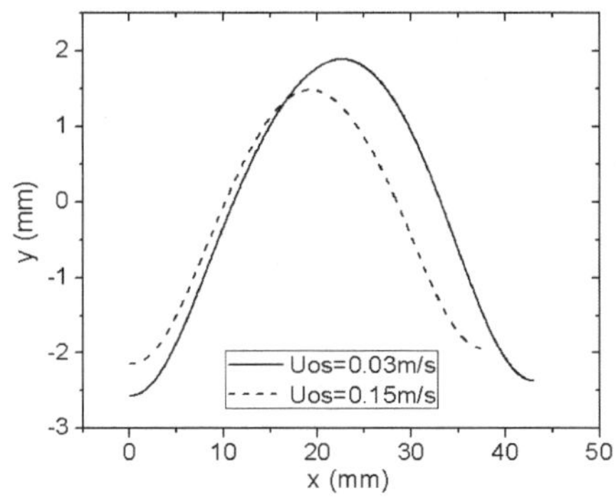

Figure 9: Mean interfacial wave shape for horizontal flow, U_{ws}=0.15m/s.

In Figure 9 the solid line presents data of oil superficial velocity and relative velocity of 0.03m/s and 0.2m/s, respectively, and the dashed line presents data of oil superficial velocity and relative velocity of 0.15m/s and 0.33m/s, therefore an increase of 33% in the relative velocity. One may see that the wavelength decreases with the increase of relative velocity. This difference in wave shape might help to explain the stability of stratified flows, *i.e.*, flows with higher relative velocity would tend to be more sensitive to disturbances at the interface or less stable.

7 Conclusions

New data of interfacial wave geometrical properties and celerity in viscous oil-water horizontal stratified flow are offered. The wave amplitude, wavelength and wave celerity are dependent on holdup and relative velocity between the water and oil phases. The wavelength tends to decrease with the increase of the relative velocity and with the oil holdup. The celerity tends to increase with the increase of the oil holdup and relative velocity. Both results indicate higher degree of hydrodynamic instability at higher relative velocities. The mean wave shape depends on the relative velocity. The new experimental data and findings can be used for improving pressure drop, holdup and stability models of liquid-liquid stratified flow.

Acknowledgements

The authors are gratefull to FAPESP (Fundação de Amparo à Pesquisa do Estado de São Paulo, process 2008/06922-9 and process 2010/03254-5) for the financial support.

References

[1] Bontozoglou, V. and Hanratty, T.J., Wave Height Estimation In Stratified Gas-Liquid Flows. *Aiche Journal*, New York, 35 (8), pp. 1346-1350, 1989.

[2] Bontozoglou, V., Weakly Nonlinear Kelvin-Helmholtz Waves Between Fluids Of Finite Depth. *International Journal of Multiphase Flow*, Oxford, V.17, N.4, P.509-518, July/Aug. 1991.

[3] Li, G.J.; Guo, L.; Chen, X.J., An experimental investigation on the interfacial waves in air-water two-phase flow within horizontal pipelines. *Chinese Journal Chemical Engineering*, China, 5 (4), pp. 316-324, Dec. 1997.

[4] Trallero, J.L.; Sarica, C.; Brill, J.P., A Study of oil/water flow patterns in horizontal pipes. *SPE Production & Facilities*, Richardson, 12 (3), pp.165-172, Aug. 1997.

[5] Elseth, G., An experimental study of oil-water flow in horizontal pipes, Ph.D thesis, Porsgrunn: Norwegian University of Science and Technology, 270p, 2001.

[6] Alkaya, B., Jayawardena, S.S., Brill, J.P., Oil Water Flow Patterns in Slightly inclined Pipes. In: *Etce/Omae200 Joint Conference: Energy For The News Millennium*, 2000, New Orleans. Proceedings. NewYork: Asme. pp. 14-17, 2000.

[7] Bannwart, A.C. et al., Flow patterns in heavy crude oil-water flow. *Journal of Energy Resources Technology*, 126 (3), pp. 184-189, New York, Sept. 2004.

[8] Brauner, N.; Rovinsky, J.; Maron, D.M., Determination Of The Interface Curvature In Stratified Two-Phase Systems By Energy Considerations. *International Journal of Multiphase Flow*, Oxford, 22 (6), pp. 1167-1185, 1996.

[9] Trallero, J.L. (1995). Oil-water flow patterns in horizontal pipes. Ph.D. Thesis - The University of Tulsa, Tulsa, 1995.

[10] Rodriguez, O.M.H.; Oliemans, R.V.A.; Mude, R.F. Stability Analysis Of Slightly-Inclined Stratified Oil-Water Flow, Including The Distribution Coefficients And The Cross-Section Curvature. *In: Proc. Of the 5th North American Conference On Multiphase Technology*, Banff. Proc. Cranfield: Brh Group, pp. 229-245, 2006.

[11] Al-Wahaibi, and P. Angeli, Experimental Studies on Flow Pattern Transitions in Horizontal Oil-Water Flow, *Proc. 6th International Conference on Multiphase Flow (ICMF 2007)*, Leipzig, Germany, July 2007.

[12] Rodriguez, I. H., Yamaguti, H. K. B., de Castro, M. S., Da Silva, M. J. and Rodriguez, O. M. H., Slip ratio in dispersed viscous oil–water pipe flow, *Experimental Thermal and Fluid Science*, 35 (1), pp. 11-19, January 2011.

[13] Bendat, J. S. and Piersol, A. G., *Random Data: analysis and measurement procedures*. John Wiley & Sons: New York, 2000.

[14] Taitel, Y. and Dukler, A. E., A model for predicting flow regime transitions in horizontal and near horizontal gas-liquid flows. AIChE Journal, 22, pp. 47-55, 1976.

[15] Rodriguez, O.M.H. and Banwart, A. C., Stability Analysis of Core-Annular Flow and Neutral Stability Wave Number, *AIChE Journal*, 54 (1), January, 2008.

[16] Rodriguez, O. M. H. and Oliemans, R.V. A., Experimental study on oil-water flow in horizontal and slightly inclined pipes, *International Journal of Multiphase Flow*, 32, pp. 323-343, 2006.

[17] Whitham, G. B. Linear and Nonlinear Waves. New York: John Wiley & Sons, 1974.

[18] P. G. Drazin, W. H. Reid, Hydrodynamic stability. Cambridge University Press, Cambridge, 1981.

[19] Wallis, G.B., One-Dimensional Two-Phase Flow. MacGraw-Hill, New York, USA, 1969.

A numerical model for partially-wetted flow of thin liquid films

K. V. Meredith, J. de Vries & Y. Xin
FM Global, Research Division, USA

Abstract

A model for simulating thin liquid film transport over solid surfaces has been developed. The film transport for mass continuity and momentum were formulated as a two-dimensional set of equations using thin-film assumptions. These equations have been implemented in an open-source CFD code (OpenFOAM). Treatments for partial-wetting phenomena have been included in the model to account for the behavior near the contact-line. A surface-tangential force along the contact line has been developed to allow for the simulation of rivulets and dry patches in two-dimensional surface flow. An approach for applying contact angle effects to the model for a real stochastic surface is outlined. Additionally, experimental measurements were made for film flow over an inclined surface for a wide range of flow rates. Using these experimental results, the model has been validated for partially wetted flow over an inclined panel. The critical flow rate of a film over a given surface was used as validation for the model. Results show that for flow rates below the critical flow rate, the partially wetted behavior of the flow was reproduced. Comparisons to experimental flow patterns and wetted-area fractions were made.
Keywords: thin-film assumption, partial wetting, contact line, critical flow rate.

1 Introduction

The goal of this research was to develop a robust water-film transport model for use in simulating the partial wetting behavior of thin film flow in practical applications over realistic surfaces. To this extent, a thin film transport model has been developed in an open source computational fluid dynamics (CFD) framework (OpenFOAM).

WIT Transactions on Engineering Sciences, Vol 70,
www.witpress.com, ISSN 1743-3533 (on-line)
doi:10.2495/MPF110201

Work on understanding thin-film flows began in 1910 with the experimental measurements of Hopf [1]. Later, Nusselt [2,3] proposed a theoretical treatment of thin-film flow and heat transfer. An excellent review of the work prior to 1964 has been offered by Fulford [4]. Recent film modeling has focused on volume of fluid (VOF) approaches [5,6] and on applying the lubrication approximation to thin-film flow [7–9]. The VOF models are computationally expensive. The majority of the lubrication approximation models are not aimed at practical problems, include a limited amount of additional physics, and are overly focused on detailed behavior near the contact line.

The phenomena that occur during the lifetime of a liquid film consist of partial wetted flow behavior and interfacial transport of mass, momentum, and energy. The subject of this work will focus on partially wetted flow behavior, i.e., surface film flows delineated by a contact line separating the dry and wet surfaces. In addition, this approach includes treatment for phenomena such as rivulets, dry patches, and isolated wet spots.

2 Modeling approach

Due to the thin nature of the liquid film (typically less than 1 mm), the flow in the direction normal to the surface can reasonably be assumed to be negligible. The transport model only needs to describe film advection in the surface-tangential direction, thus greatly simplifying the model. In addition, due to the 'thinness' of the film, the diffusive transport of mass/momentum/energy in the surface-normal direction will dominate and the surface-tangential diffusion will be small in comparison. In other words, advection can be treated in the wall-tangential direction and diffusion processes treated in the wall-normal direction. This is known as the thin-film assumption and has been used successfully by a number of authors [10–13]. Tangential diffusion effects may prove to be important along the contact line (e.g. at the edges of rivulets), and this feature may need to be added in the future.

The thin-film assumption allows integration of the three-dimensional transport equations in the wall-normal direction from the wall surface to the film surface to obtain a set of equations for the film thickness and the wall-tangential components of velocity.

2.1 Mass continuity

The integrated mass continuity equation is defined as

$$\frac{\partial \rho\delta}{\partial t} + \nabla_s \cdot [\rho\delta\mathbf{U}] = S_{\rho\delta} \tag{1}$$

where ρ is the liquid density, δ is the film thickness, $\mathbf{U}$ is the film velocity, ∇_s is the vector differential operator tangential to the surface, $\left(\frac{\partial}{\partial x}, \frac{\partial}{\partial y}\right)^{\mathsf{T}}$. $S_{\rho\delta}$ is the mass source per unit wall area due to impingement, splashing, evaporation, absorption into the solid, and film separation. For the purposes of this study, where the focus was on film flow in the absence of spray interactions and phase change, the source term is reduced to

$$S_{\rho\delta} = 0 \tag{2}$$

2.2 Momentum transport

The momentum equation, integrated over film height, is

$$\frac{\partial \rho\delta\mathbf{U}}{\partial t} + \nabla_s \cdot [\rho\delta\mathbf{U}\mathbf{U}] = \underbrace{-\delta\nabla_s p}_{\text{pressure based}} + \underbrace{\mathbf{S}_{\rho\delta\mathbf{U}}}_{\text{stress based}} \tag{3}$$

where $\mathbf{U}$ represents the mean, tangential velocity of the film. The momentum source terms are split into pressure based (tangential gradients in wall-normal forces) and stress based (forces tangential to wall).

The pressure term, p, is comprised of forces in the wall-normal direction and consists of capillary effects (p_σ), hydrostatic pressure head (p_δ), and local gas-phase pressure (p_g).

$$p = p_\sigma + p_\delta + p_g \tag{4}$$

The stress term, $\mathbf{S}_{\rho\delta\mathbf{U}}$, includes the viscous shear stresses $(\boldsymbol{\tau}_g, \boldsymbol{\tau}_w)$, gravity body force $(\rho\mathbf{g}_t\delta)$, and stress related to contact-angle force $(\boldsymbol{\tau}_\theta)$.

$$\mathbf{S}_{\rho\delta\mathbf{U}} = \boldsymbol{\tau}_g + \boldsymbol{\tau}_w + \rho\mathbf{g}_t\delta + \boldsymbol{\tau}_\theta \tag{5}$$

The specific definitions and submodels for the pressure terms tangential momentum source terms in Equations 4 and 5 are given in Section 3.

3 Source terms and submodels

Relevant source terms in Equations (4) and (5) are described in detail. Where appropriate, a background section is included describing the theory behind the submodel for that particular source term.

3.1 Capillary pressure

The capillary pressure, p_σ, required by Equation (4), is the pressure component due to surface tension based on the curvature of the film surface and represented by the Laplace–Young boundary condition at the air-fluid interface [7, 14]. Using the Laplacian of film thickness to estimate the curvature, the term for the pressure contribution of surface tension is given as

$$p_\sigma = -\sigma{\nabla_s}^2\delta \tag{6}$$

where σ is the surface tension and ${\nabla_s}^2\delta$ approximates the curvature of the liquid surface. This relationship is valid only for surfaces with slight curvature [14].

3.2 Hydrostatic pressure

The hydrostatic pressure, p_δ, required by Equation (4), is the pressure component due to hydrostatic pressure head, and is given by

$$p_\delta = -\rho\left(\mathbf{n}\cdot\mathbf{g}\right)\delta \tag{7}$$

where $\mathbf{g}$ is the gravity force vector and $\mathbf{n}$ is the surface normal vector. This pressure term is eliminated for purely vertical surfaces ($\mathbf{n}\cdot\mathbf{g} = 0$).

3.3 Gravity body force

The gravity body force term, required by Equation 5, is represented by $\rho\mathbf{g}_t\delta$ where $\mathbf{g}_t$ denotes the gravity components tangential to the wall. This term is zero for horizontal surfaces, and is the main driving force for flow over vertical and inclined surfaces.

3.4 Shear stress

The shear stress terms $\boldsymbol{\tau}_g$ and $\boldsymbol{\tau}_w$, required by Equation (5), represent the shear at the film-gas interface and the film-wall interface respectively.

In most cases for film flow in the presence of a quiescent gas, $\boldsymbol{\tau}_g \ll \boldsymbol{\tau}_w$ because gas-phase velocities are typically very low and also because the viscosity of the gas is low compared to that of the liquid. For the purposes of this study, $\boldsymbol{\tau}_g$ is assumed negligible.

The film-wall stress term, $\boldsymbol{\tau}_w$, is modeled based on an assumed velocity profile in the wall-normal direction. Assuming laminar flow and taking the velocity at the wall to be zero and the velocity gradient at the gas interface to be zero, the expression for the velocity distribution through the film can be shown to be

$$\mathbf{u}(z) = \frac{3\mathbf{U}}{\delta}\left(z - \frac{z^2}{2\delta}\right) \tag{8}$$

where $\mathbf{U}$ represents the mean film velocity, z represents the surface-normal coordinate, and δ represents the film thickness. Employing the quadratic velocity profile assumption, the shear stress at the wall is found as

$$\boldsymbol{\tau}_w = -\mu\frac{3\mathbf{U}}{\delta} \tag{9}$$

Experimental data confirms that this parabolic velocity profile remains valid even for wavy film flows [15]. A film can exhibit wavy behavior without necessarily being turbulent, and is not fully turbulent until $\mathrm{Re}_f = \Gamma/\mu > 900$ [4], which is well above the range of interest for this study. Here, Γ is defined as the mass flow rate per unit width of inlet (kg/m-s).

3.5 Partial wetting behavior

Simulating partial wetting behavior involves treatment of the contact line (the delineation between wet/dry regions). Along the contact line delineating the separation of wet and dry regions of the flow there is a surface-tangential force that limits the film from spreading. This force only exists for partially wetting fluids (i.e. $\theta_E > 0$, where θ_E represents the contact angle). The contact angle defines the ability of the liquid to flood the surface. This force is a major cause of partially wetted flow behavior such as rivulets, dry spots, and film rupture. Thus, the effect of contact angle must be included in the film transport model in order to accurately predict partial wetting behavior.

The tangential surface forces due to interfacial tension, ($\mathbf{F}'_s$), are expressed as

$$\mathbf{F}'_s = (\sigma + \sigma_{sl}) - \sigma_{so} \tag{10}$$

where σ is the liquid-air interfacial tension, σ_{sl} is the solid-liquid interfacial tension, and σ_{so} is the solid-air interfacial tension.

Since only σ in Equation (10) is easily measured experimentally, it is helpful to reduce this surface-tangential force to a function of σ and θ_E (also a measurable quantity) by employing Young's Law [16, 17]. Young's Law states that the forces along the contact line are balanced at equilibrium as

$$\sigma_{so} - (\sigma \cos\theta_E + \sigma_{sl}) = 0 \tag{11}$$

Combining Equations (10) and (11) yields what is termed here as the 'contact-angle force'.

$$\mathbf{F}'_s = \sigma(1 - \cos\theta_E) \tag{12}$$

$\mathbf{F}'_s$ represents the surface-tangential force normal to the contact line (still in the tangential plane) per unit width of contact line.

The contact-angle force is applied in a straight forward fashion to the momentum equation, as is shown here:

$$\boldsymbol{\tau}_\theta = \beta \frac{\sigma(1 - \cos\theta_E)}{\Delta_{cl}} \mathbf{n}_{cl} \tag{13}$$

First, $\mathbf{F}'_s$ is divided by the width of the computational cell, Δ_{cl} in the direction normal to the contact line. Secondly, $\mathbf{F}'_s$ is multiplied by the unit vector normal to the contact line in the surface-tangential plane. Finally, an empirical parameter β is introduced to account for discrepancies between the theory and observed behavior of real surfaces. Values for β can be determined by experimentally measuring the critical flow rate (i.e. the flow rate at which transition occurs from rivulet flow to continuous film flow occurs) for a given surface, and then adjusting β to match this behavior in the model.

3.6 Specification of contact line

The contact line is defined as the interface between the wet regions and dry regions in the computation. In this context, it is helpful to think in terms of a critical film thickness. For a given contact angle, a film will exhibit a minimum or critical film thickness, δ_c, below which the film flow will not exert enough force to overcome the contact angle force. Analytical expressions can be derived for δ_c on horizontal, vertical, and arbitrary inclination angle surfaces. In the model, a region of the flow is considered wet when denoted by an ω value of 1, and considered dry when denoted by an ω value of 0. Determination of ω depends on the critical film thickness and the approach varies for hydrophobic surfaces versus hydrophilic surfaces.

Hydrophobic surfaces For surfaces that are non-absorbent of water, when $\delta > \delta_c$, then that region of the film is considered 'wet', while for $\delta < \delta_c$ the surface is considered 'dry'.

Hydrophilic surfaces For certain porous materials (such as corrugated cardboard) or surfaces with micro-grooves that tend to attract water, once the surface has been 'wetted' it is very difficult to return it to the 'dry' state, even when $\delta < \delta_c$. To account for this in the model, a 'memory' is introduced that records whether that region of the surface was previously wetted. If a region of the film was previously wetted, then δ must reach a value much lower than δ_c (e.g. $\delta < 0.001\ \delta_c$) in order to be considered dry again.

To account for the scenario where the surface is wetted by means other than film flow (e.g. spray impingement) and the film thickness is still lower than δ_c, the wetting criterion has been relaxed to be, for example, $\delta > \frac{1}{2}\delta_c$.

In practice, this phenomenologically based treatment does a very acceptable job of replicating experimentally observed partial wetting behavior.

3.7 Specification of contact angle

One of the major issues with the above approaches is determining the contact angle. The contact angle for advancing films or flowing rivulets has usually been found to be different than equilibrium contact angles measured from the droplet method [15, 18]. The contact angle has even been found to be dependent on flow rate, rivulet width, and heat transfer rate [19]. For water on cast acrylic (Plexiglas®), the surface used in this study, the contact angle has been observed by the authors to typically be in the range of 70°. Thus, in the model a normal distribution of contact angle is used based on an experimentally observed mean value and standard deviation. For the results shown here, the standard deviation was chosen as 10° with a mean of 75°. At each computational cell a random contact angle is chosen from this normal distribution and is fixed for the duration of the calculation. The result of this approach is a contact angle that varies over the

WIT Transactions on Engineering Sciences, Vol 70, © 2011 WIT Press
www.witpress.com, ISSN 1743-3533 (on-line)

surface in a random manner. This gives rise to nonuniform flow and leads to the formation of rivulets and dry regions.

4 Experimental setup

The experiments used a test apparatus specially designed to generate uniform water flows. The main body of the apparatus consists of a vertical cast acrylic plate (0.61 m wide and 1.22 m long). On the upper end of the plate, a perforated copper tube (1.27 cm diameter, 0.51 m long) was positioned laterally to discharge water through sixty holes (1 mm diameter) onto the cast acrylic panel. Since the discharge of the individual stream created wavy flow, the water discharged from tube applicator was forced through a diffusive medium attached to the upper edge of the cast acrylic panel, creating a smooth initial flow. Water flow rates were measured via a flow meter (Micro Motion® 3700 MVD). Regular heated tap water (43 °C) was used for all experiments. All images were taken using an IR camera (SC655, 50 Hz, 640×480). All thermal images were corrected for perspective projection using a 4 point transformation in Matlab. The tubular water applicator and the entire apparatus were installed on a Aluminum frame (80/20 Inc.), which allowed variation of the panel inclination angle between 5° and 90°.

5 Results

The model has previously been validated against the Nusselt solution theory and experimental data for film thickness and velocity of continuous thin film flow over an inclined surface [20]. The results for this work are focused on partially wetted flow behavior.

Experiments and simulations were performed for partially wetted flow conditions over a range of flow rates and inclination angles. A uniform grid spacing of 3.8 mm was used for the calculations. The simulations are compared here to the experimentally measured partially wetted flow conditions. First, qualitative visual comparison is shown for the flow behavior, followed by comparison based on the measured wetted area fraction.

Figure 1(a–d) shows instantaneous experimental images of flows on a 5° sloped surface. The wetted surface is displayed as black, while the dry surfaces are shown as white. At the lower flow rates, partial wetting of the surface is observed, yielding a rivulet based flow. As flow rate increases past a critical value, the surface becomes flooded. The transition from rivulet flow to continuous film flow occurs between $\Gamma = 108$ g/m/s and $\Gamma = 174$ g/m/s. In the flooded condition, considerable 'necking' of the flow is observed. The extent of necking decreases with increasing flow rate.

Simulation results for the 5° incline condition are shown in Figure 1(e–h). The model shows considerably less meandering behavior for the rivulet flow. Still, similarities are seen in the wetting behavior of the liquid film. The lower flow rate simulations predict rivulet flow, while the higher flow rate simulations

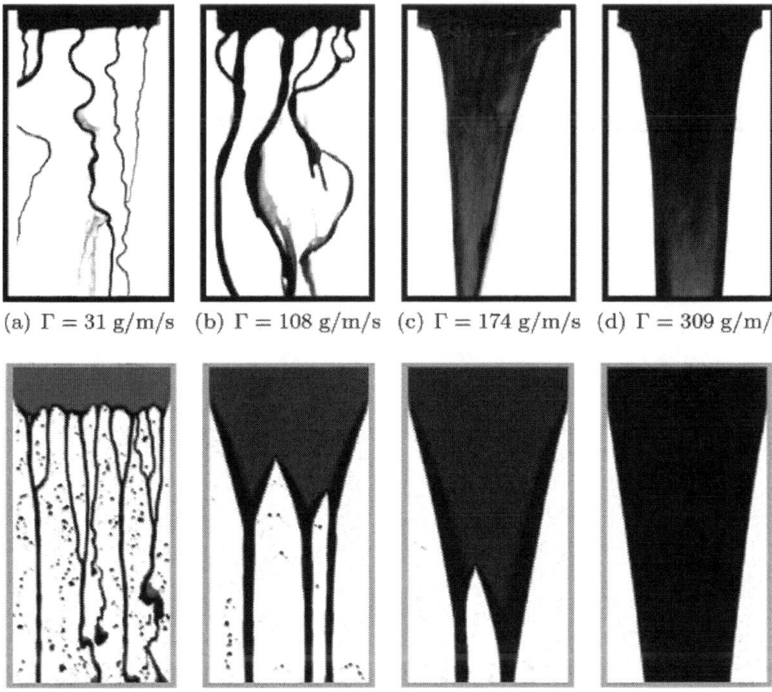

(a) $\Gamma = 31$ g/m/s (b) $\Gamma = 108$ g/m/s (c) $\Gamma = 174$ g/m/s (d) $\Gamma = 309$ g/m/s

(e) $\Gamma = 50$ g/m/s (f) $\Gamma = 150$ g/m/s (g) $\Gamma = 200$ g/m/s (h) $\Gamma = 400$ g/m/s

Figure 1: Experimental film flow (a–d) compared with simulated film flow (e–h) (5° incline, $\beta = 0.4$).

predict flooded conditions. Even the general trend of the reduction in necking as flow rate increases is captured in the model, if only qualitatively. The simulation results show isolated droplets that form as a rivulet passes. This behavior was observed experimentally also, however the post-processing of the experimental images removed the droplets from visibility.

Figure 2(a–d) shows experimental images of flows on a 90° sloped surface for a range of flow rates. The transition from rivulet flow to flooded film flow occurs between Γ values of 125 and 212 g/m/s. Simulation results for the 90° incline condition are shown in Figure 2(e–h). The transition from partially wetted flow to flooded flow occurs between $\Gamma = 150$ g/m/s to $\Gamma = 300$ g/m/s.

Several possible causes exist for the discrepancies seen between the qualitative visual comparisons of the experimental images and simulated results. First, the simulations used a steady inlet condition, with film thickness and velocity introduced at the top of the panel uniformly in space and time. No perturbations of the flow were simulated at the inlet. This behavior differs from the experimentally observed fluctuations in film thickness at the inlet. Another cause for the discrepancy is possibly the thin film assumption. The simulations do not account

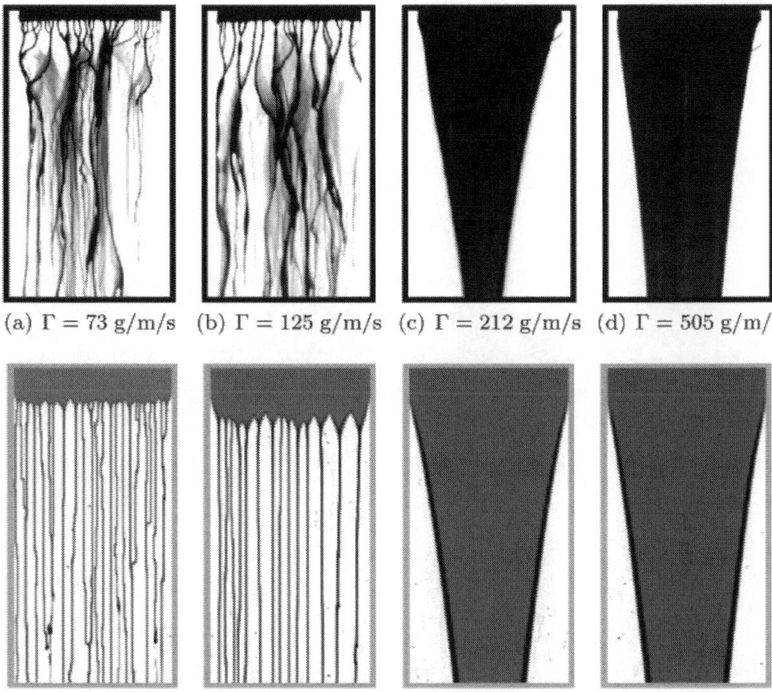

(a) $\Gamma = 73$ g/m/s (b) $\Gamma = 125$ g/m/s (c) $\Gamma = 212$ g/m/s (d) $\Gamma = 505$ g/m/s

(e) $\Gamma = 50$ g/m/s (f) $\Gamma = 150$ g/m/s (g) $\Gamma = 300$ g/m/s (h) $\Gamma = 400$ g/m/s

Figure 2: Experimental film flow (a–d) compared with simulated film flow (e–h) (90° incline, $\beta = 1.0$).

for any surface-normal components of the flow. Experimentally, observations show that such surface-normal velocities do seem to occur. Another possible discrepancy is the assumed viscous shear stress that is based on a quadratic velocity profile. This assumption holds for continuous films, but may break down when the flow is constrained to narrow rivulets. Also, local variations in surface roughness and/or curvature of the panel are not captured in the simulations. Finally, thermocapillary instabilities may play a role in determining the flow characteristics. The experimental flow was set to a temperature higher than that of the panel to assist in image acquisition with an IR camera. Although cast acrylic is not a very conductive material ($k = 0.56$ W/m-K), small thermal gradients may give rise to thermocapillary instabilities and enhance the meandering behavior of the flow. Experimental observation has shown markedly different behavior for a hot film on a cold panel versus a cold film on a hot panel [21, 22].

A more quantitative comparison was made between the experimental and simulated results. The wetted area fraction (WAF) is the fraction of the total solid panel surface area that is covered by liquid. The WAF was digitally extracted from

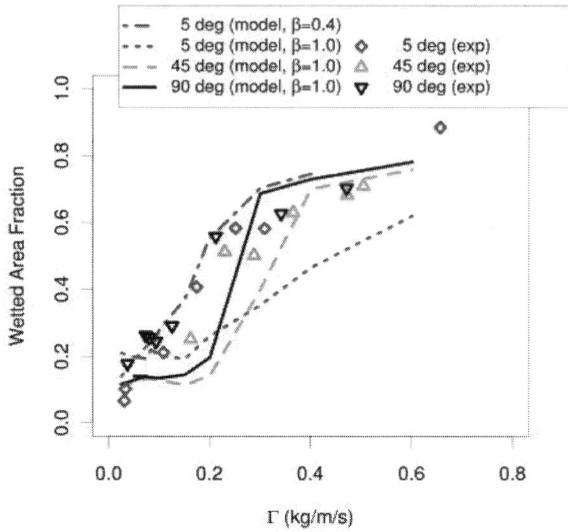

Figure 3: Wetted area fraction as a function of flow rate.

the experimental results using a threshold method in Matlab. Figure 3 shows this comparison to the fractions computed from the simulation results.

At the lower flow rates (rivulet flow), the wetted area fraction is between 0.1 and 0.2 for both model and experiments. As the flow rate increases, the model predicts a slightly more rapid transition to flooded film flow for the 45° and 90° film cases. The 5° incline results demonstrate a much more gradual transition to fully flooded conditions. A simulated result for the 5° incline case with a value of $\beta = 0.4$ is shown in Figure 3 for comparison.

6 Summary and conclusions

The partial wetting behavior of thin-film flow was simulated for a wide range of flow rates and for three inclination angles. Although a few of the more subtle behaviors of the flow are not captured (e.g. meandering rivulets), the visual comparisons to the experimental flow patterns are encouraging. In particular, the rivulet flow transitioning to flooded film flow behavior of the model is very reasonable. The predictions for wetted area fraction are also encouraging. The wetted area fraction predictions were shown to be sensitive to the β parameter. However, values for $\beta = 1$ yielded a good match to the experimental values for the 45° and 90° cases.

More work is needed in simulating realistic inlet boundary conditions for the film. Ideally the inlet flow rate would be allowed to vary as a function of space and time. Also, the sensitivity of the simulation results to contact angle and grid spacing needs to be studied.

Acknowledgement

This work was funded by FM Global as part of the Strategic Research Program for Fire and Suppression Modeling.

References

[1] Hopf, L., Turbulenz bei einem flusse. *Annalen der Physik*, **337(9)**, pp. 777–808, 1910.

[2] Nusselt, W., Die oberflachenkondensation des wasserdampfes. *Zeitschrift des Vereines Deutscher Ingenieure*, **60**, pp. 541–546 and 569–575, 1916.

[3] Nusselt, W., Heat transfer in the trickle cooler. *Zeitschrift des Vereines Deutscher Ingenieure*, **67**, pp. 206–210, 1923.

[4] Fulford, G.D., The flow of liquids in thin films. Academic Press, volume 5 of *Advances in Chemical Engineering*, pp. 151–236, 1964.

[5] Tong, A.Y. & Wang, Z., A numerical method for capillarity-dominant free surface flows. *Journal of Computational Physics*, **221(2)**, pp. 506 – 523, 2007.

[6] Saha, A.A. & Mitra, S.K., Effect of dynamic contact angle in a volume of fluid (vof) model for a microfluidic capillary flow. *Journal of Colloid and Interface Science*, **39(2)**, pp. 461–480, 2009.

[7] Diez, J.A. & Kondic, L., Computing three-dimensional thin film flows including contact lines. *Journal of Computational Physics*, **183(1)**, pp. 274–306, 2002.

[8] Frassy, J., Lecot, C., Murariu, M., Delattre, C. & Soucemarianadin, A., Transient spreading of films on heterogeneous surfaces. *Proceedings of the COMSOL Users Conference 2007 Grenoble*, 2007.

[9] King, A.A., Cummings, L.J., Naire, S. & Jensen, O.E., Liquid film dynamics in horizontal and tilted tubes: dry spots and sliding drops. *Physics of Fluids*, 2007.

[10] Bai, C.X. & Gosman, A.D., Mathematical modeling of wall films formed by impinging sprays. *Society of Automotive Engineers*, **(SAE 960626)**, 1996.

[11] Stanton, D. & Rutland, C., Multi-dimensional modeling of heat and mass transfer of fuel films resulting from impinging sprays. *SAE Technical Paper Series*, 1998.

[12] O'Rourke, P.J. & Amsden, A.A., A particle numerical model for wall film dynamics in port-fuel injected engines. *SAE Paper 961961*, 1996.

[13] O'Brien, S. & Schwartz, L., Theory and modeling of thin film flows. *Encyclopedia of Surface and Colloid Science*, p. 5283, 2002.

[14] Landau, L.D. & Lifshitz, E.M., *Fluid Mechanics, Second Edition: Volume 6 (Course of Theoretical Physics)*, 1984.

[15] Trela, M., A semi-theoretical model of stability of vertical falling liquid films. *Chemical Engineering Science*, **49(7)**, pp. 1007–1013, 1994.

[16] Hartley, D. & Murgatroyd, W., Criteria for the break-up of thin liquid layers flowing isothermally over solid surfaces. *International Journal of Heat and Mass Transfer*, **7(9)**, pp. 1003–1015, 1964.
[17] de Gennes, P.G., Borochard-Wyart, F. & Quere, D., *Capillarity and Wetting Phenomena*. Springer, 2003.
[18] Semiczek-Szulc, S. & Mikielewicz, J., Experimental investigation of contact angles of rivulets flowing down a vertical solid surface. *International Journal of Heat and Mass Transfer*, **21(12)**, pp. 1625–1625, 1978.
[19] Ponter, A.B. & Aswald, K.M., Minimum thickness of a liquid film flowing down a vertical surface–validity of Mikielewicz and Moszynskfs equation. *International Journal of Heat and Mass Transfer*, **20(5)**, pp. 575–576, 1977.
[20] Meredith, K.V., Xin, Y.B. & de Vries, J., A numerical model for simulation of thin-film water transport over solid fuel surfaces. *Fire Safety Science*, **submitted**, 2011.
[21] Zhang, F., Zhao, X., Geng, J., Wu, Y.T. & Zhang, Z., A new insight into Marangoni effect in non-isothermal falling liquid films. *Experimental Thermal and Fluid Science*, **31**, pp. 361–365, 2007.
[22] Zhang, F., Wu, Y.T., Geng, J. & Zhang, Z.B., An investigation of falling liquid films on a vertical heated/cooled plate. *International Journal of Multiphase Flow*, **34(1)**, pp. 13–28, 2008.

Section 6
Energy applications

A techno-economical view on energy losses at hydropower dams (case study of Karun III Dam and Hydropower Plant)

M. Jorabloo[1], M. Abdolahpour[2], R. Roshan[2] & H. Sarkardeh[2]
[1]*Department of Engineering, Water Engineering Group, Islamic Azad University, Iran*
[2]*Hydraulic Structures Division, Water Research Institute, Iran*

Abstract

In order to support the future economic developments in the world, one of the most important objectives is the reduction of energy losses. Apart from this objective, hydropower dams which are considered a renewable energy source encounter this problem. Karun III Dam and Hydropower Plant is constructed on the Karun River in the province of Khuzestan, Iran, with a capacity of 2000 MW. The Dam is a double arch concrete type, 205m high from the foundation and 185m high from the river bed. In the present work by using the physical model of Karun III Dam and measuring pressures in all parts of conveyance members, energy losses for each part were calculated separately. Experiments were conducted in three different elevations and with maximum design discharge. By calculating the cavitation index along the related hydraulic structures, probability of cavitation was studied. Discharge coefficient was also estimated in the operation conditions. Experiments showed that when the reservoir elevation was increased in constant discharge; the discharge coefficient was increased respectively. Also it was observed that increasing the reservoir elevation causes head losses to increase in different parts of conveyance members except the anti-vortex structure which was installed at the top of the intake. Moreover, energy losses at the Y-Branch member were observed more than other parts.
Keywords: Karun III Dam and Hydropower Plant, energy losses, physical model, cavitation.

WIT Transactions on Engineering Sciences, Vol 70, © 2011 WIT Press
www.witpress.com, ISSN 1743-3533 (on-line)
doi:10.2495/MPF110211

1 Introduction

Dams are constructed for different purposes. The most important targets could be the supply of drinking water, industrial, irrigation, power generation, prevention of floods, providing hydraulic head and also recreation purposes. Meanwhile, use of dams is a safe and green way for power generation [1]. It is clear that the occurrence of energy losses in hydraulic structures is inevitable. Minimizing the losses can lead to an increase of energy generation in hydropower plants. Some of the important structures in view of energy losses are intakes, tunnels, transitions and Y-branches. In the flow path two types of losses are expected: friction losses and local losses. Friction losses or longitudinal losses caused by passing the flow over the surface roughness. Darcy-Weisbach presented the following equation to calculate the head losses along the pressure conduits [2]:

$$h_f = f\frac{L}{D}\frac{V^2}{2g} \quad (1)$$

where h_f is longitudinal head losses, f is Darcy-Weisbach Coefficient, D is the diameter, L is the length and V is flow velocity in the tunnel.

Since, the head losses due to suddenly changes occur in the flow pass, the following equation calculates the head of energy losses [2]:

$$h_f = k\frac{V^2}{2g} \quad (2)$$

where K is local head loss coefficient.

To calculate power generation in a hydropower plant the following equation can be used:

$$P = \gamma QH \quad (3)$$

where P is the generated power, γ is the specific gravity of the water, Q is flow rate and H is water head.

According to the Bernoulli Equation, the discharge coefficient can be calculated. By summarizing related equations, discharge coefficient is as follows:

$$C_d = \frac{V}{2g(h+h_f-\frac{P}{\gamma})} \quad (4)$$

$$K = \frac{1}{C_d^2} - 1 \quad (5)$$

where C_d is discharge coefficient, h is the water depth over the intake (the distance between the centerline of the intake and water surface).

Cavitation is a phenomenon which may cause damage and cavity in hydraulic structures at high velocities. Sometimes in a hydraulic system, because of increasing velocity and existing small roughness at flow walls, pressure decreases. Decreasing local pressure may cause vapor pressure and creation of bubbles. Produced bubbles, distribute in the flow and transport farther, a place with high pressure. In this time, they will explode and produce noise and huge impact stresses into the walls. Due to the impact of exploded bubbles, walls may face serious damage that is known as "Cavitation phenomenon". In some cases the bubbles explode at the site of creation because of unstable flow regimes and leads to destruction just in place.

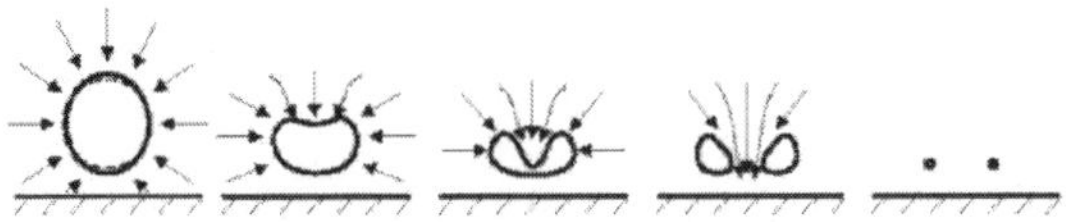

Figure 1: Schematic view of cavitation process.

Existence of swirling flows can lead to a decrease in the turbine efficiency, introducing vibration in the system, increasing hydraulic losses at the entrance of power intakes, blocking the trashrack and tunnels because of air suction and floating materials from the reservoir, and finally reduction of turbine life [3]. In this regard, using the anti-vortex structures in dams is recommended. Based on Sarkardeh et al. [4], vortices are divided into three classes. Vortices Class C are observed as a weak rotation of flow at water surface and in addition to water surface rotation a drop may also be observed in water surface. In vortices Class B the rotation of flow is extended down to the intake itself and in stronger position, dragging debris and trash into the intake is expected. In Class A vortices, some air bubbles are entrained from water surface and are transported down to the intake and in the strongest vortices, a stable air core is formed in the center of the vortex and air is entrained into the tunnel steadily.

Investigation on hydropower energy losses in the various structures has been considered by some researchers. In the present work by using the physical model of Karun III Dam, head losses at all parts due to power generation from reservoir to turbine, discharge coefficient in the presence of anti-vortex walls. Moreover, cavitation probability was investigated at the all parts of the conveyance structures. Experiments were conducted in three different elevations and with maximum design discharge.

2 Experimental works

All experiments were carried out on the physical model of the Karun III Dam and Hydropower Plant. The Dam is constructed on the Karun River in the province of Khuzistan, Iran, with the capacity of 2000 MW. The Dam is a double arch concrete type, 205 m high from the foundation and 185 m high from the river bed. The power plant consisted of 8 units, each unit supply the capacity of 250 MW. The physical model of Dam and its hydropower plant were constructed to the scale of 1:33.3. This scale was selected considering the presented criteria by Anwar et al. and Jain et al. to prevent scale effects [5, 6].

Hydraulic model of Karun III Dam power intakes were included: parts of the dam reservoir, intakes, power tunnels, braches (Figures 2 and 3). The model has two power tunnels with 38 cm diameter (power tunnel 1) (Figure 3). Design discharge of each intake in the model was 110 lit/s. Considering the design discharge, dimensions of thc intakes were selected 104 cm height and 52 cm width. Maximum and minimum depth of water over the intake center line was 1.80 m and 0.60 m respectively. Experiments showed that at 0.90 m depth, stable vortices of Class B were formed at the intakes. Since scale effects in the physical model may reduce the type of the vortex so formation of vortex Class A is probable in the prototype.

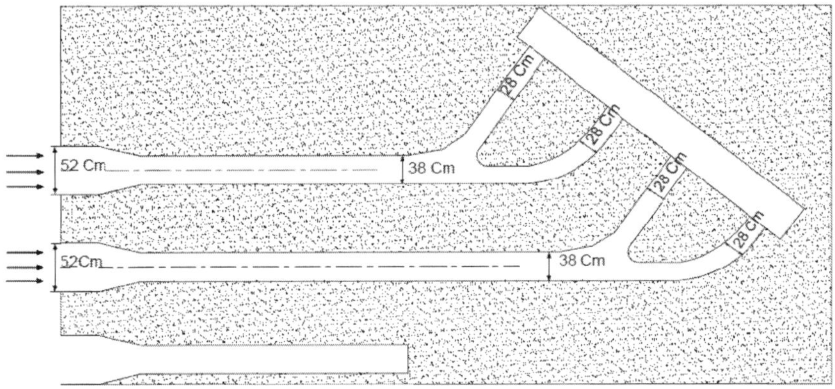

Figure 2: Plan view of convenience structures from the reservoir to the turbine of Karun III Dam.

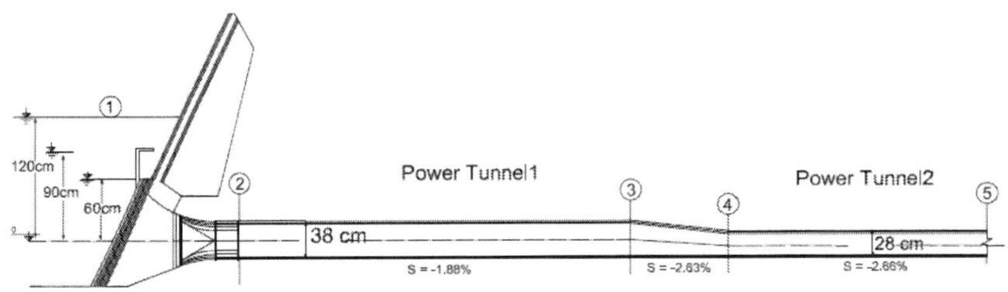

Figure 3: Side view of the convenience structures from reservoir to turbine of Karun III Dam.

To measure the flow discharge in the model, four rectangular weirs were calibrated and used for each unit. Water level of the reservoir was measured by a point gauge with accuracy of 0.1 mm. All experiments were carried out in the three different depths of 0.60 m, 0.90 m and 1.20 m from the intake center line with a flow rate equal to 110 lit/s in the model at each intake. In order to investigate the real condition, all the experiments had been done in the presence of designed anti-vortex walls [7].

3 Results and discussions

Experiments showed that vortices were formed on the top of the intakes at a depth of 0.90 m above the intake center line, since strong and stable vortices were observed in the model operation at depth less than 0.90 m using of anti-vortex structure was necessary. Anti-vortex structures can be used for elimination or reduction of vortex in dams. Previous researches by Sarkardeh et al. [7] showed that groove vertical anti-vortex walls have a better performance in reducing vortex rather than continuous vertical anti-vortex walls. Therefore, groove vertical anti-vortex walls were selected as the final alternative.

3.1 Measuring discharge coefficient

In order to calculate discharge coefficient in power intakes of Karun III Dam with installing anti-vortex walls, pressure values were measured in a section immediately after the intake entrance (Figure 3, Section 2). In the next step, discharge coefficients were calculated by Equation (4) at different depths of 0.60 m, 0.90 m and 1.20 m above the intake center line (Figure 4).

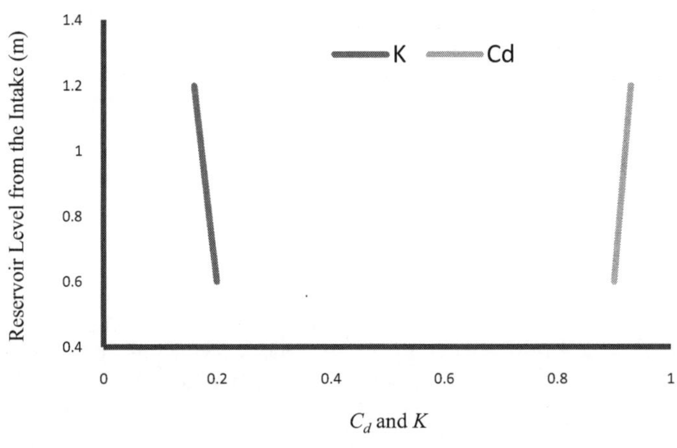

Figure 4: C_d and K of the intake in presence of anti-vortex structure.

Results showed that by increasing the water level in the reservoir, discharge coefficient was increased, but this increase was not considerable. So it could be said that anti-vortex walls have no meaningful effect on the discharge coefficient in different water levels.

According to conducted experiments at three different levels of reservoir elevation (0.60, 0.90 and 1.20 m), the pressure values (by using about 400 piezometers) showed that the probability of cavitation was not exist along the different convenience structures.

3.2 Calculating head losses due to power generation

One of the most important and effective parameters in hydropower dam design is the maximum use of the power of the water over the dam. So calculation of hydraulic head losses in the related structures has a great importance [8]. To calculate head losses along the structures, experiments were carried out with three different water levels, constant discharge and with presence of anti-vortex walls. Five different sections were used for measuring the pressure values (Figure 3). According to the Figure 5, it can be concluded that by increasing the water surface depth, head losses will increase respectively.

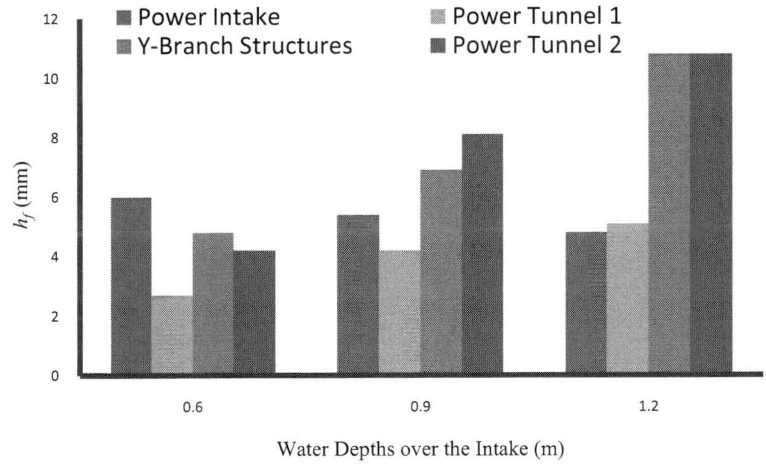

Figure 5: Head loss in different structures and reservoir levels.

This increasing trend is similar in different structures except power intake (in presence of anti-vortex device). Decreasing trend of head loss in intake inlet could be caused by the effect of anti-vortex wall in different intake levels. Maximum head loss is related to the depth of 1.20 m of the reservoir at the Y-branch structure and power tunnel 2.

Economical analysis of the project is very important in the engineering design process, especially in dam design. By using Equation 3, the amount of energy could be calculated in this project. Figure 6 shows the power loss due to power generation in the Dam.

As can be seen from Figure 6, the total head loss will increase by increasing the reservoir elevation. This figure could show to the engineers how hydraulic head losses could lead to a great amount of power lost in the hydropower plants in the prototype.

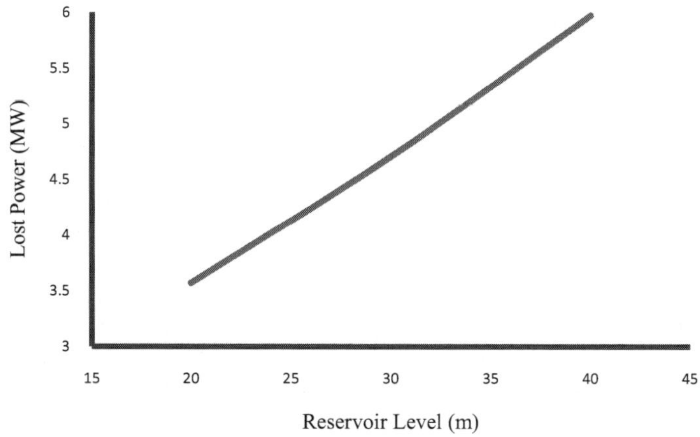

Figure 6: Total power losses in different reservoir levels of Karun III Dam.

4 Summary and conclusion

In the present study by using the physical model of Karun III Dam and Hydropower Plant, energy losses were calculated due to the energy generation process. The hydraulic model consisted of an anti-vortex structure, power intake, transition structure, power tunnel with 38 cm diameter in the model, Y-branch structure and finally power tunnel with 28 cm diameter in the model. Experiments were performed in three different reservoir levels and in maximum design discharge. Experiments showed that the Y-branch structure has most energy loss among the other structures. Occurrence of cavitation is a very important factor in safety of hydraulic structures. In this regard, the pressures were measured in different points of structures. The probability of cavitation was ruled out of possibility.

Acknowledgement

The authors would like to thank from the Water Research Institute (WRI) for their kind co-operation and using data.

References

[1] ASCE, Committee on Hydropower Intakes of the Energy Division, (1995), "*Guidelines for Design of Intakes for Hydroelectric Plants*", published by ASCE, New York.
[2] White, F.M. (1991), "*Fluid Mechanics*", McGraw-Hill.
[3] Knauss, J. (1987), "Swirling Flow Problems at intakes", Balkema.

[4] Sarkardeh, H., Zarrati, A.R., Roshan, R., (2010), "*Effect of intake head wall and trash rack on vortices*", Journal of Hydraulic Research Vol. 48, No. 1, pp. 108–112.
[5] Anwar, H.O., Weller, J.A. and Amphlett, M.B. (1978), "*Similarity of Free Vortex at Horizontal Intake*", Journal of Hydraulic Research, No. 2, pp 95-105.
[6] Jain, A.K., Raju, K.G.R and Grade, R.J. (1987), "*Vortex Formation at Vertical Pipe Intake*", Journal of Hydraulic Engineering, Vol. 100, No. 10, pp 1427-1445.
[7] Sarkardeh, H., Safavi, K., Karaminejad, R., H.M. Azamathulla, (2009), "*Prevention of Vortex Formation at Karun III Power Tunnels*", 5th National Civil Engineering Conference AWAM '09), Malaysia.
[8] Sarkardeh, H. (2007), "*Hydraulics of Power Tunnels and its Effect on Power Generation*", M.Sc. Seminar, Amirkabir University of Technology (Tehran Polytechnic).

Hydrate formation in natural gas pipelines

M. Naseer & W. Brandstätter
Department of Petroleum Production and Processing,
University of Leoben, Austria

Abstract

Gas hydrates pose a problem to the flow assurance programs in the oil and gas industry. This paper highlights the mechanisms of hydrate formation in natural gas pipelines. Computational Fluid Dynamics (CFD) was used to gain insight in to the mechanisms of water vapour condensation and consequent water accumulation in the lowered sections of a gas pipeline. The pipeline temperature profile, condensation of water vapour at walls, hydrate formation and hydrate slurry rheology were modelled with user defined functions (UDF) integrated into the CFD-software Fluent. It was found that the uphill sections of gas pipelines are natural sites for water accumulation and hydrate formation. It is demonstrated that CFD is a valuable tool to understand complex physical phenomena occurring in multiphase flow problems.
Keywords: natural gas, pipelines, condensation, hydrate formation, uphill sections, viscosity, nucleation, sagging sections, kinetics, macroscopic.

1 Introduction

The growth of energy demand is constantly pushing the gas producers to operate at higher pressures. In the start of the twentieth century, the natural gas industry started facing the problem of unexpected pipeline blockage. It was shown in 1934 by HammerSchmidt that the blockages appearing to be ice were actually gas hydrates [1]. The gas hydrate formation in oil and gas pipelines is an undesirable situation, undermining the flow assurance programs in addition to posing threat to the personnel and equipment. Clathrate hydrates are non-stoichiometric mixtures of water and natural gas in which the gas molecules are trapped in polygonal crystalline structure made of water molecules [2]. The water molecules arrange themselves in an orderly fashion around the gas molecules, thus entrapping them.

WIT Transactions on Engineering Sciences, Vol 70, © 2011 WIT Press
www.witpress.com, ISSN 1743-3533 (on-line)
doi:10.2495/MPF110221

These gas hydrates have the appearance of ice but can form well above the freezing point of water at high pressures.

Hydrate formation in natural gas pipelines is a complex phenomenon to capture in its entirety. The sequence of events leading to hydrate formation in gas pipelines involves water vapor condensation, accumulation of water at lowered sections of the pipeline, nucleation and growth of hydrate particles ultimately blocking the pipeline. The site of pipeline blockage may not be the same as that of hydrate deposition because just at the beginning of summer, the hydrate just released from the pipewall along the length may migrate downstream and deposit at sagging sections of the pipeline to cause the blockage [3]. Experiments on natural gas hydrate under actual operating conditions in pipelines is difficult to emulate due to scale of the problem and high pressures involved. However, different experimenters have focused on various aspects of the hydrate formation. Bishoni et al. [4] presented the formation and decomposition kinetics of gas hydrates at a conceptual level, highlighting the different phases of hydrate formation and growth. Riberio et al. [5] provided a critical review of literature regarding hydrate formation kinetics, highlighting the modeling efforts by different researchers. In an effort to emulate actual hydrate formation conditions, industrial scale flow loop experiments were setup by ExxonMobil Upstream Research Company [6] and Tulsa University [7] and Davies et al. [8] validated their hydrate formation model against the experimental data. While the hydrate formation model just mentioned was for the oil dominated flowlines, the hydrate formation may be quite different in natural gas flowlines.

2 Theory

Natural gas pipelines always have some moisture content as the gas from the formations reach the well head saturated with water [3]. During the cold season, the ground temperatures are lower resulting in the temperature drop of the pipeline content. If the temperature of the gas falls below the saturation temperature corresponding to the dew point of water, condensation of water vapors will start. In straighter sections of the pipeline, the condensate starts to accumulate at the base of the pipe due to gravity. The accumulation may increase at the base of the sagging/ uphill sections of the pipe. If enough water is present and the thermodynamic conditions are suitable, hydrate nucleation processes are initiated. To analyse hydrate formation on a theoretical basis in principle two choices exist: one-dimensional (1-D) or three-dimensional (3-D) models. While 1-D CFD models may be suitable for very long pipeline sections, the effects of water condensation and accumulation can only be modelled using the 3-D CFD methods.

3 CFD model

As an application example the pipeline section depicted in figure 1 was selected. The computational domain consists of approximately 84000 polyhedral control

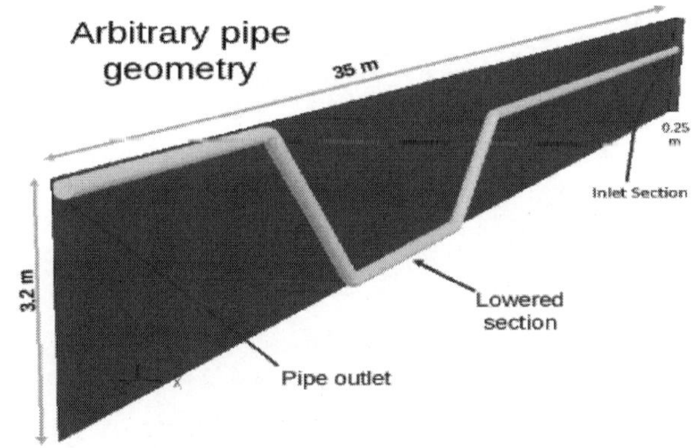

Figure 1: Natural gas pipeline.

volumes. The grid was refined during the computational run time to resolve the accumulation phenomena.

The "Mixture Model" of Fluent [9] was used to represent the interpenetrating continua consisting of mixture components.

Table 1: Model parameters.

No.	Model parameter	Value
1.	Pipe diameter	0.25m
2.	Inlet velocity	2m/s
3.	Water vapor volume fraction	0.2
4.	Inlet temperature	315K
5.	Operating pressure	6.3Mpa

Table 1 depicts different parameters of the pipeline section. The Pressure Implicit Split Operator (PISO) algorithm was used to couple the pressure and velocity fields. The K-Epsilon model was included to model the effects of turbulence.

3.1 Mathematical framework

The mixture model treats 'n' different phases as interpenetrating and models the fluid flow with conservation equations of mass, momentum and energy. The continuity, momentum and energy equations for the mixture in differential form [9] are given below.

Continuity:

$$\frac{\partial}{\partial t}(\rho_m) + \nabla \cdot (\rho_m \vec{v}_m) = 0 \tag{1}$$

Momentum:

$$\frac{\partial}{\partial t}(\rho_m \vec{v}_m) + \nabla \cdot (\rho_m \vec{v}_m \vec{v}_m) = \nabla p + \nabla \cdot \left[\mu_m \left(\nabla \vec{v_m} + \nabla \vec{v}_m^T\right) + \rho_m \vec{g},\right.$$
$$+ \vec{F} + \nabla \cdot \left(\sum_{k=1}^{n} \alpha_k \rho_k \vec{v}_{dr,k} \vec{v}_{dr,k} \right) \tag{2}$$

Energy:

$$\frac{\partial}{\partial t} \sum_{k=1}^{n} (\alpha_k \rho_k E_k) + \nabla \cdot \sum_{k=1}^{n} (\alpha_k \vec{v}_k (\rho_k E_k + p)) = \nabla \cdot [k_{eff} \nabla T] + S_E \tag{3}$$

where $\vec{v}_m$ is the mass averaged velocity, ρ_m and μ_m are volume fraction averaged mixture density and viscosity respectively, $\vec{v}_{dr,k}$ is the drift velocity of individual phases, E_k is the energy of phase k, α_k represents the volume fraction of phase k, p represents pressure, k_{eff} is the effective thermal conductivity, T represents the temperature and S_E represents energy source. In order to model the drag of the primary phase on the secondary phases, a drift flux model available in the Fluent was used where, the slip velocity was modelled as provided by Minnan et al. [9] and the drag function correlation of Schiller and Neumann [9, 10] was used.

3.2 Model setup

For the sake of this study, a certain composition of natural gas was considered at the inlet to the pipe. The molar concentrations were used as input to the program 'CSM Hyd' [2] and a pressure-temperature equilibrium curve for the hydrate formation was obtained. Based on the best fit through the equilibrium data, an equation was obtained representing the threshold temperature for hydrate formation as

$$T = 8.5274 \times \ln(P) + 270.86 \tag{4}$$

Where T is Temperature in Kelvin and P is the pressure in MPa.

To calculate the saturation temperature according to the partial pressure of vapor, the following equation was used.

$$T_{sat} = 16.335 \times \ln(P) + 167.08 \tag{5}$$

In order to accurately model the condensation in Fluent, a relationship developed by De Schepper et al. [11, 12] was employed.

$$M_{water} = 0.1 \times \alpha_{vapour} \times \rho_{vapour} |T_{sat} - T_{wall}| / T_{sat} \tag{6}$$

Modelling the temperature profile of the pipeline content along the length of the buried pipeline is a challenging task because it varies with the seasonal changes and also with the local changes in the conditions in the vicinity of the pipeline. A simplified expression was used [13] to describe the pipe temperature profile along the length.

$$T(x) = T_{Gr} + (T_o - T_{Gr}) \exp\left(\frac{-k\pi Dx}{0.28QC_p}\right) \tag{7}$$

where T_{Gr} is the ground temperature, x is the distance along the pipe from the inlet, D is the pipe diameter, k is thermal conductivity, Q is the volumetric flow rate and C_p is the specific heat capacity.

The kinetic modeling of hydrate nucleation and growth requires the definition of a driving force. While different researchers have presented their versions of kinetic models, there is a general lack of consensus among them. The applicability of these models to dynamic flow systems is limited by assumptions regarding thermodynamics at the molecular level and to the specifics of the experimental setup. Most notable among them is the work by Kashchiev and Firoozabadi who considered the difference between chemical potential of the species as the driving force [14].

An authentic kinetic model can serve as bridge between the microscopic phenomenon of crystallization and its macroscopic effects. The assumption is made here that there is an onset of instantaneous nucleation due to high number of nucleation sites (on the pipe wall due to roughness) and high supersaturation of the condensed liquid water film on the pipe wall with hydrocarbon gas at high pressure. Based on the general stoichiometric molar relationship for hydrate formation, the threshold mass fraction (considering water and hydrocarbon gas) in a cell was set to be approximately 0.866. This threshold value was imposed through the UDFs.

When hydrate formation conditions are reached in a computational cell, the hydrate formation is modelled by mass transfer equations for both the phases i.e. water and gas by the following simplified relationship:

$$M_{hyd} = (\rho_{water} \times \alpha_{water}) + (\rho_{gas} \times \alpha_{gas}) \tag{8}$$

To incorporate the rheological effects of the increase of suspension in the solution, the following relationship was used [10, 15]:

$$\mu_{slurry} = \alpha_{water} \left(1 - \alpha_{hyd}\right)^{-2.5} \tag{9}$$

4 Results and discussion

The effect of water condensation at the pipe walls and accumulation at the bottom of the pipe section is depicted in figure 2. At the onset of condensation of water vapor, the condensed water starts to accumulate at the bottom section of the pipe i.e. at the 6 o'clock position and it flows along the mean flow direction. This effect can be clearly observed in figure 2. It is asserted that the effect of accumulation can only be highlighted in three dimensional simulations and can be completely overlooked in the one dimensional models. Furthermore, the accumulation of free

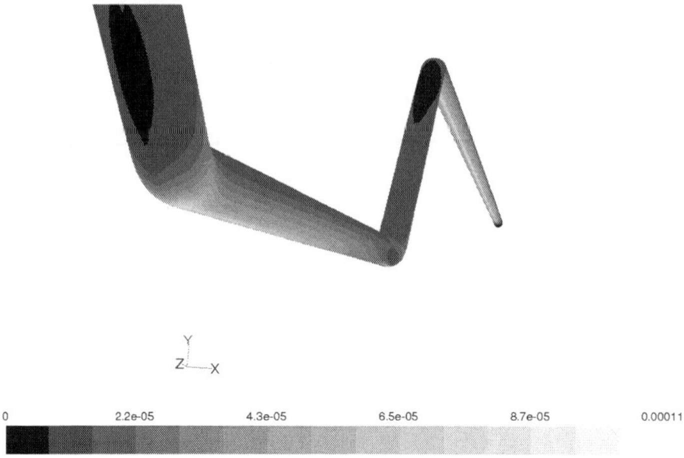

Figure 2: Water phase contours showing condensation at wall and accumulation at the bottom of the pipe.

water at the base of the uphill section of the pipe is displayed. It is concluded that there are two main mechanisms of water accumulation in pipelines due to gravity i.e. accumulation at bottom section of the pipe and accumulation at the base of uphill/ sagging sections of the pipe. The accumulation phenomena is a compound effect taking place at the base of uphill section of the pipe for aforementioned reasons and is clearly visible in figure 2.

Presence of free water is a key factor in the hydrate formation initiation as argued by Sloan [2]. The simulation shows that the base of the sagging sections of pipelines are most susceptible to hydrate nucleation as these regions serve as stagnation points/ traps for any condensed water in gas flows. The hydrate formation at these sections can start in winters. Multiple hydrate deposits along the length of the pipe result in additional pressure drop. The real problem may start at the beginning of summers, when the hydrate loosens up at the pipeline walls (not completely dissociated) and starts traveling [3]. The sloughed off hydrate then flows along the mean flow direction until it reaches lowered sections and gets trapped as packed deposits.

The region of high hydrate volume fraction in accumulated water is shown in figure 3. The high hydrate volume fraction region is marked by the iso-surfaces of hydrate concentration which can be clearly distinguished from the water iso-surfaces. It is evident that the water stagnation points are most susceptible to hydrate formation. The hydrate formation acts as additional resistance to the flow as shown by the velocity vectors in the cut-plane through $x = 0$. The velocity vectors actually drawn at $x = 0$ iso-plane are offset by some distance for clearer visibility. Stagnation zone with high hydrate volume fraction regions can be observed. In real pipelines the difference in velocities between the mean flow and the hydrate deposits would result in a drag force between phases leading to pressure drop undermining the flow assurance.

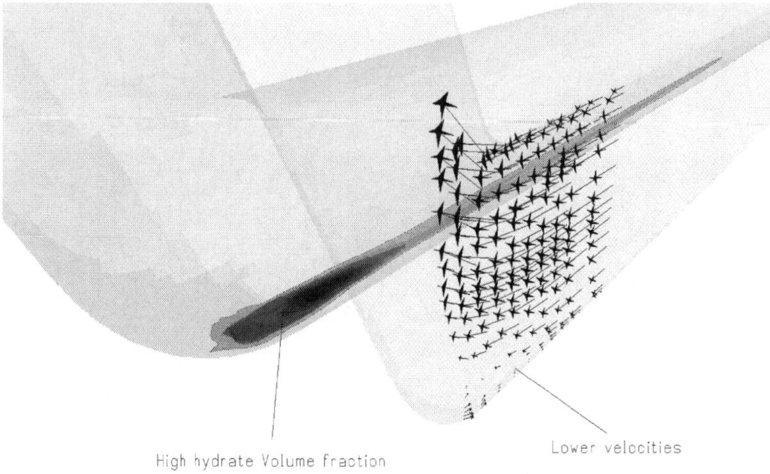

Figure 3: Velocity vectors at x=0 plane drawn at an offset showing the effect of high hydrate concentration.

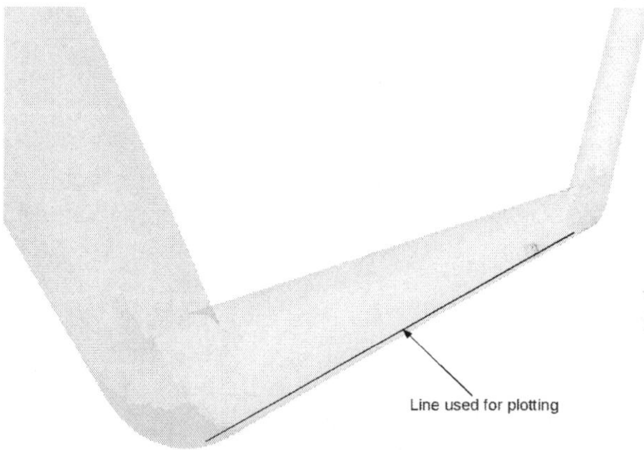

Figure 4: Line used for the plots.

In order to plot different parameters, a line just at the bottom of the lowered section of the pipe in between the sections 22m and 28m (measured from the inlet) was used as shown in figure 4.

Figure 5 shows the variation of viscosity and water volume fraction at the base of the lowered (straight) section. The gradual increase in viscosity mixture defined as $\mu_m = \sum_{k=1}^{n} \alpha_k \mu_k$ goes hand in hand with increase in the water volume fraction with a sudden spike indicating high hydrate volume fraction influencing the mixture viscosity. The gradual slope in the curve before 27m highlights the accumulation phenomena of water (pileup). The sudden increase in the viscosity

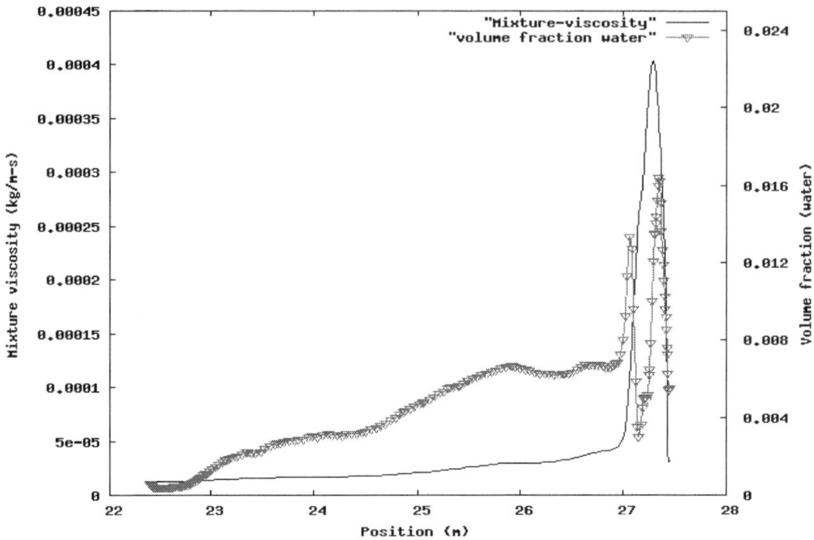

Figure 5: Mixture viscosity and water volume fraction along the length of lowered section of the pipe.

beyond 27m is indicative of the formation of a new phase drastically effecting the viscosity of the mixture.

The water volume fraction depicted in figure 5 shows a gradual increase initially followed by a sharp increase, a sharp decrease and again a sharp increase. The sharp decrease belongs to the region of high hydrate volume fraction and is acting as a 'double sided tidal barrage' between down stream and upstream water. It is interesting to note the effect of hydrate deposition on the water film moving in the mean flow direction. At the upstream side of the deposit, it is acting like a wall, resulting in increase in water volume as depicted in the figure and at the downstream side, the result of water accumulation is due to the backflow of water film adjacent to the deposit. The backflow is assisted by loss in the dynamic pressure in the region shadowed by the deposit and gravity effects at the uphill section.

The mixture velocity $\vec{v_m} = \sum_{k=1}^{n} \alpha_k \rho_k \vec{v_k} / \rho_m$ at the bottom of the pipe section is shown in figure 6. The increase in the velocity between 22–22.5m section is due to mean flow from the downhill section impinging the wall resulting in increase in the velocity. Beyond 22.5m the mixture velocity drops due to friction at the walls. In the section between 23–26m, the mixture velocity oscillates with a mean value of about 0.5m/s. The "fish-hook" curve beyond 27m shows the drastic drop in the mixture velocity due to high hydrate volume fraction.

The variation of static and dynamic pressure is depicted in figure 7. The drastic increase of static pressure at about 27m (and decrease of dynamic pressure) indicates the effects of high hydrate deposition.

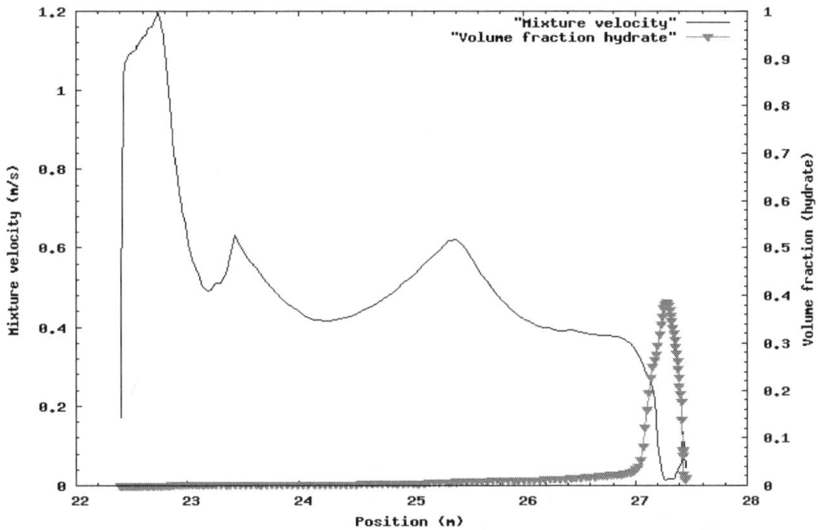

Figure 6: Mixture velocity and hydrate volume fraction along the length of lowered section of the pipe.

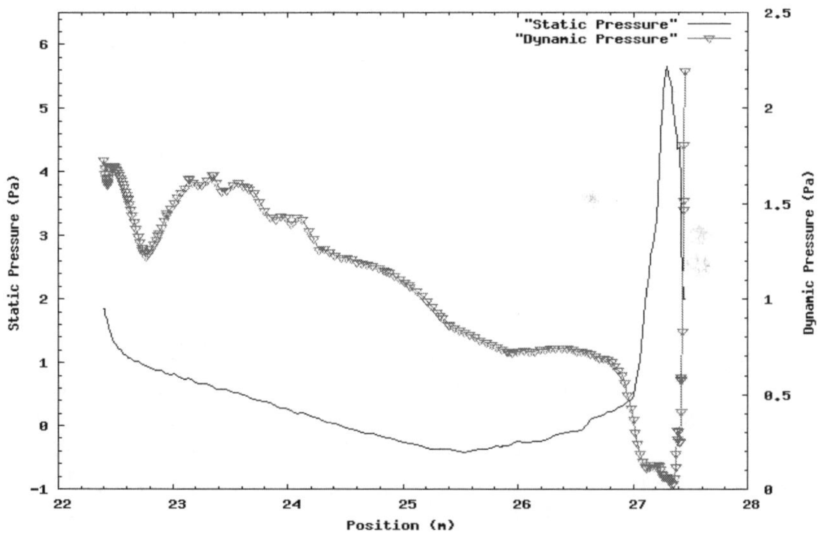

Figure 7: Pressure variation along the length of lowered section of the pipe.

5 Conclusions and future work

A 3D-CFD model for hydrate formation in gas pipelines has been developed. The results achieved so far are in agreement with existing knowledge about the condensation and accumulation of liquid water regions due to gravity in hilly terrain pipelines. Furthermore critical regions for hydrate formation could be identified. Overall very plausible theoretical explanations about the complex physical phenomenon taking place could be found.

Unfortunately no direct comparison with experimental results due to lack of data so far could be made. However, this will be subject to a forthcoming paper. Measurements for model validation in a real hilly terrain pipeline are currently ongoing. Future work is also aimed at integrating a correct kinetic model to simulate the effect of hydrate nucleation and growth phenomena on a macroscopic scale.

References

[1] J.J. Carroll, *Natural Gas Hydrate: A Guide for Engineers*. Elsevier Science, 2003.

[2] E.D. Sloan & K.A. Koh, *Clathrate Hydrates of Natural Gases*, 2008.

[3] W.M. Deaton & E.M. Frost, Jr., Gas hydrates and their relation to the operation of natural-gas pipe lines. Technical report, United States Department of the Interior, Bureau of Mines, July 1946.

[4] P.Raj Bishoni & V. Natarajan, Formation and decomposition of gas hydrates. *Fluid Phase Equilibria*, **117**, pp. 168–177, 1996.

[5] C.P. Ribeiro Jr. & P.L.C. Lage, Modelling of hydrate formation kinetics: State-of-the-art and future directions. *Chemical Engineering Science*, **63(8)**, pp. 2007–2034, 2008.

[6] UBC, https://circle.ubc.ca/bitstream/handle/2429/1133/5434.pdf.

[7] UT, http://www.tuhfp.utulsa.edu/.

[8] S.R. Davies, J.A. Boxall, L.E. Dieker, A.K. Sum, C.A. Koh, E.D. Sloan, J.L. Creek & Z.-G. Xu, Predicting hydrate plug formation in oil-dominated flowlines. *Journal of Petroleum Science and Engineering*, **72(3-4)**, pp. 302–309, 2010.

[9] *Fluent 6.3 Users Guide*.

[10] B.V. Balakin, A.C. Hoffmann & P. Kosinki, Experimental study and computational fluid dynamics modeling of deposition of hydrate particles in a pipeline with turbulent water flow. *Chemical Engineering Science*, 22 November 2010.

[11] S.C.K. De Schepper, G.J. Heynderickx & G.B. Marin, Modeling the evaporation of a hydrocarbon feedstock in the convection section of a steam cracker. *Computers & Chemical Engineering*, **33(1)**, pp. 122–132, 2009.

[12] A. Alizadehdakhel, M. Rahimi & A.A. Alsairafi, Cfd modeling of flow and heat transfer in a thermosyphon. *International Communications in Heat and Mass Transfer*, **37(3)**, pp. 312–318, 2010.

[13] Y.F. Makogon, *Hydrates of Hydrocarbons*. PennWell Books, 1997.

[14] D. Kashchiev and A. Firoozabadi, Nucleation of gas hydrates. *Journal of Crystal Growth*, **243(3-4)**, pp. 476–489, 2002.

[15] P. Mills, Non-Newtonian behaviour of flocculated suspensions. *J. Phys. Lett.*, **46**, pp. L301–L309, 1985.

Flow patterns in inclined gas-liquid annular duct flow

F. A. A. Mendes[1], O. M. H. Rodriguez[1], V. Estevam[2] & D. Lopes[3]
[1]*Department of Mechanical Engineering, Engineering School of São Carlos, University of São Paulo (USP), Brazil*
[2]*PETROBRAS/E&P-ENGP/TPP/EE, Brazil*
[3]*PETROBRAS/UN-RIO/ENGP/EE, Brazil*

Abstract

The submergible centrifugal pump (SCP) is one of the most common artificial lifting techniques employed in the Brazilian offshore scenario. However, free gas in the suction of the SCP is one of the most important limitations in the design of the pumping system. Gas-liquid flow in annular-ducts is found in the petroleum industry associated with gravitational gas separators that are applied with the SCP technique for oil production in directional wells. The main goal of this study is to investigate the behavior, i.e., flow patterns and maps, of gas-liquid flow in an annular duct at several inclinations in a setup with dimensions which are closer to real offshore application. High-speed video recording together with an objective technique based on the time-frequency pressure-signature analysis were applied for the flow pattern characterization. The experimental data were collected in the apparatus consisted of an inclinable test section, 0 to 90 degrees, with inner and outer diameters of 75 mm and 111 mm, respectively, hence the annular channel possessing an 18 mm gap. The total length of the test section was of 10.5 m. Air, water and oil at near atmospheric pressure constituted the gas and liquid phases. The air, water and oil superficial velocities were in the range of 0.02–30 m/s, 0.02–5 m/s and 0.005–0.5 m/s, respectively. Comparisons between data and flow pattern maps available in the literature for annular duct allowed the discrimination of regions and flow situations for which improvements of the transition modelling are required.

Keywords: two phase flow, annular duct, flow patterns maps, time-frequency pressure signature, directional well, inclined flow.

WIT Transactions on Engineering Sciences, Vol 70, © 2011 WIT Press
www.witpress.com, ISSN 1743-3533 (on-line)
doi:10.2495/MPF110231

1 Introduction

The multiphase flow in annular-duct can be found frequently in the oil industry; however, little is known about the phenomenology of this flow type. This knowledge is important for the design of submergible centrifugal pump (SCP). The SCP is one of the most common artificial lifting techniques employed in the Brazilian offshore scenario. In addition, free gas in the suction of the SCP is one of the most critical limitations in the design of the pumping system. Gas-liquid flow in annular-ducts is found in the petroleum industry associated with gravitational gas separators that are applied with the SCP technique for oil production in directional wells.

Kelessidis and Dukler [1] presented a study on upward-vertical two-phase flow in an annular duct. The authors traced flow patterns maps and established the transitions of flow patterns experimentally. Hasan and Kabir [2] proposed a phenomenological model for transitions in upward vertical and inclined annular duct flow and used data from the literature to validate the model. Caetano *et al.* [3] carried out an experimental and theoretical study of the effect of eccentricity in an upward-vertical two-phase annular-duct flow. The only quoted authors who studied the influence of inclination on flow patterns was Hasan and Kabir [2], however the model was validated only with data of upward-vertical flow. The studies [4] and [5] show results for horizontal two-phase annular-duct flow. In the first study, the authors were looking for the effect of slope on flow patterns, while in the second study the air-water flow patterns were classified and compared in two different sections test. In both studies, they used small tubes (annular channel gap of about 1 mm), which is not compatible with the dimensions used by the oil industry. Furthermore, the authors have not proposed any model for flow pattern transition. An objective identification and characterization of flow patterns based on statistical analysis of time-frequency pressure fluctuation signal was proposed by Drahos *et al.* [6] and Matsui [7]. Drahos *et al.* [6] used an analysis of the PDFs (Probability density function) and PSDs (Power Spectrum Density) to identify and characterize the flow patterns in a gas-liquid two-phase flow in a horizontal pipe. Matsui [7] applied statistical tools for analyzing the PDF of the differential pressure signals in vertical gas-liquid two-phase pipe flow.

Blanco *et al.* [8] conducted a thorough review on the study of gas-liquid flow in annular duct and proposed flow patterns maps for the annular-duct flow in horizontal and vertical inclinations. However, those authors don't validated the proposed model with data related to dimensions similar to those found in the oil industry because of the lack of available data in the literature. The main purpose of this study is to expand the existing experimental database on two-phase flow in annular ducts and validate the phenomenological model proposed by Blanco *et al.* [8].

2 Experiments

2.1 Experimental apparatus

The experimental data were obtained from the experimental apparatus illustrated by fig. 1. This experimental apparatus is mounted in Thermal-Fluids Engineering Laboratory (NETeF) of the Engineering School of São Carlos of the University of São Paulo. The test section (TS) has an annular geometry and is made of borosilicate glass. It has 10.5 m of length and is sustained by an inclinable truss beam capable of going from horizontal to 90 degrees of inclination. There is a flow development section of 1.5 m and 111 mm of internal diameter before the annular duct of 7.5 m and 111 mm of outside diameter (Do) and 75 mm of internal diameter (Di), fig. 2. A pipe section identical to the development section was positioned after the annular test section.

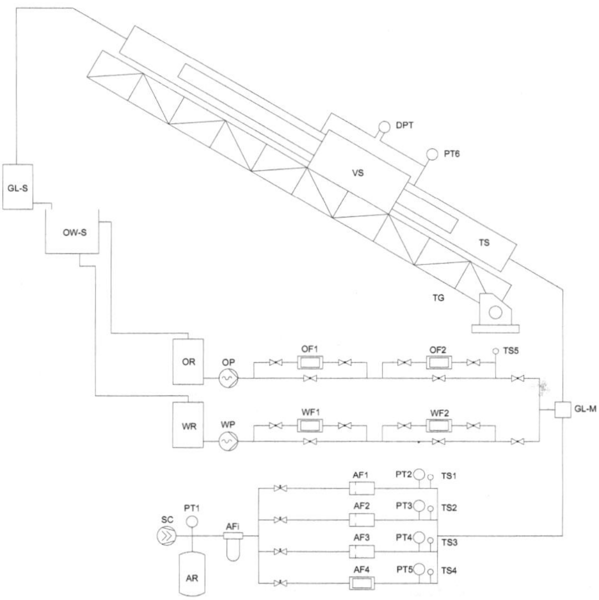

Figure 1: Schematic of experimental apparatus.

The air is supplied by a Worthington screw compressor of 50 kW model ROLLAIR 40 (SC), whose flow control is done by needle valves positioned before the flow meters. Three orifice plates (AF1, AF2, AF3) and an Oval positive displacement flow meter model Gal 50 (AF4) are used for measuring the air flow. The tap water used in test is pumped from a water tank (WR) by a Weatherford progressive cavity helical pump of 11kW model WHT (WP). To measure the water flow it is used a Badger Meter positive displacement flow meter OGT (WF1) and an Oval vortex water flow meter model EX Delta. The oil (density 828 kg/m³, viscosity 0,22 Pa.s and superficial tension oil-ar $52x10^{-3}$ N/m at 25°C) is pumped by a Weatherford progressive cavity helical pump of 11kW model WHT (OP). Two Oval positive displacement flow meters are used to measure the oil flow: MIII (OF1) and Flowpet EG (OF2). The gas-liquid mixture is formed in the gas-liquid mixer (GL-M), it is separated in the gas-liquid separator (GL-S) and the air is vented to the atmosphere. The liquid returned by gravity to tank (WR) or (OR). Novus pressure transducers TP-150 with range of 0-5 bar, accuracy 0,5% (PT2, PT3, PT4, PT5,PT6) were used to measure pressure and IOPE temperature sensor model TW-TC/2 (TS1, TS2, TS3, TS4, TS5), range -20–140°C and accuracy ±0,5°C were used to measure temperature. Tab. 1 summarizes the measurements instruments used in this study.

Table 1: Measurement instruments.

Flow meter	Range at 1 atm and 25°C (l/min)	Accuracy
AF1	57–262	±1% RD
AF2	172–794	±1% RD
AF3	524–2400	±1% RD
AF4	1–20	+1%~ -5% RD
OF1	0,05–8	±1% RD
OF2	2,5–106	±1% RD
WF1	1–35	±1% RD
WF2	30–1300	±1% RD

An Optron is high speed camera model CamRecord 600 is positioned in front of the visualization section (VS) and used to capture images of the flow. Xenon lamps of 35 W and 6000 K are positioned so as to provide sufficient lighting to obtain the images, fig. 3. A Validayne differential pressure transducer, model DP-15 (DPT) is used to acquire the two-phase flow pressure signature. At the entrance of the acquisition system it is installed a low-pass filter of 20 Hz to eliminate the signal noise of the pressure transducer. The remote control and acquisition system consisted of a PC equiped with a National Instruments acquisition board model PCI-6224. The water pump is controled with the assistance of a CAM system. The inclination angle of the test section (TS) is measured with a Bosh angle meter DNM 60L, accuracy ±0,2°.

2.2 Experimental procedure

The test section (TS) was set to the desired angle. The liquid superficial velocity was established by the pump control system using a PID controller implemented in *LabView*™ platform. The air injection was controlled manually through the needle valves. After five minutes, which was enough time to ensure steady state, the signal provided by the pressure transducer was saved and the image acquisition performed for a period of 60 s. The flow-pattern transition boundaries were determined *a posteriori* by visual observations (subjective technique) and the signal analysis (objective technique) based on the Fourier transform (FFT) and the Probability density function (PDF) of the pressure signature.

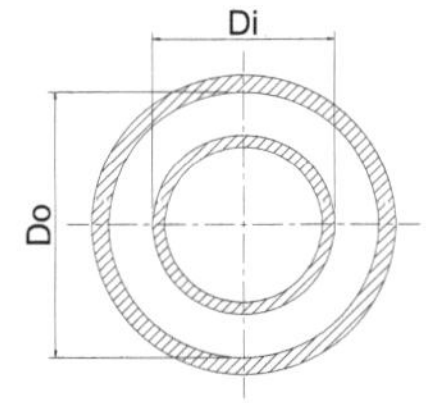

Figure 2: Schematic of the test section.

Figure 3: High-speed camera setup.

3 Results and discussion

3.1 Subjective analysis

In the subjective analysis the movies of the flow were carefully analyzed in slow motion, which allowed the identification and characterization of the flow patterns. Fig. 4 illustrates the horizontal water-air wavy stratified flow pattern obtained at an acquisition rate of 1000 fps and resolution of 800 x 500 pixels. The superficial velocities were set at J_G=10.26 m/s and J_L=0.25 m/s m/s. The four snapshots show an interfacial wave crossing the visualization section.

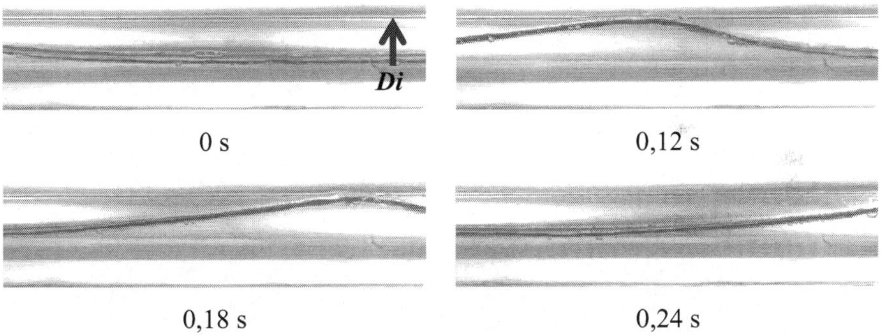

Figure 4: Lateral view of wavy stratified flow (J_G=10.26 m/s e J_L=0.25 m/s); Di stands for internal diameter.

Fig. 5 illustrates an elongated bubble observed from above during plug flow. One can see the deformed bubble due to the annular geometry. The bubble takes on a form that resembles of a flatworm. The lateral view of a “flatworm” bubble crossing the visualization section during plug flow can be seen in fig. 6. The camera was set to an acquisition rate of 1000 fps and a resolution of 800 x 500 pixels. The air and water superficial velocities were J_G=0,25 m/s and J_L=3,52 m/s.

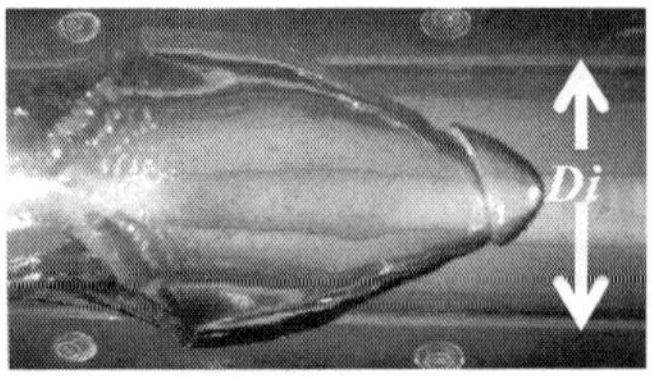

Figure 5: Top view of an elongated "flatworm" bubble in plug flow.

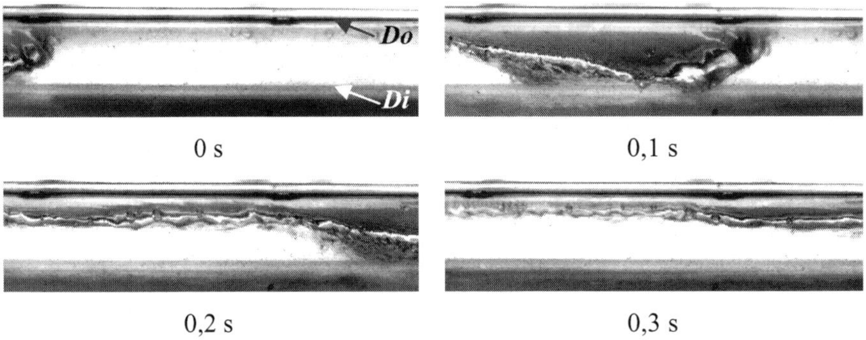

Figure 6: Lateral view of a "flatworm" bubble in plug flow; J_G=0.25 m/s and J_L=3.52 m/s.

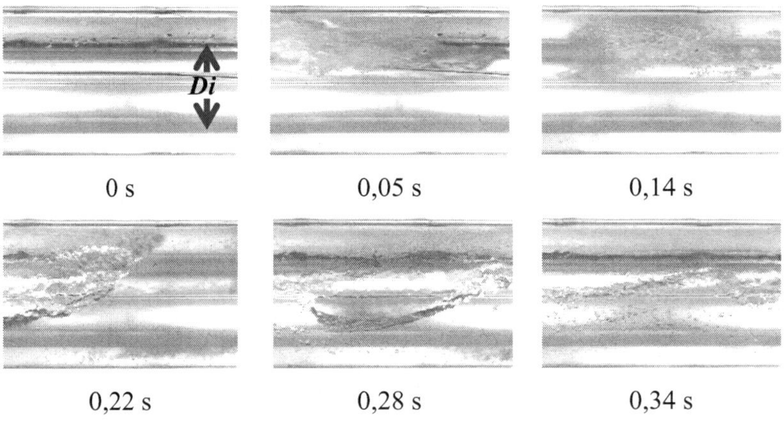

Figure 7: Lateral view of slug flow; J_G=2 m/s and J_L=1.25 m/s.

Fig. 7 shows a sequence of images of the slug flow pattern, J_G=2 m/s and J_L=1,25 m/s, with the camera set at an acquisition rate of 1000 fps and 800 x 500 pixels. The first snapshot (t = 0 s) shows a smooth stratified flow pattern. The following five snapshots show the aerated water slug crossing the visualization section. Notice that at t = 0.34 s the flow resembles an aerated wavy stratified flow. A picture of the dispersed bubble flow pattern is shown by Fig. 8. The

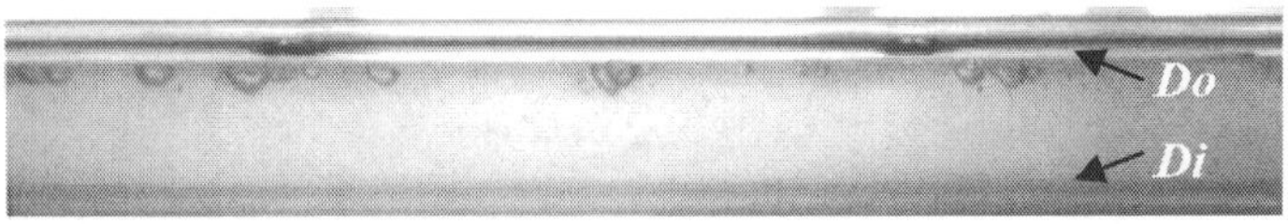

Figure 8: Lateral view of dispersed bubble flow, J_G=0.05 m/s e J_L=4.38 m/s.

sequence of images was carried out at a rate of 1250 fps and a resolution of 1280 x 256 pixels. The superficial velocities were J_G=0.05 m/s e J_L=4.38 m/s.

3.2 Objective analysis

The Validyne differential pressure transducer was mounted with diaphragm 20 (pressure range of -860 to 860 Pa) to identify the stratified flow and diaphragm 32 (pressure range of -14 to 14 kPa) for all the other flow patterns. Data were acquired at a rate of 5 kHz, with an accuracy of 0.25% FS and analyzed by a homemade program implemented in *LabView*™.

3.2.1 Horizontal flow

Air-water. Figs. 9–12 illustrates the acquired signal in time and frequency domains related to air-water smooth stratified flow, wavy stratified flow, slug flow and dispersed bubble flow, respectively. The basic difference between slug flow (fig.11) and plug flow (fig.12) is the observed intermittence of the signal in time. In addition, it is clear from the PSD graph that for plug flow the frequency peak is in between that of stratified flow and dispersed bubble flow. In stratified flow the dominant frequency is around 1 Hz, while in plug flow it is around 2 Hz and in dispersed bubble flow it is above 4 Hz. In the time domain, the smoothness of the signal is a clear discriminator between smooth (fig. 10) and wavy (fig. 11) stratified flow. In addition, the PSD of the smooth stratified flow signal presents a narrower frequency range than that of wavy stratified flow. The slug flow signal is characterized by strong pulsation in time (fig. 12). It indicates the passage of the piston of liquid that is formed by Kelvin-Helmholtz instability. The main feature of the signal related to dispersed bubble flow is the existence of a well defined high frequency oscillation (fig. 13).

Air-oil. One can see in Figs. 14 and 15 the signals related to smooth and wavy stratified flow in the tests with air and oil. In the frequency domain, the smooth stratified flow signal shows essentially a single low oscillation frequency (fig. 14), whereas the signal related to wavy stratified flow (fig. 15) presents a higher and broader distribution of frequencies in the PSD graph. The intermittent flow is divided in slug and plug flow. The signal related to plug flow (fig. 17)

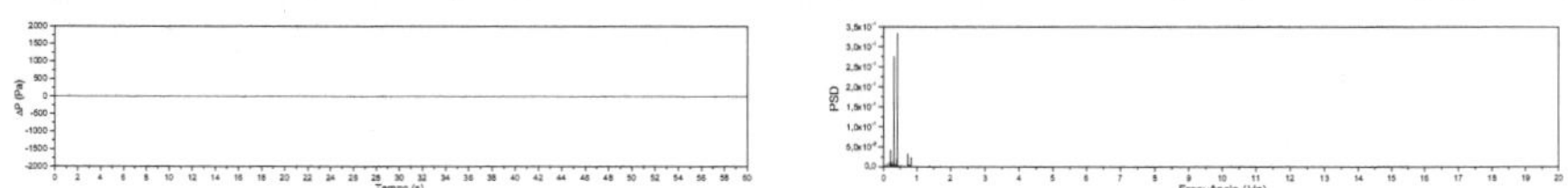

Figure 9: Smooth stratified flow (J_G=0.05 m/s; J_L=0.05 m/s).

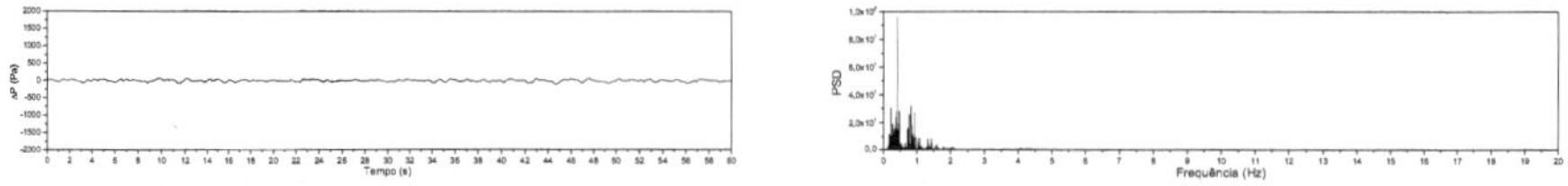

Signal PSD

Figure 10: Wavy stratified flow (J_G=5.00 m/s; J_L=0.20 m/s).

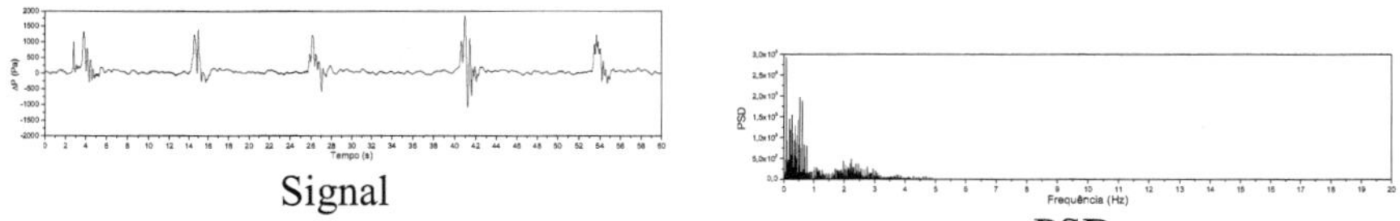

Signal PSD

Figure 11: Slug flow (J_G=3 m/s; J_L=2 m/s).

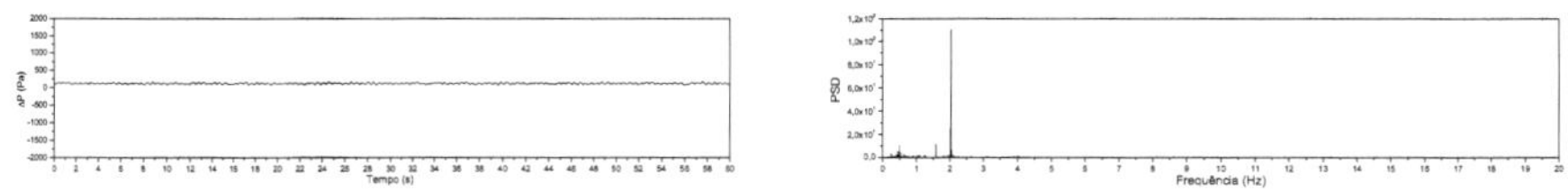

Signal PSD

Figure 12: Plug flow (J_G=0.1 m/s; J_L=1 m/s).

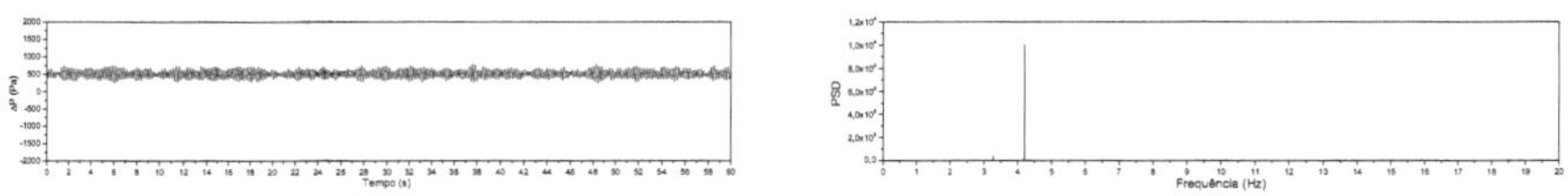

Signal PSD

Figure 13: Dispersed bubble flow (J_G=0.05 m/s; J_L=4.00 m/s).

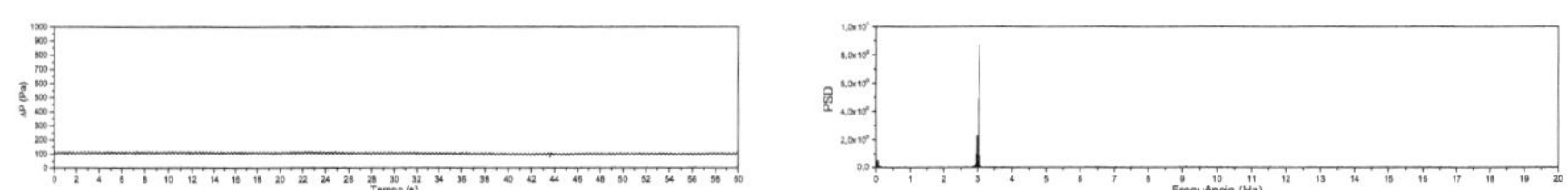

Signal PSD

Figure 14: Smooth stratified flow (JG=0,1 m/s; JL=0,03 m/s).

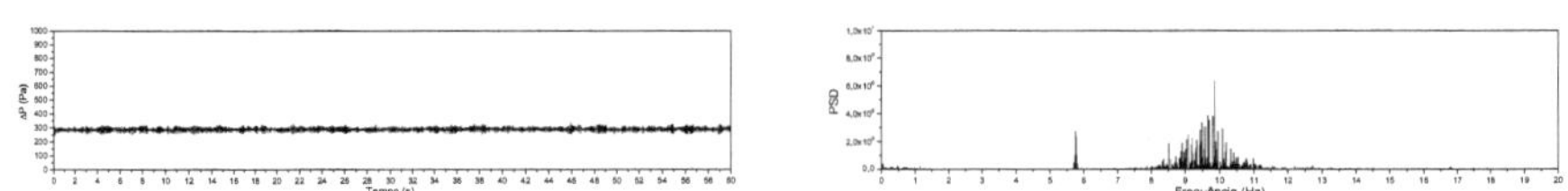

Signal PSD

Figure 15: Wavy stratified flow (J_G=15,00 m/s; J_L=0,06 m/s).

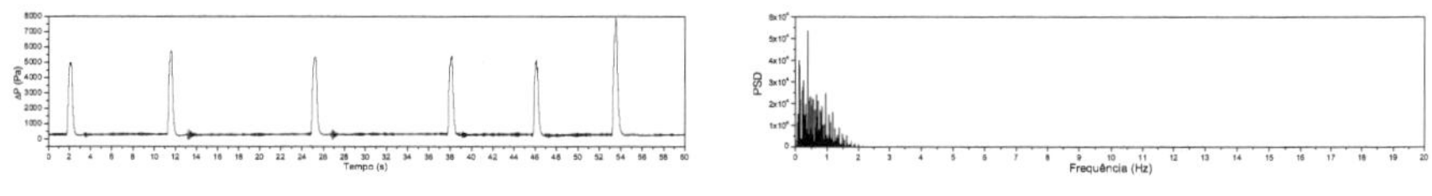

Signal PSD

Figure 16: Slug flow (J_G=10 m/s; J_L=0,2 m/s).

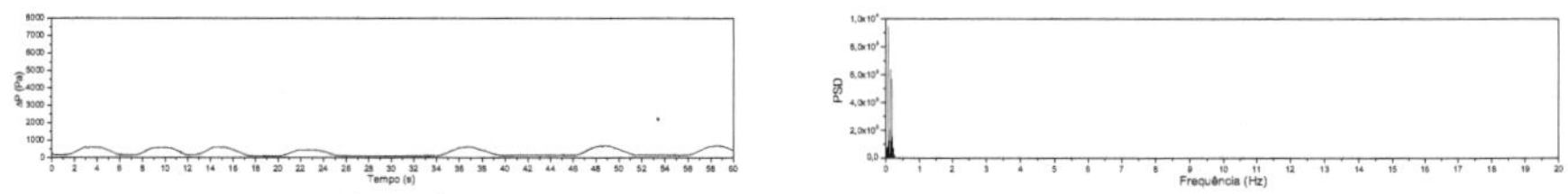

Signal PSD

Figure 17: Plug flow (J_G=0,5 m/s; J_L=0,2 m/s).

presents smoother oscillations and the PSD shows a narrow and quite low-frequency distribution. The main features of signal related to slug flow (fig. 16) is the abrupt oscillation in time and a broader PSD distribution moving away of zero.

3.2.2 Inclined flow (45°)

Figs. 18-20 illustrate the time-domain signals in upward inclined air-water bubble, slug and churn flow, respectively. The pressure signature related to bubble flow is smoother than that of slug or churn flow. In addition, the PSD indicates the presence of high frequency oscillations around 10 Hz only for bubble flow. In churn flow, fig. 20, the signal is characterized by high amplitude. It is worth noting that the signal related to churn flow can reach values that overcome the value of the hydrostatic pressure of water (10.4 kPa). This feature of the signal obtained in churn flow is likely due to a frictional pressure loss of downward liquid film flow added with the high void fraction.

3.3 Flow patterns maps

In modeling horizontal annular duct, Blanco *et. al.* [8] chose to use a model proposed by Rodriguez *et al.* [9] to define the stratified-flow transition boundaries and the approach proposed by Taitel and Dukler [10] to define all the other transitions. Fig. 20 illustrates the horizontal air-water flow pattern map drawn by Blanco *et al.* [8] and the flow pattern map obtained experimentally in this work. Some inconsistency can be observed when comparing data with predictions. The flow map of Blanco *et al.* [8] predicts the transition from stratified to intermittent flow for liquid superficial velocities below 0.2 m/s, but according to the data such transition occurs at liquid superficial velocity around 1 m/s. The stratified wavy flow was observed up to gas superficial velocity of 20 m/s in the experiments, whereas the flow map proposed by Blanco *et al.* [8] predicts annular flow.

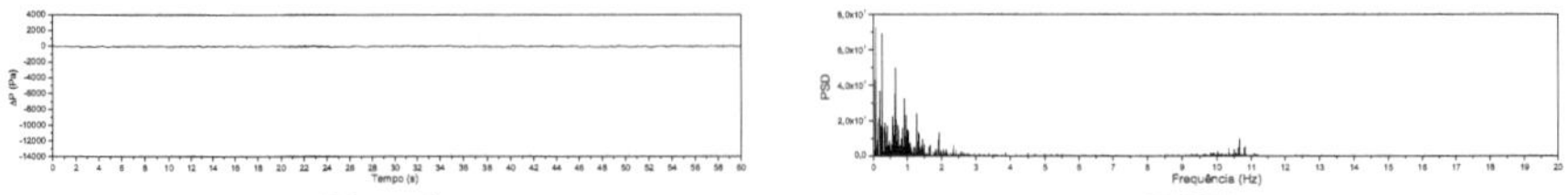

Signal PSD

Figure 18: Bubble flow (J_G=0.05 m/s; J_L=0.5 m/s).

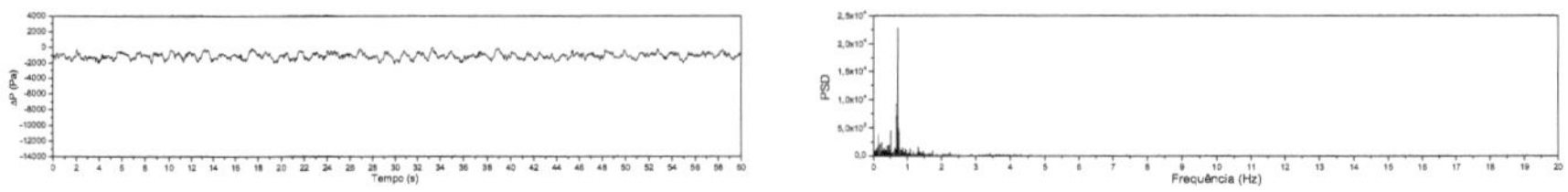

Signal PSD

Figure 19: Slug flow (J_G=0.5 m/s; J_L=0.05 m/s).

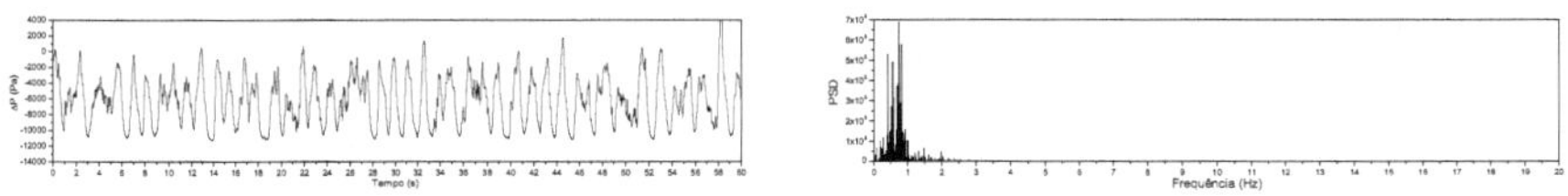

Signal PSD

Figure 20: Churn flow (J_G=7 m/s; J_L=0.1 m/s).

In fig. 22 one can notice the air-water flow pattern map with the test section inclined at +1°. It should be pointed out that stratified flow was no longer observed in the test section at one degree of inclination from horizontal for both air-water and air-oil flows; in the region of smooth stratified flow it was observed plug flow and in the region that was occupied by wave stratified flow it was observed slug flow, respectively. It is interesting to see the similarity between fig. 22 and 25. The horizontal air-oil flow pattern map in annular duct experimentally obtained can be seen in fig. 24. The stratified-flow transition boundary occurs for oil superficial velocities below 0.08 m/s, whereas in the air-water flow pattern map (fig. 21) the stratified flow occurs for liquid superficial velocities below 1 m/s. The transition to wave stratified flow is observed for gas superficial velocities above 1 m/s in the air-water flow pattern map, but in the

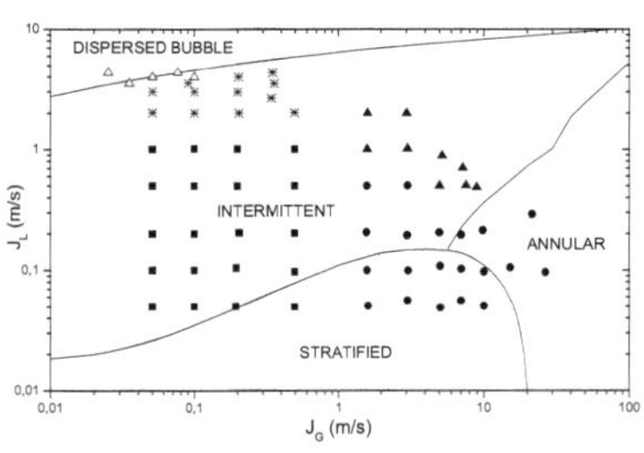

Figure 21: Horizontal air–water flow pattern map.

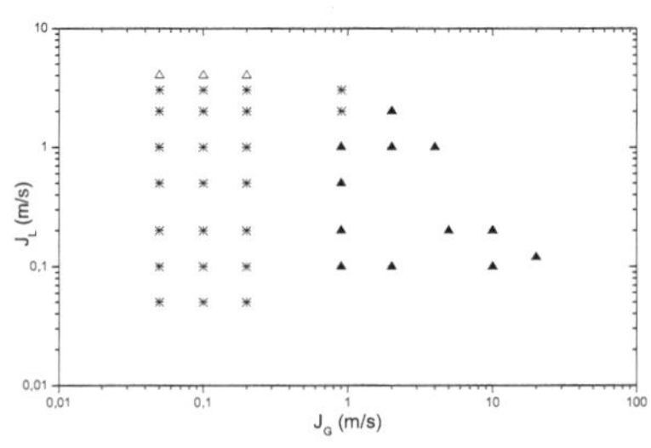

Figure 22: Air–water flow pattern map at +1°.

case of air-oil the boundary transition from smooth stratified flow to wave stratified flow occurs for gas superficial velocities above 0.1 m/s. The slug flow is observed at gas superficial velocities above 0.5 m/s in both air-oil and air-water flows.

As one can see in fig. 23, the methodology proposed by Blanco *et al.* [8] for upward gas-liquid annular duct flow at several inclinations was able to predict with quite good accuracy the various observed flow patterns. The flow patterns bubble, slug, churn and dispersed bubble were observed with the test section inclined at 45°. The plug flow observed in the air-water flow pattern map at one degree of inclination from horizontal was not observed in the air-water flow pattern map at +45°. The dispersed bubble flow occurs for liquid superficial velocities above 4 m/s for the flow pattern map at +1° and liquid superficial velocities above 2 m/s for the flow pattern map at +45°.

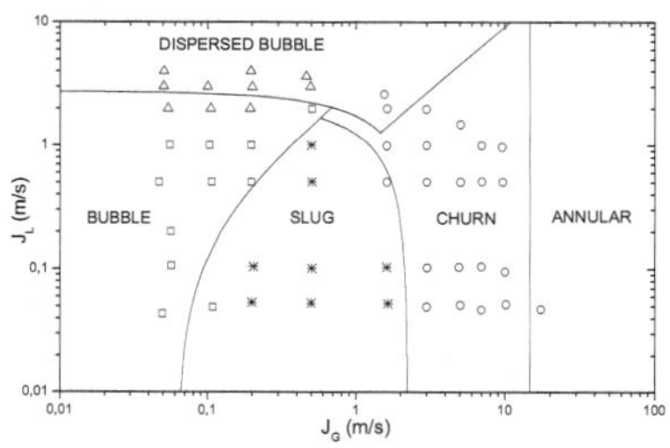

Figure 23: Air–water flow pattern map at +45°.

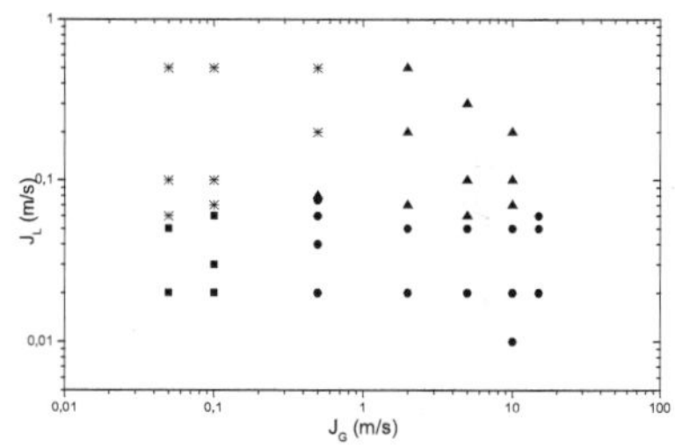

Figure 24: Horizontal air–oil flow pattern map.

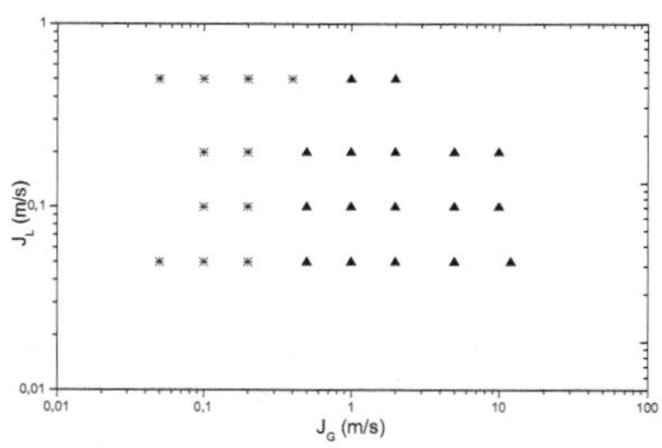

Figure 25: Air–oil flow pattern map at +1°.

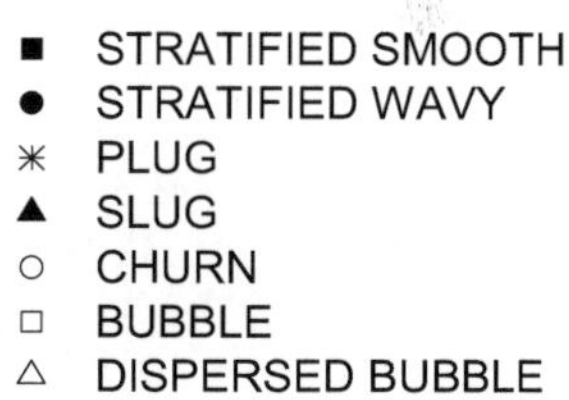

4 Conclusion

Air-water and air-oil flow patterns for horizontal and inclined flow in a large annular duct were identified and characterized in this study. The experimental results were obtained in a 10.5 m length test section of borosilicate glass with inner diameter of 75 mm and outer diameter of 111 mm. The inclinations of the test section were 0°, +1° and +45° in relation to horizontal. The ranges of superficial velocities of water, oil and air were 0,044 to 4,373 m/s, 0,01 to 0,5 m/s and 0,025 to 26,78 m/s, respectively. The flow patterns smooth stratified, wavy stratified, intermittent (plug and slug) and dispersed bubbles were identified at zero degrees of inclination (horizontal). With the test section inclined at one degree, bubbles, intermittent and dispersed bubbles were identified. In the tests where the test section was inclined at 45°, bubbles, slug, churn and dispersed bubbles were observed. The transition to wave stratified flow is observed for gas superficial velocities above 1 m/s in the air-water flow pattern map, but in the case of air-oil the boundary transition from smooth stratified flow to wave stratified flow occurs for gas superficial velocities above 0.1 m/s. The slug flow is observed at gas superficial velocities above 0.5 m/s in both air-oil and air-water flows. The plug flow observed in the air-water flow pattern map at one degree of inclination from horizontal was not observed in the air-water flow pattern map at +45°. The dispersed bubble flow occurs for liquid superficial velocities above 4 m/s for the flow pattern map at +1° and liquid superficial velocities above 2 m/s for the flow pattern map at +45°. The results of this study help to assess the potentiality of using a simple pressure-signature objective technique as a tool in identifying flow pattern transition boundaries in air-water and air-oil flows in a big annular duct. Another finding is the clear necessity of developing a phenomenological model capable of accurately generate flow pattern maps for horizontal gas-liquid flow in annular duct with dimensions similar to that used in the petroleum industry.

Acknowledgement

The present study was financially supported by PETROBRAS, whose guidance and assistance are gratefully acknowledged.

References

[1] Kelessidis, V.C. and Dukler, A.E., Modeling flow pattern transitions for upward gas-liquid flow in vertical concentric end eccentric annuli, *International Journal of Multiphase Flow*, 15(2), pp. 173-191, 1989.

[2] Hasan, A.R. and Kabir, C.S., Two-phase flow in vertical and inclined annuli, *International Journal of Multiphase* Flow, 18(2), pp. 279-293, 1992.

[3] Caetano, E.F., Shoham, O. and Brill, J.P., Upward vertical two-phase flow through an annulus. Part I: Single-phase fiction factor, Taylor bubbles rise velocity and flow pattern prediction, *Journal of Energy Resources Technology*, 114, pp. 1-13, 1992.

[4] Wongwises, S. and Pipathattakul, M., Flow pattern, pressure drop and void fraction of two-phase gas-liquid flow in an inclined narrow annular channel. *Experimental Thermal and Fluid Science*, 30(4), pp. 345-354, 2006.

[5] Ekberg, N.P., Ghiaasiaan, S.M., Abdel-Khalik, S.I., Yoda, M. and Jeter, S.M., Gas-liquid two-phase flow in narrow horizontal annuli, *Nuclear Engineering and Design*, 192(1), pp. 59-80, 1999.

[6] J. Drahos, J. Cermak and K. Selucky, Characterization of Hydrodynamic Regimes in Horizontal Two-Phase Flow Part II: Analysis of Wall Pressure Fluctuations, *Chem. Eng. Process.*, vol. 22, pp. 45-52, 1987.

[7] G. Matsui, Automatic identification of flow regimes in vertical two-phase flow using differential pressure fluctuations, *Nuclear Engineering and Design*, vol. 95, pp. 221-231, 1986.

[8] Blanco C.P., Albieri T.F. and Rodriguez O.M.H., Revisão de modelos para transição de padrão de escoamento gás-líquido em duto anular vertical e horizontal, *ENCIT 2008*, Belo Horizonte, Brazil, 2008.

[9] Rodriguez, O.M.H.; Mudde, R.F.; Oliemans, R.V.A., Stability analysis of slightly-inclined stratified oil-water flow, including the distribution coefficients and the cross-section curvature. In: North American Conference on Multiphase Technology, 5, 2006, Banff, Canada. Proceedings. Banff, Canada: BHR Group Limited. p.229-245.

[10] Taitel, Y. and Dukler, A. E., A model for predicting flow regime transitions in horizontal and near horizontal gas-liquid flow, A.I.Ch.E. Journal, 22, pp. 47-55, 1976.

Section 7
Flow in porous media

Modelling effective permeability of fracture networks in permeable rock formations by singular integral equations method

A. Pouya[1], M. N. Vu[1,2] & D. Seyedi[2]
[1]Université Paris-Est, Laboratoire Navier (ENPC/IFSTTAR/CNRS), Ecole des Ponts ParisTech, Marne la Vallée, France
[2]BRGM/RNSC, Orléans, France

Abstract

In this paper, theoretical and numerical formulations of plane steady-state fluid flow in a fractured porous rock are used to investigate its effective permeability. If the far field inflow is uniform, the theoretical solution shows that the pressure field in the matrix is a function of the discharge in the fracture network. A numerical resolution based on singular integral equations is employed to derive the general problem of many intersected fractures in order to obtain the pressure field in anisotropic matrix. This solution allows computing the flux in the fractures which is the key issue for upscalling the equivalent permeability. This paper presents in detail the method for deriving the equivalent permeability from this solution. This method is applied to two real cases: an Excavation Damage Zone (EDZ) around a deep underground gallery and a geological rock formation presenting several families of fractures. The results of the both cases show that the developed method provides an easy and efficient way to determine the equivalent permeability of the fractured porous rock medium. This equivalent permeability can be implemented in analytical and numerical tools for continuous media towards estimating the flow characteristics in the rock formation.

Keywords: facture network, porous rock, steady-state flow, effective permeability.

WIT Transactions on Engineering Sciences, Vol 70, © 2011 WIT Press
www.witpress.com, ISSN 1743-3533 (on-line)
doi:10.2495/MPF110241

1 Introduction

In the context of important industrial applications such as underground nuclear waste disposal in claystone, geological CO_2 storage or hydrocabure reservoir, the numerical modelling of fluid flow in fractured porous materials is intensively investigated. More specifically, in damaged zones, cracks and geological faults generally have a great effect on the effective permeability of porous media or rock masses. The interest of this question also extends to other fields, such as hydrogeology for water resources management in aquifers, etc.

Different approaches are used to study the problem of flow in porous media containing fractures with mass exchange between matrix and fracture, a review of which can be found in Sanchez-Vila et al. [16], Goméz-Hernandez and Wen [7] and Renard and de Marsily [15]. However, these works are essentially based on some numerical-empirical modelling. Pouya and Fouché [11] presented some rigorous theoretical-numerical results based on homogenisation theories for the equivalent permeability of heterogeneous or fractured media. The self-consistent scheme is investigated by Dormieux and Kondo [5] and Barthélémy [1] for upscalling the effective permeability of cracked porous media. In their method, cracks are assimilated to ellipsoidal inclusion obeying to a Darcy's law of flow. Following a different way, based on theory of Cauchy integral, Liolios and Exadaktylos [9] studied mathematically the problem of plane steady-state flow in an infinite isotropic porous media with multiple non-intersecting fractures. In this approach, the fracture is considered as zero thickness discontinuity line and complex numbers are used. Recently, Pouya and Ghabezloo [14] used a direct approach to obtain a general potential pressure solution function of flux in the fracture network for anisotropic matrix containing intersecting curvilinear fractures. By continuing this work, Pouya and Vu [13] employed the singular integral equation method to establish a fast numerical calculation that can be used to derive the general solution and to show the fluid flow around many intersecting fractures.

Using the recent theoretical and numerical advances, this paper presents a method to evaluate the effective permeability of the permeable rock formation containing a dense fracture network. The numerical solution of pressure and discharge in the fracture, and then, the equivalent permeability using the formulas of average velocity and of average pressure gradient. The numerical calculation is applied to two cases. The first one is an EDZ around a deep underground gallery where the fractures are curved surfaces with a well defined shape and are almost regularly distributed in this zone. The second is a fractured geological formation with several families of fractures in which the fractures are plane surfaces and their characteristic parameters (orientation, length and position) are defined by a stochastic law. This approach provides an easy and efficient tool to determine the effective permeability of fractured porous rock formation.

2 Mathematical formulation

An infinite homogeneous body Ω containing a fracture network is considered (fig.1). Fracture number m is denoted Γ_m. It is generally presented by a smooth curve $\underline{z}^m$ of the curvilinear abscise s. The intersection points of two or more fractures and the extremities of fractures ending in the matrix constitute a set of singular point S.
Fluid velocity $\underline{v}(\underline{x})$ in the matrix is given by Darcy's law:

$$\forall \underline{x} \in \Omega - \Gamma \qquad \underline{v}(\underline{x}) = -\mathbf{k}(\underline{x}).\underline{\nabla} p(\underline{x}) \tag{1}$$

where $\mathbf{k}$ is the matrix permeability and $p(\underline{x})$ the pressure field. The flow through the discontinuity is usually expressed in terms of the Poiseuille's law:

$$\forall s \in \Gamma \qquad q(s) = -c(s).\partial_s p \tag{2}$$

where, s is the abscise along the fracture and c is the fracture's conductivity.

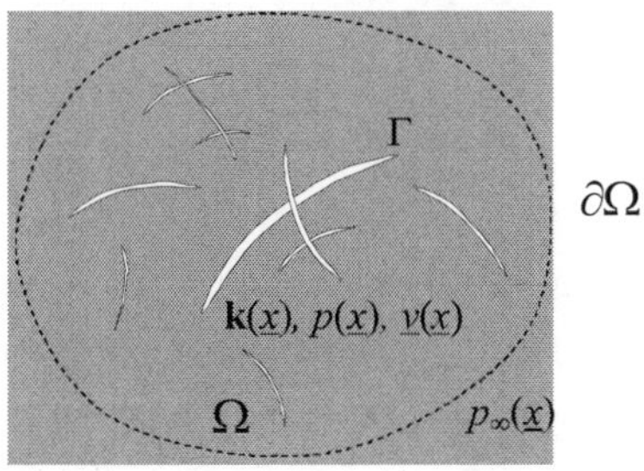

Figure 1: Rock mass containing fractures.

Mass conservation in the matrix reads:

$$\forall \underline{x} \in \Omega - \Gamma \qquad \underline{\nabla}.\underline{v}(\underline{x}) = 0 \tag{3}$$

The fracture-matrix mass exchange law on the fracture excluding singular points is obtained by considering masse balance in a portion ds of the fractures (fig. 2a)

$$\forall s \in \Gamma \qquad [\![\underline{v}(\underline{z})]\!].\underline{n}(s) + \partial_s p(s) = 0 \tag{4}$$

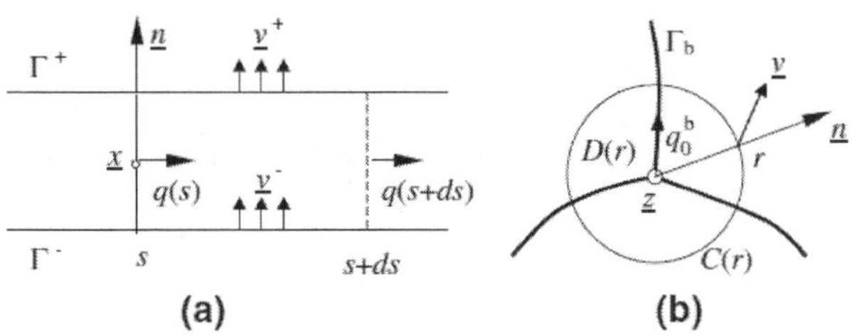

Figure 2: a Mass exchange between matrix and fracture. b Mass balance in a disc around an intersection point (Pouya and Ghabezloo [14]).

At the intersection points of fractures (fig. 2b), Pouya and Ghabezloo [14] analyzed the masse balance in a small disc centred on the singular point and deduced the following relationship:

$$\forall \underline{z} \in S \qquad \underline{\nabla}.\underline{v}(\underline{x}) + \left(\sum_{b} q_0^b \right) \delta(\underline{x} - \underline{z}) = 0 \tag{5}$$

with q_0^b is the outgoing flow on the branch of fracture b and δ is the Dirac distribution. Nevertheless, a deeper mathematical analysis allows us to go further than eqn. (5) (Pouya and Vu [13]) and shows that:

$$\sum_{b} q_0^b = 0 \tag{6}$$

In order to calculate the effective permeability, it is sufficient to impose a constant pressure gradient at infinity: $p_\infty(\underline{x}) = \underline{A}.\underline{x}$. The previous problem was derived rigorously by Pouya and Ghabezloo [14] and a general solution of potential was given:

$$p(\underline{x}) = p_\infty(\underline{x}) + \frac{1}{2\pi\kappa} \sum_{\mathrm{m}} \int_{\Gamma_{\mathrm{m}}} q^{\mathrm{m}}(s) \frac{\underline{x} - \underline{z}^{\mathrm{m}}(s)}{\left\| \sqrt{\mathbf{k}^{-1}}.\left[\underline{x} - \underline{z}^{\mathrm{m}}(s) \right] \right\|^2} . \mathbf{k}^{-1} . \underline{t}^{\mathrm{m}}(s)\, ds \tag{7}$$

where, $\underline{x}$ is a current point in the material, $\underline{z}^{\mathrm{m}}(s)$ is the point the Γ_m at the curvilinear abscise s, $q^m(s)$ and $\underline{t}^m(s)$ are respectively the discharge along the fracture and the unit vector tangent to the fracture at this point, and κ is the square root of determinant of $\boldsymbol{k}$: $\kappa = \sqrt{|\mathbf{k}|}$.

3 Numerical calculation

When field point $\underline{x}$ is located in the fractures; we obtain a singular integral equation of unknown pressure field in the fracture networks by replacing eqn. (2) into eqn. (7). This equation was resolved numerically by collocation method (Bonnet [3]) which consists of enforcing exactly eqn. (7) at finite number of points called collocation point.

Curvilinear fractures are approximated by a series of small linear segments and are discretized by E elements numbered n and denoted E_n ($1 \le n \le E$). Thus, eqn. (7) is written as follows:

$$p(\underline{x}) = p_\infty(\underline{x}) + \sum_{n=1}^{N} I_n \tag{8}$$

with:

$$I_n(\underline{x}) = \int_{E_n} q^{\mathrm{n}}(s) \frac{\underline{x} - \underline{z}^{\mathrm{n}}(s)}{\left\| \sqrt{\mathbf{k}^{-1}}.\left[\underline{x} - \underline{z}^{\mathrm{n}}(s) \right] \right\|^2} . \mathbf{k}^{-1} . \underline{t}^{\mathrm{n}}\, ds \tag{9}$$

Two types of elements are distinguished: extremity elements and current elements. Linear interpolation of pressure is used for current elements. General theoretical results of velocity field singularity around a fracture tip lead us to choose an interpolation function for the flux q(s) corresponding to a variation as $s^{1/2}$ where s is the distance on the fracture line to the extremity point. Therefore, interpolation function for pressure on the left extremity elements (tip at $s=0$), current element, and right extremity elements (tip at $s=L$) are respectively expressed as following:

$$p(s) = p_1 + \frac{p_2 - p_1}{\sqrt{L}^3}\sqrt{s}^3$$

$$p(s) = p_1 + \frac{p_2 - p_1}{L}s \tag{10}$$

$$p(s) = p_2 - \frac{p_2 - p_1}{\sqrt{L}^3}\sqrt{L-s}^3$$

At first, a collocation point $\underline{x}$ is selected per element which verifies $I_n(\underline{x}) = 0$. And then, the set of extremity points of fracture network is added as supplementary collocation points. It should be noted that no matter how the fractures intersect and how the discretization is made, the number of nodes N is always smaller than number of collocation points M. Otherwise, this method of selection of collocation points always leads to the number of equations M larger than the number of nodal pressure unknown. It is interesting that all elementary integrals $I_n(\underline{x})$ in eqn. (9) can be calculated analytically by using the variable interpolations in eqn. (10), which provides a very fast calculation method.
Enforcing eqn. (7) in M collocation points leads to following matrix equation:

$$\mathbf{H}.P=Y \tag{11}$$

where, $P(N\text{x}1)$ is the column of nodal pressure unknown, $Y(M\text{x}1)$ is the column of infinite pressure field at collocation points and $\mathbf{H}(M\text{x}N)$ is computed from the assembly operation after computing all elementary integrals. The approximated solution can be found by the least squares fitting method. Pouya and Vu [13] have well shown the validity of this method by comparing the numerical solution of a single straight fracture in an infinite homogeneous media with the known closed-form solution of the same problem.

Once the eqn. (11) is solved, the nodal pressure $P\{p_n\}$ $(1 \leq n \leq N)$ are known, using eqn. (2), we can deduce the flux in each element. In addition, in the post-processing stage, eqn. (11) shall be employed again to compute directly the field pressure in porous matrix.

For a homogeneous rock mass Ω, containing a fractures network Γ_m that is considered as the limit case of thin permeable layers, the average velocity and pressure gradient in Ω is given, respectively, by:

$$\underline{V} \equiv \frac{1}{\Omega}\left[\int_{\Omega} \underline{v} d\Omega + \sum_{m} \int_{\Gamma_m} q\underline{t} ds\right]; \quad \underline{G} \equiv \frac{1}{\Omega}\int_{\Omega} \underline{\nabla} p d\Omega \tag{12}$$

Let us to suppose that a linear pressure $p(\underline{x}) = \underline{A}.\underline{x}$ is applied at the infinity. The pressure field in matrix is known, especially on the boundary $\partial\Omega$, hence, the vector $\underline{G}$ can be deduced such as:

$$\underline{G} \equiv \frac{1}{\Omega}\int_{\partial\Omega} p.\underline{n}ds \tag{13}$$

where, $\underline{n}$ is the unit outward normal on $\partial\Omega$. Besides, Pouya and Fouché [11] demonstrated that $\frac{1}{\Omega}\int_{\Omega} \underline{v}d\Omega = -\mathbf{k}.\underline{G}$. Hence, the linearity of all velocities and flux with respect to $\underline{G}$ implies that there exists a tensor $\mathbf{k}^f$ that satisfies the following relationship:

$$\frac{1}{\Omega}\sum_{m}\int_{\Gamma_m} q\underline{t}ds = -\mathbf{k}^f.\underline{G} \tag{14}$$

Thus, under boundary condition of linear pressure, the global permeability tensor of Ω domain is given by $\mathbf{k}^p=\mathbf{k}+\mathbf{k}^f$. In the case of weak fracture density, the Mori-Tanaka estimation of $\mathbf{k}^p$ is determined by neglecting the interaction between fractures i.e. assuming that the flux q in the left-hand side of eqn. (14) is deduced from the results obtained for a single fracture in an infinite body which is given in Pouya and Ghabezloo [14]. However, in the case of high fracture density (or dense fracture networks), the integral of flux on the fracture network, i.e. the term described in left-hand side of eqn. (13) could be computed numerically for a fracture network discretized by E element.

In the next section, we shall take up some real examples for calculating the effective permeability in using the recent advance of theoretical development and numerical tools which are described above.

4 Application

4.1 Effective permeability of EDZ

Excavation of underground galleries generates cracks and fractures in a zone around the gallery called Excavation Damaged Zone (EDZ). The effect of fractures on the hydraulic properties of EDZ can be estimated through a numerical method taking into account the complex geometry and density of discontinuities. Three zones around the gallery (fig. 3) can be distinguished: the zone crossed by fractures; damaged or micro-fractured zone beyond fractured zone and the zone of intact rock beyond EDZ. The first zone is considered here to calculate its equivalent permeability. The fractures appearing in this zone has more or less regular geometry with the same shape, spacing, extensions that are repeated along the axis of gallery.

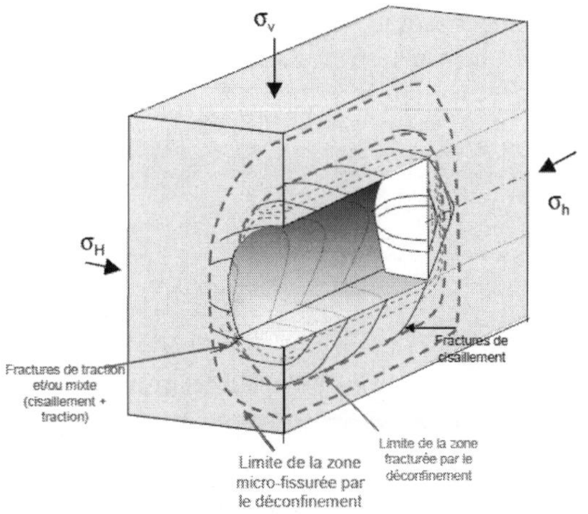

Figure 3: Representation of the EDZ around the gallery and the shape of the fracture (ANDRA 2008 [18]).

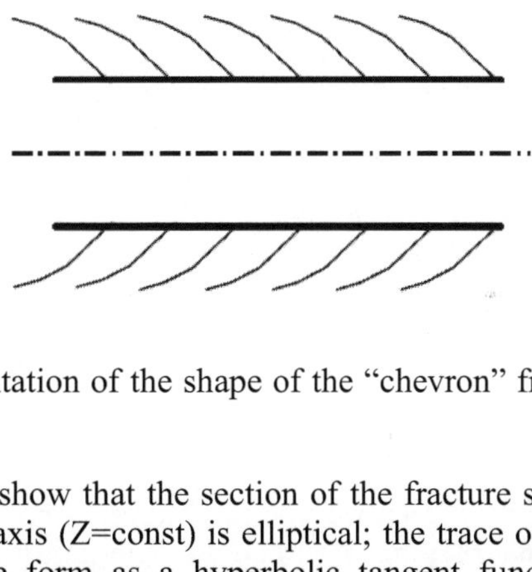

Figure 4: Representation of the shape of the "chevron" fractures in a vertical plane.

Field observations show that the section of the fracture surfaces in the plane orthogonal to gallery axis (Z=const) is elliptical; the trace of fracture in vertical plane has a curvature form as a hyperbolic tangent function; the extension fracture in the horizontal fracture is more or less constant. This observation leads to take the following expression (eqn. 15) to represent the equation of the fracture surface (Pouya et al. [17]).

$$\frac{X^2}{a^2}+\frac{Y^2}{b^2\tanh(\lambda Z)}=1 \tag{15}$$

The parameters a, b, λ depend on the orientation of the gallery. And then, the values: $a=b=4$, $\lambda=0.57$ are determined by comparing the trace of theoretical surface to observed fracture in field. In addition, the last parameter d=60 cm of spacing between fractures is fixed by the field measurement.

The above geometrical model is used to evaluate the effective permeability of EDZ (fig. 5) with the isotropic matrix permeability k = 10^{-12}m/s and the hydraulic conductivity of fracture c=10^{-11}m^2/s. The chevron fracture form is discretized by series of small linear segments.

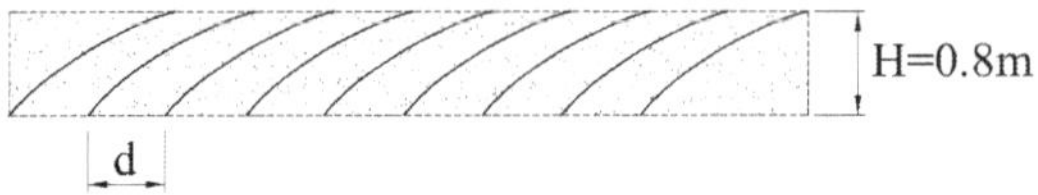

Figure 5: Representative segment of EDZ in calculating the effective permeability.

The linear pressure $p(\underline{x}) = \underline{A}.\underline{x}$ is imposed at infinity. By solving the problem for two vectors of pressure gradient A_1=(1,0) and A_2=(0,1), two average velocity vectors $\underline{V}^1$, $\underline{V}^2$ and two average pressure gradient vectors $\underline{G}^1$, $\underline{G}^2$ are obtained. The equivalent permeability $\mathbf{k}^p$ is then deduced from the equation $\underline{V} = -\mathbf{k}^p.\underline{G}$. Launching the numerical program for the EDZ zone, the following effective permeability matrix is found:

$$\mathbf{k}^p = \begin{bmatrix} 9.09 & -5,39 \\ -5,26 & 8.19 \end{bmatrix} \text{ x } 10^{-12}(\text{m/s})$$

This result shows that the effective permeability matrix is almost symmetric according to the results announced by Pouya and Fouché [11]. The principal directions of this permeability tensor can be calculated easily. They are found to be approximately parallel and orthogonal to the fractures surfaces. Otherwise, permeability predicted in this way is smaller than that given by the Mori-Tanaka estimation based on theoretical results of Pouya and Ghabezloo [14]. The difference can be explained by the fact that the Mori-Tanaka estimation does not take into account the fractures interaction.

4.2 Fracture network in a permeable rock

The study of fluid flow in fractured permeable rocks needs, first, a geometrical model of fractures that may be described the deterministic models (Kolditz and Clauser [8]) or by stochastic approach (Cacas et al. [4]; Billaux [2]; Gervais [6] and Maleki et al. [10]). The stochastic simulation of joint network is based on a hierarchical probabilistic model that realistically reproduces fracture connectivity using minimal data such as the number of fracture sets, the fracture length, spacing and density. In practice, the stochastic distribution is often used for modelling fluid flow in hydrogeology or for natural geological reservoir since their large flexibility. In this work, the study introduced by Maleki et al. [10] is adopted. A square domain is characterized by two corners (X_{min},Y_{min}) and (X_{max},Y_{max}). Three fracture families are defined each one by a density ρ (number of fracture per area unit), the half of fracture length r = L/2 and the orientation θ (angle between fracture line and x-axis, $0 \leq \theta < \pi$). The fracture line is completely determined by these parameters and the coordinates (x_0,y_0) of its center. A uniform distribution of the fracture center points and of the angle θ; and an exponential law of length distribution are supposed.

In order to study the effect of fracture density on the effective permeability of fractured porous media, three families are generated in a domain $10x10m^2$ of rock formation. Several configurations are considered for which all the fractures parameters are constant except for the density that increases (fig. 6). Herein, we introduce the dimensionless parameter $\lambda = c / 2\pi k\xi$, with $\xi = 1m$; and then we use the numerical procedure to compute the equivalent permeability of fractured porous rock with homogeneous permeability k for matrix formation.

(a) (b)

Figure 6: Fracture network in porous rock formation: (a) non-percolated network, (b) percolated network.

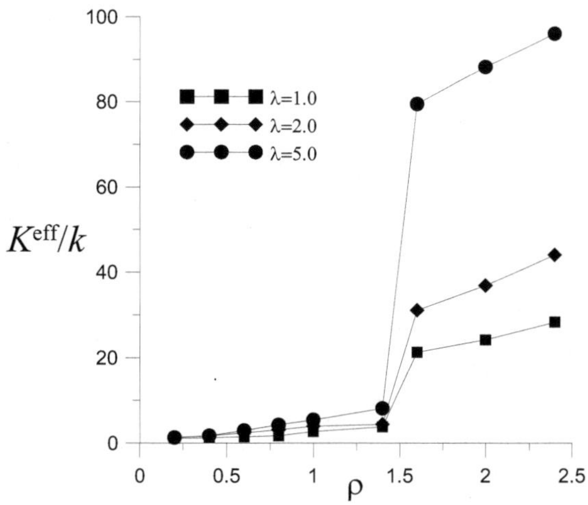

Figure 7: Determinant of effective permeability matrix varies with fracture density for different values of λ .

Figure 7 depicts the evolution of the determinant of the effective permeability tensor $K^{eff} = \left|\mathbf{k}^{p}\right|$ with the fracture density for different value of λ.

The equivalent permeability increases first linearly with the fracture density ρ, for small values of the density. A sharp increase is observed for a critical density ρ≈1.5 that represents the percolation threshold. The dimensionless density $\bar{\rho} = \pi \bar{R}^2 \rho$, where $\bar{R}$ is the mean half-length of the fractures, is found to be around 1.2. After percolation threshold, the effective permeability function of density increases with a different pace as observed also by Maleki et al. [10].

5 Conclusion

The effective permeability of fractured porous rocks is studied numerically using analytical and numerical developments based on the singular integral equations. This method that is clearly presented in this paper allows taking into account the real geometry of the fractures and of their interactions effects. The geometrical shape of individual fractures may be introduced precisely in the model as it has been done for the case of the EDZ studied here above. In the case of great number of fractures, the geometry of the fractures network can be described and generated numerically in the model by stochastic laws. The parameters of these stochastic laws are deduced from observation data. The numerical program that has been developed using this method provides an efficient tool for quick evaluation of the effective permeability of fractured porous rocks. The results presented in this paper to illustrate this method show also the important effect of percolation which modifies significantly the magnitude of effective permeability.

Acknowledgements

This work is co-funded by IFSTTAR and BRGM.

Reference

[1] Barthélémy, J.-F., Effective permeability of media with a dense network of long and micro fractures. *Transport in Porous Media*. **76**, pp.153–178, 2009.

[2] Billaux, D., *Hydrogéologie des milieux fracturés. Géométrie, connectivité et comportement hydraulique*, Thèse de doctorat, Ecole Nationale Supérieure des Mines de Paris, 1990.

[3] Bonnet, M., *Equations intégrales et éléments de frontière*, CNRS Editions/Eyrolles : Paris, 1995.

[4] Cacas, M. C., Ledoux, E., De Marsily, G. 1 Tillie, B., Modeling Fracture Flow with A Stochastic Discrete Fracture Network: Calibration and Validation: 1. The Flow Model. *Water Resources Research*, **26**(3), pp. 479-489, 1990.

[5] Dormieux, L. & Kondo D. 2004. Approche micromécanique du couplage perméabilité–endommagement. *C.R. Mecanique*, **332**, pp. 135-140, 2004.

[6] Gervais F., *Modélisation géométrique d'un réseau de fractures dans un massif rocheux stratifié. Application aux carrières marbrières de Comblanchien (Côte d'Or, France)*, PhD dissertation. Ecole Nationale Supérieure des Mines de Paris, 1993.

[7] Goméz-Hernández, J. J. & Wen, X. H., Upscaling hydraulic conductivities in heterogeneous media: an overview. *Journal of Hydrology*, **183**, pp. ix–xxxii, 1996.

[8] Kolditz, O. & Clauser, C., Numerical Simulation of Flow and Heat Transfer in Fractured Crystalline Rocks: Application to the Hot Dry Rock Site in Rosemanowes (U.K.). *Geothermics*, **27**, pp. 1-23, 1998.

[9] Liolios, P.A. & Exadaktylos, G. E., A solution of steady-state fluid flow in multiply fractured isotropic porous media. *International Journal of Solids and Structures*, **43**, pp. 3960–3982, 2006.

[10] Maleki, K., Pouya, A., Courtois, A. & Su, K., Modélisation numérique du couplage entre l'endommagement et la perméabilité des argiles raides. *16ème Congrès Français de Mécanique*, Nice, 2003.

[11] Pouya, A. & Fouché, O., Permeability of 3D discontinuity networks: new tensors from boundary-conditioned homogenization. *Advance Water Resources*, **32**, pp. 303-314, 2009.

[12] Pouya, A., Bourgeois, E., Haxaire, A. & Poutrel, A., Continuum equivalent model for the fractured EDZ around underground galleries in claystone. *Congrès ANDRA*, Nante, 2010.

[13] Pouya, A. & Vu, M. N., Modeling steady-state flow in fractured porous rocks by singular integral equations method. *In Proceedings of the 13th International Conference of the International Association for Computer Methods and Advances in Geomechanics*, Melbourne, 2011.

[14] Pouya, A. & Ghabezloo, S., Flow around a crack in a porous matrix and related problems. *Transport in Porous Media*, **84(2)**, pp. 511-532, 2010.

[15] Renard, P. & de Marsily, G., Calculating equivalent permeability: a review. *Advance Water Resources*, **20 (5–6)**, pp. 253–278, 1997.

[16] Sánchez-Vila, X., Girardi, G.P. & Carrera, J., A synthesis of approaches to upscaling of hydraulic conductivities. *Water Resources Research*, **31(4)**, pp. 867–882, 1995.

[17] Pouya, A., Bourgeois, E., Haxiare, A., *Elaboration des modèles mécaniques et hydromécaniques homogénéisés de l'EDZ au champ proche d'un ouvrage de stockage*, Rapport ANDRA, C.RP.0LCP.10.0003, 2010.

[18] ANDRA 2008, *Synthèse de la réunion du 15-11-2007 sur la conceptualisation des réseaux de fractures dans l'EDZ « initiale »*, C.NT.AEAP.08.0002, 2008.

Modelling multiphase flow inside the porous media of a polymer electrolyte membrane fuel cell

T. Berning & S. K. Kær
Department of Energy Technology, Aalborg University, Denmark

Abstract

Transport processes inside polymer electrolyte membrane fuel cells (PEMFC's) are highly complex and involve convective and diffusive multiphase, multispecies flow through porous media along with heat and mass transfer and electrochemical reactions in conjunction with water transport through an electrolyte membrane. We will present a computational model of a PEMFC with focus on capillary transport of water through the porous layers and phase change and discuss the impact of the liquid phase boundary condition between the porous gas diffusion layer and the flow channels, where water droplets can emerge and be entrained into the gas stream.
Keywords: fuel cell modelling, multi-phase flow, GDL/channel interface.

1 Introduction

Polymer electrolyte membrane fuel cells (PEMFC's) combine hydrogen with oxygen from air in order to generate electricity with the only by-products being water and waste heat. The overall reactions are as follows:

Anode:	$2\ H_2$	$\rightarrow$	$4\ H^+ + 4\ e^-$
Cathode:	$O_2 + 4\ H^+ + 4\ e^-$	$\rightarrow$	$2\ H_2O$
Combined:	$2\ H_2 + O_2$	$\rightarrow$	$2\ H_2O$

A schematic of a single PEMFC is given in Figure 1. Depending on the detailed operating conditions the product water can be in the liquid or gas phase. Above reaction occur inside the catalyst layers (CL's) and this means that the product water has to travel though the porous CL's and the porous transport

WIT Transactions on Engineering Sciences, Vol 70,
www.witpress.com, ISSN 1743-3533 (on-line)
doi:10.2495/MPF110251

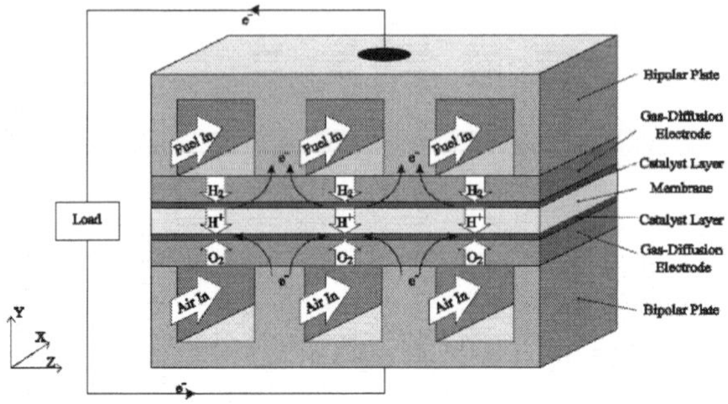

Figure 1: PEMFC schematic.

layers (PTLs) towards the flow channels where it can be picked up by the reactant.

The primary mechanism that drives the liquid water through the porous sublayers is capillary action, as was shown by Sinha and Wang [1], which is a diffusion-like mechanism that drives the water from regions of high concentrations to regions of low concentrations. The capillary pressure itself is a function of the material parameters of the porous material (contact angle, permeability, porosity) and of the liquid water volume fraction, also called "saturation". A frequently used equation to define the capillary pressure is the Leverett function, but recently there has been experimental effort to measure the capillary pressure as function of the saturation. Numerical models based on CFD typically use the Leverett function (described below) and may vary the coefficients in order to account for the pore-size distribution (PSD).

A common problem in these models has been the description of the liquid phase boundary condition at the interface between the porous gas diffusion layer (GDL) and the gas flow channel. A frequent approach was to specify a capillary pressure of zero at this interface, which corresponds to having a negligible amount of liquid phase at this interface. This was done by Berning and Djilali [2], among others. Alternatively, an arbitrary non-zero liquid saturation was prescribed by Natarajan and Nguyen [3] which ended up dominating the amount of liquid saturation inside the porous media. A more elaborate interfacial coverage model was proposed by Meng and Wang [4] in order to relate the liquid saturation at the interface to the cell potential and flow conditions in the channel. However, specifying any fixed value, either a value for the saturation or a capillary pressure, will not necessarily capture essential physics, as was pointed out by Gurau et al. [5]. This group in turn considered pending droplets, which are fed by one or more capillaries, and claim that the pressure downstream of the capillaries is equal to the pressure in the attached droplets which they feed. This means that equilibrium is assumed between the droplet and the pore that is feeding it. Such equilibrium must be questioned as it is an experimental observation that droplets continue to grow at a lower channel gas stream, which means that the capillary pressure of the pore feeding the droplet must always be larger than the capillary pressure of the sessile droplet. Weber and Newman [6] assumed a zero liquid water flux if the gas phase pressure exceeds the liquid phase pressure while the capillary pressure is set to zero otherwise.

From a modelling perspective it is desirable to de-couple the transport phenomena inside the porous media from the channel, i.e. to prescribe a boundary condition where the GDL is not affected by the channel flow. The channel flow in turn will always be disturbed by the growing droplets, but these phenomena are very difficult to simulate in a model. Recently our group has suggested an approach of prescribing a boundary condition by applying a liquid phase pressure gradient across the interface at the last layer of the GDL [7]. This pressure gradient is derived from the Hagen-Poiseuille equation and is based on the experimental observation that liquid phase typically enters the channel through the same, "active" pores of the GDL [8]. It was found experimentally that the liquid water level inside the porous layers could be changed by perforating the GDL. In this work we want to investigate the effect of the boundary condition that our group has suggested on the predicted water content inside the porous media. A rendering of the liquid water transport is shown in Figure 2 [9].

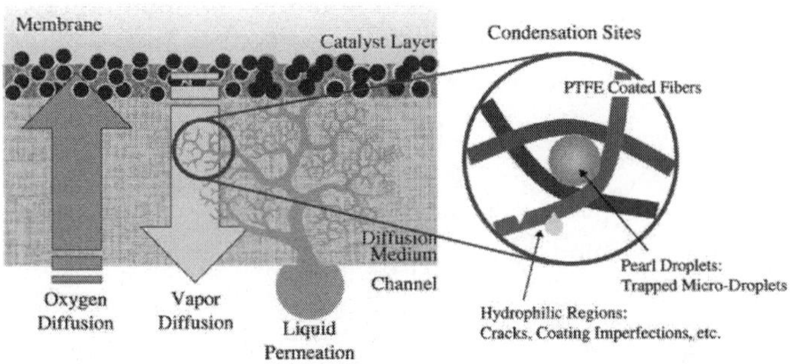

Figure 2: Rendering of the water transport model [9].

2 Model description

For the sake of brevity we do not list in detail all modelling equations employed in our comprehensive three-dimensional, non-isothermal CFD model. Instead we want to limit the list of equations to the description of liquid water flux through the porous media and across the GDL/channel interface.

As was previously shown, the three-dimensional momentum equations for the liquid phase in our multi-fluid model can be reduced to Darcy's law for the gas phase and liquid phase [7]:

$$\mathbf{u}_g = \frac{k_{rel,g} K}{\mu_g} \nabla p_g \tag{1}$$

$$\mathbf{u}_l = \frac{k_{rel,l} K}{\mu_l} \nabla p_l = \frac{k_{rel,l} K}{\mu_l} \nabla p_g - \frac{k_{rel,l} K}{\mu_l} \nabla p_{cap} \tag{2}$$

where the capillary pressure is defined as:

$$p_{cap} = \sigma \cos\theta \left(\frac{\varepsilon}{K}\right)^{0.5} J(S) \tag{3}$$

In above equations $\mathbf{u}_g$ and $\mathbf{u}_l$ are the three-dimensional velocity vectors, k_{rel} is the relative permeability of each phase, K is the dry permeability of the porous medium, μ is the dynamic viscosity and p is the pressure. Moreover, σ denotes the surface tension between the gas phase and the liquid phase (here 0.0625 *N/m*) and θ is the effective contact angle between the liquid phase and the porous medium. Here and in the following the subscripts "*g*" and "*l*" denote the gas phase and liquid phase, respectively. It is also important to realize that the terms including the permeability and the viscosity represent a flow resistance in Darcy's law. Thus the resistance for the liquid phase may be denoted:

$$R_{Darcy,l} = \frac{\mu_l}{k_{rel,l} K} \tag{4}$$

The relative permeability of the liquid phase depends on the volume fraction of the liquid phase inside the porous media, termed liquid saturation *s*, according to:

$$k_{rel,l} = s^n \tag{5}$$

where *s* is the liquid saturation and *n* is a prescribed exponent. In this work we apply a value of n=3.

The inverse of the property $(\varepsilon/K)^{0.5}$ is known as the characteristic pore-radius of the porous medium in question:

$$r_c^{-1} = \left(\frac{\varepsilon}{K}\right)^{0.5} \tag{6}$$

The Leverett *J*-function is typically of cubic nature, and our group has previously argued that it must be closely related to the integrated pore-size distribution (PSD), starting from the largest pores.

$$J(S) = 1.417S - 2.120S^2 + 1.263S^3 \tag{7}$$

A schematic of the liquid pressure as function of the saturation level for two different porous media is shown in Figure 3. For hydrophobic porous media the liquid pressure exceeds the gas phase pressure and the capillary pressure is negative.

Eqn. (3) also shows that different porous media have different capillary pressure functions. In a fuel cell there are porous layers with different material properties adjacent to each other. Because of the continuity of the pressure fields

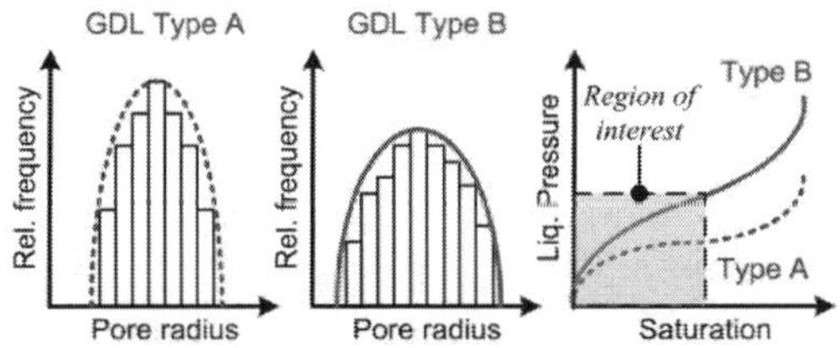

Figure 3: Expected liquid pressure curves for two types of porous media.

this leads to jump conditions for the liquid saturation across these interfaces. These jump conditions occur for example at the interface between the catalyst layer and the micro-porous layer as well as between the micro-porous layer and the GDL. Figure 4 shows schematically the capillary pressure curves for two different porous media and the arising jump in the liquid water saturation caused by the continuity in the capillary pressure across such an interface.

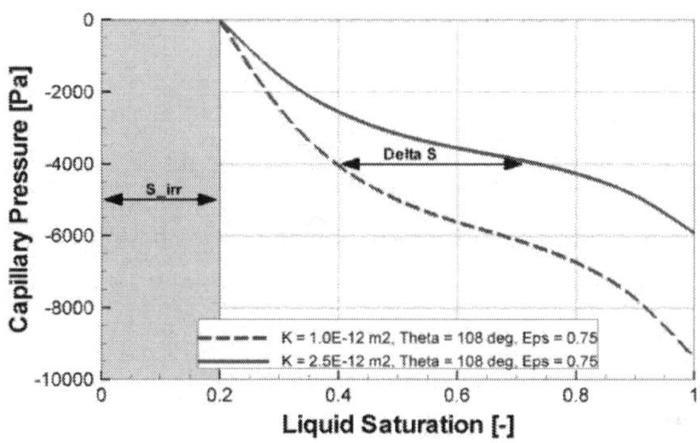

Figure 4: Capillary pressure versus saturation for two different material types.

While the above equations are sufficient to describe the capillary liquid water transport through the various porous layers in a fuel cell including the interface conditions, an important question still concerns the boundary condition for the liquid phase at the interface between the porous GDL/PTL and the gas flow channel.

In contrast to specifying fixed values for the capillary pressure or saturation at the GDL/channel interface our group has suggested to specify the liquid pressure gradient based on the Hagen-Poiseuille equation which relates the liquid phase velocity to the pressure gradient according to:

$$\mathbf{U}_l = -\frac{1}{8}\frac{r_c^2}{\mu_l}\nabla p_l = -\frac{1}{8}\left(\frac{K}{\varepsilon}\right)\frac{1}{\mu_l}\nabla p_l \tag{8}$$

In this equation $\mathbf{U_l}$ denotes the intrinsic velocity which is related to the superficial velocity in the through-plane direction according to:

$$V_l A_{pores} = v_l A_y \tag{9}$$

where V_l is now the intrinsic velocity component in the y-direction of the liquid phase described by the Hagen-Poiseuille equation and v_l is the superficial velocity component that corresponds to the velocity field that CFX-4 calculates.

Combining the last two equations yields:

$$v_l = \frac{A_{pores}}{A_y} V_l = \frac{n_{pores} \times \pi \times r_c^2}{A_y} V_l = -n''_{pores} \frac{\pi}{8} \left(\frac{K}{\varepsilon} \right)^2 \frac{1}{\mu_l} \frac{\partial p_l}{\partial y} \tag{10}$$

In above equation the number of pores in the cell area in y-direction A_y is related to the number of pores per unit area according to:

$$n_{pores} = n''_{pores} A_y \tag{11}$$

where n_{pores} is the number of "active" pores in the control volume under consideration, and a control volume is a volume element in the numerical grid of the CFD model. The area of the control volume in the y-direction is denoted A_y, and we are only considering those volume elements located inside the GDL adjacent to the flow channels.

Eqn. (10) denotes again the liquid flow velocity as function of a resistance and the driving pressure gradient, and the resistance is in this case:

$$R_{H.-P.} = \frac{1}{n''_{pores}} \frac{8}{\pi} \left(\frac{\varepsilon}{K} \right)^2 \mu_l \tag{12}$$

Hence we have described two resistances, the Darcian resistance which depends on the liquid saturation and the Hagen-Poiseuille resistance which depends on the number of active sites per unit area. This resistance is only invoked at the GDL/channel interface in order to capture the experimental observation that the liquid phase enters the channel through a number of open, "active" pores that are the same over a wide range of operating conditions. A comparison of the order of magnitude of the Darcian resistance to the Hagen-Poiseuille resistance is shown in Figure 5. The porosity in this case is $\varepsilon = 75\%$ and the dry permeability of $K = 1$ Darcy $= 10^{-12}$ m^2. According to Eqn. (5) the Darcy resistance also depends on the exponent in the relative permeability, while the overall resistance in the Hagen-Poiseuille expression depends on the number of active sites for unit area which is in the order of 10^6 m^{-2}.

In our code the Hagen-Poiseuille resistance is only implemented for the last layer of fibres between the GDL and the flow channel. In the interior we consider Darcian resistance only. When we consider the flow exiting through the last pore we need to estimate the length of these pores on order to obtain the expected pressure drop. The carbon fibres of the GDL typically have a diameter of 7–10 μm, and so we estimated the pore length for the Hagen-Poiseuille flow to

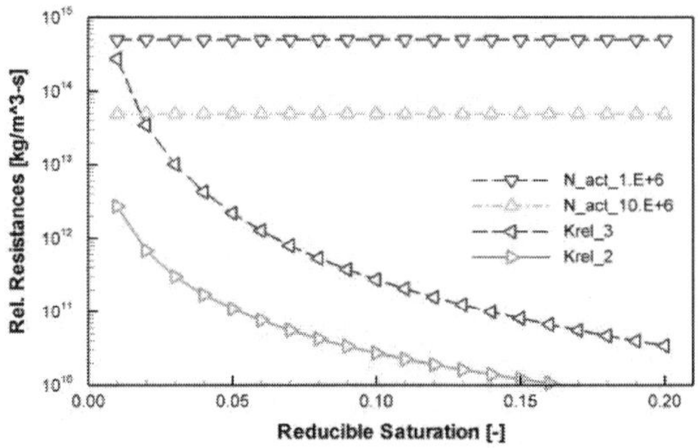

Figure 5: Comparison between the Hagen-Poiseuille and Darcian resistance.

be $l_p = 4\ \mu m$. In a pore network model this properties typically call the "throat length".

In our CFD code we need to combine both resistances due to the fact that the cell height of the control volumes in our numerical grid at the GDL/channel interface is larger than the outlet pore height, which we may leave adjustable. Hence we need to account for a Darcian resistance and a resistance at the outlet pores in the same control volume, where the superficial velocity as calculated by the CFD code is the same in both resistance regions, but the pressure drops are additive and combine to yield the pressure gradient as calculated by the CFD code.

Denoting the height of the control volume dy_c yields for the Darcian part:

$$v_l = \frac{1}{R_{Darcy,y}} \frac{\Delta p_{l,Darcy,y}}{(dy_c - l_p)} \tag{13}$$

The height of the control volume depends on the grid size and typically ranges around 20 – 30 μm. The Hagen-Poiseuille equation applies to the pore height l_p and yields:

$$v_l = \frac{1}{R_{H-P,y}} \frac{\Delta p_{l,H-P,y}}{l_p} \tag{14}$$

The pressure drops are additive to yield the pressure drop over the entire control volume according to:

$$\Delta p_{l,y} = \Delta p_{l,Darcy,y} + \Delta p_{l,H-P,y} \tag{15}$$

Finally it holds for the superficial velocity:

$$v_l = \frac{1}{R_{\text{int},y}} \frac{\Delta p_{l,y}}{dy_c} \tag{16}$$

where $R_{int,y}$ is now the combined resistance at the GDL/channel interface as it is implemented in our CFD model. This can be expressed as function of the partial resistances due to Darcy and Hagen-Poiseuille:

$$R_{\text{int},y} = \ldots = (1 - l_p / dy_c) R_{Darcy} + (l_p / dy_c) R_{H-P} \tag{17}$$

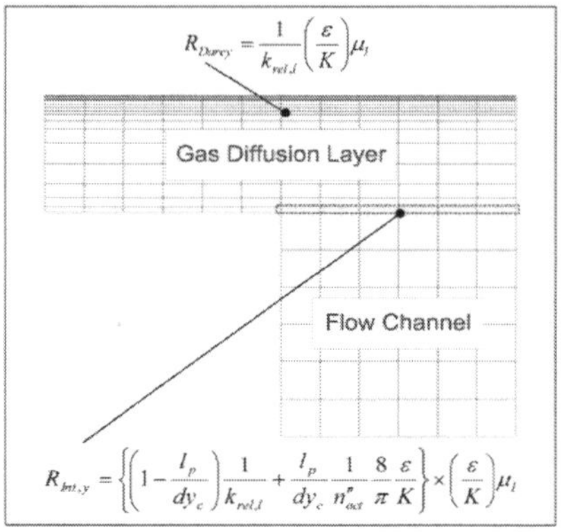

Figure 6: Local resistances inside the porous layers and numerical grid.

Figure 6 shows schematically the numerical grid of the cathode porous layers and gas flow channel and the resistance terms implemented. As described above the resistance due to the Hagen-Poiseuille equation is only implemented at the control volume of the GDL that are adjacent to the gas flow channel.

3 Model results

Clearly the complete model is much more complex than described above as it also contains multi-species flow at the cathode and anode side, heat transfer including phase change, electrochemical reactions and water transport through the electrolyte membrane. However, in this study we want to focus on the description of the liquid phase transport through the porous media by capillary action.

The model was applied to study the behaviour of a PEMFC operating at a current density of 0.4 A/cm^2. The cell temperature is 353 K and the operating pressure is 1.5 *atm* (total) at the outlet. Both the air at the cathode and the hydrogen at the anode are humidified to 75% RH at the inlet, and the stoichiometric flow ratio is 3 at both anode and cathode side. In a previous study we have found that the specific surface area of the electrolyte phase in the CLs is very important for the overall water balance [10]. In this study we have employed a value of 5.0×10^4 m^2/m^3, which is comparatively low and ensures that the cathode side becomes flooded, which is essential for this study. Because the primary focus of the current study is the role of the boundary condition for the liquid phase at the GDL/channel interface, the main parameter that was varied in this study is the number of active sites per unit area n_{act}''.

The important material properties applied inside the porous media are summarized in Table 1. The fact that all porous regions were assigned different material properties leads to a jump condition for the liquid saturation according to Figure 4. The effective contact angle that is used to calculate the capillary pressure was $\theta = 120°$ for all porous regions. Of note is that this value only applies to the hydrophobic pores while the hydrophilic pore fraction is included in the irreducible saturation.

Table 1: Material parameters of the different porous regions.

Porous Region	Porosity [-]	In-plane permeability [m^2]	Through-plane permeability [m^2]	Irreducible saturation [-]
CL	75%	1.E-12	1.E-12	0.3
MPL	75%	1.E-12	1.E-12	0.2
GDL	75%	20 E-12	10 E-12	0.2

For the sake of brevity we can only show a few results. Figures 7–9 show the predicted liquid saturation inside the porous media at the cathode side at mid-channel (the total channel length investigated is 25 *mm*). The results have been mirrored in the postprocessor for better visualization. The difference in these cases is only the variation in the number of active pores per unit area n_{act}''. Figure 7 is for the case of a non-modified boundary condition, where the pore length l_{pore} has been set to zero in our code which means that the standard Darcy resistance has not been modified. Figures 8 and 9 are for the case of $n_{act}'' = 10\times10^6$ and $n_{act}'' = 1\times10^6$, respectively. The visualization results as published by Zhang et al. [8] show that these are realistic values. The results indicate no discernible effect between the cases of having no additional resistance and having $n_{act}'' = 10\times10^6$, *i.e.* 10 active sites per mm^2. For an active site density of 1×10^6 there is a noticeable increase in liquid saturation, *i.e.* the boundary condition has an impact on the predicted modelling results. Of note is that the number of active sites per unit area was observed to be in exactly this region, i.e. around 1–10 mm^{-2} [8].

Figures 10–12 show the predicted liquid saturation inside the porous GDL in a cut close to the channel interface. Hence the liquid volume fraction under the channel area and under the land area of fuel cell can be seen. Again the main differences can be observed between the case of $n_{act}'' = 10\times10^6$ and $n_{act}'' = 1\times10^6$, hence in the physically realistic regime. Reducing the number of active sites per unit area leads to an increase in flow resistance for the liquid phase in the through-plane direction. Hence the water spreads out more in the in-plane direction. In any case we observe that the boundary condition as we have implemented it does have an impact on the modelling results. Gerteisen et al. [11] have found that drilling holes into the GDL to release the liquid water can stabilize the cell performance.

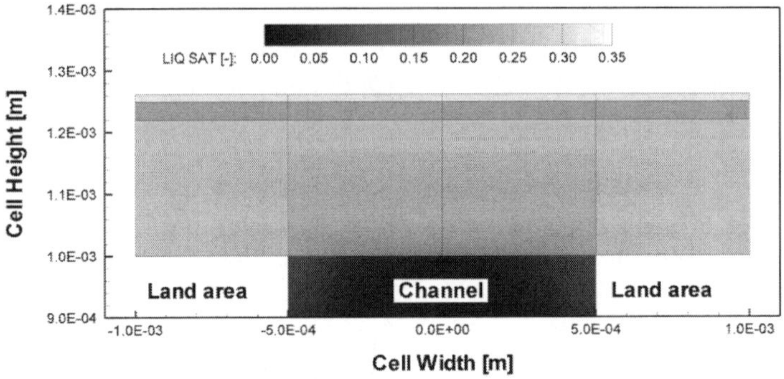

Figure 7: Predicted liquid saturation inside the cathode side porous media at mid-channel for the case of the "standard" boundary condition (l_{pore}=0 μm).

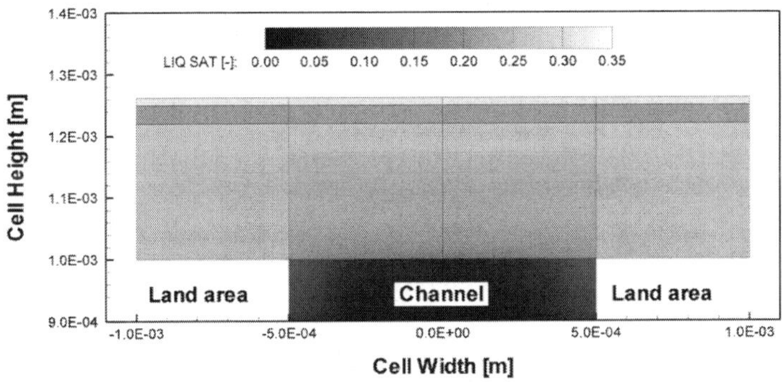

Figure 8: Predicted liquid saturation inside the cathode side porous media at mid-channel for the case of n_{act}"= 10 mm^{-2}.

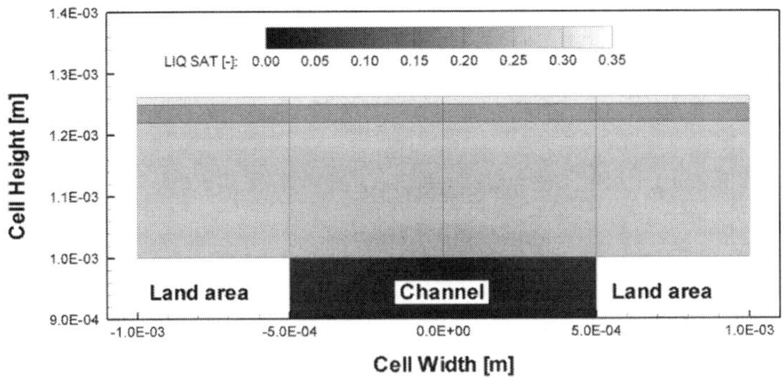

Figure 9: Predicted liquid saturation inside the cathode side porous media at mid-channel for the case of n_{act}"= 1 mm^{-2}.

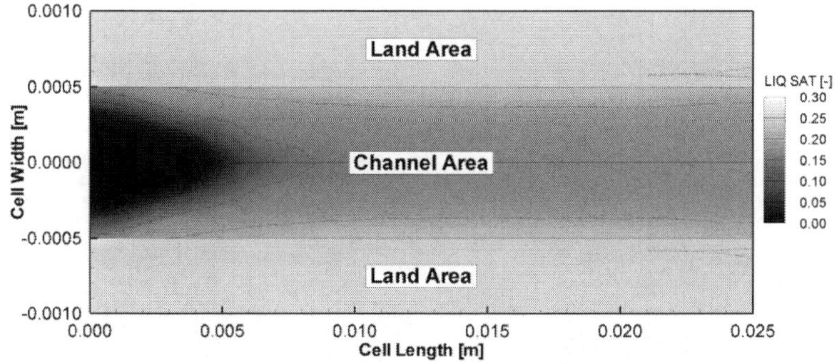

Figure 10: Predicted liquid saturation inside the cathode GDL close to the channel interface for the case of the “standard” boundary condition (l_{pore}=0 μm).

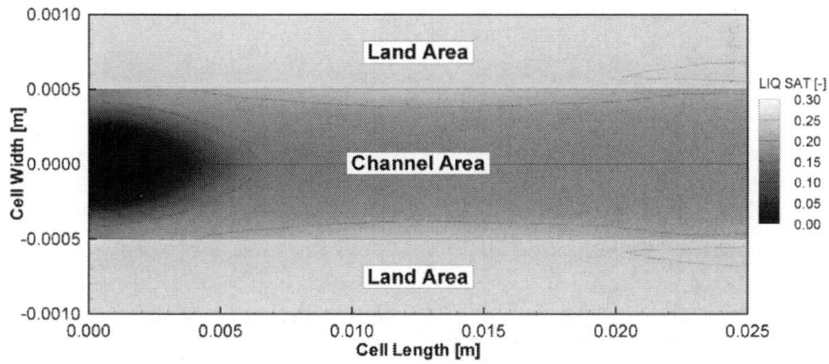

Figure 11: Predicted liquid saturation inside the cathode GDL close to the channel interface for the case of n_{act}”= 10 mm^{-2}.

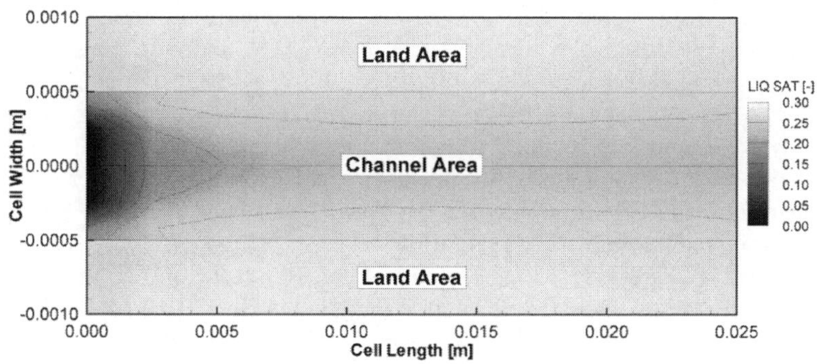

Figure 12: Predicted liquid saturation inside the cathode GDL close to the channel interface for the case of n_{act}”= 1 mm^{-2}.

4 Conclusions

We have proposed a novel boundary condition for the liquid phase at the GDL/channel interface of a PEM fuel cell. This boundary condition is based on the experimental observation that there are only a few pores per mm^2 that release the liquid water, and this means that the frequently use Darcy approach may not be applicable for this flow regime. Instead, we have utilized the Hagen-Poiseuille equation to describe the pressure drop across the outlet pores. This was implemented as an additional resistance for the liquid phase at the channel/GDL boundary. First simulations indicate that this boundary condition has a significant impact on the predicted liquid saturation inside the porous media. This is in good accord with experimental observations. More work is needed to fully understand and quantitatively verify this effect.

References

[1] Sinha, P. K. & Wang, C.-Y.: Pore-network modeling of liquid water transport in gas diffusion layer of a polymer electrolyte fuel cell, Electrochimica Acta, 52 (2007) 7936-7945.

[2] Berning, T. & Djilali, N.: A 3D, multiphase, multicomponent model of the cathode and anode of a PEM fuel cell, J. Electrochem. Soc. 150 (2003) A1589-A1598.

[3] Natarajan, D. & Nguyen, T. V.: Three-dimensional effects of liquid water flooding in the cathode of a PEM fuel cell, J. Power Sources 115 (2003) 66.

[4] Meng, H. & Wang, C.-Y.: Model of two-phase flow and flooding dynamics in polymer electrolyte fuel cells, J. Electrochem. Soc. 152 (2005) A1733-A1741.

[5] Gurau, V., Zawodzinski, T. A. & Mann J. A.: Two-phase transport in PEM fuel cell cathodes, J. Fuel Cell Science and Technology 5 (2008) 021009.

[6] Weber, A. Z. & Newman, J.: Effects of microporous layers in polymer electrolyte fuel cells, J. Electrochem. Soc. 152 (2005) A677-A688.

[7] Berning, T., Odgaard, M. & Kær, S.: A computational analysis of multi-phase flow through the porous media of a PEMFC cathode using the multi-fluid approach, J. Electrochem. Soc. 156 (2009) B1301.

[8] Zhang, F. Y., Yang, X. G. & Wang, C.-Y.: Liquid water removal from a polymer electrolyte fuel cell, J. Electrochem. Soc. 153 (2006) A225-A232.

[9] Nam, J. H. & Kaviany, M.: Effective diffusivity and water-saturation distribution in single- and two-layer PEMFC diffusion medium, Int. J. Heat Mass Transfer 46 (2003) 4595-4611.

[10] Berning, T., Odgaard, M. & Kær, S.: Water Balance Simulations of a PEM Fuel Cell Using a Two-Fluid Model, ECS Trans. 33 (2010) 1503-1513.

[11] Gerteisen, D., Heilmann, T. & Ziegler, C.: Enhancing liquid water transport by laser perforation of a GDL in a PEM fuel cell, J. Power Sources 177 (2008) 348.

Capillary trapping of supercritical CO_2 in porous media at the pore scale

T. Suekane, T. Izumi & K. Okada
Department of Mechanical Engineering,
The University of Tokushima, Japan

Abstract

Carbon dioxide capture and storage in geological formations is recognized as a promising method for decoupling fossil fuel use and carbon emissions. Carbon dioxide injected into geological formations is expected to be trapped by several mechanisms against buoyancy. In this paper, we focused on capillary rapping. Trapped bubbles in packed beds of glass beads were visualized by means of micro-focused X-ray CT at pore scale for supercritical CO_2 and water systems at reservoir condition and for nitrogen and water systems at laboratory room condition. When the diameter of glass beads is the same, distribution of volume of trapped bubbles is similar for each condition. At the pressure of 8.5MPa and the temperature of 45°C, which corresponds to just above a critical point, the morphology of the interface between the supercritical CO_2 and water suggests that CO_2 is non-wetting to glass beads. Residual gas saturation increases with a decrease in a diameter of glass beads, because the capillary pressure is higher with respect to buoyancy. Residual gas saturation can be expressed as a function of the reverse Bond number for each condition. Distribution of trapped bubble volume is identical with each other for each condition at same diameter of glass beads. The large trapped bubbles contribute to the residual gas saturation.
Keywords: capillary trapping, residual gas saturation, supercritical CO_2, micro-focused X-ray CT, carbon dioxide capture and storage.

1 Introduction

The geological storage of anthropogenic greenhouse gases to mitigate climate change is recognized as a means to reduce emissions to the atmosphere and the

WIT Transactions on Engineering Sciences, Vol 70, © 2011 WIT Press
www.witpress.com, ISSN 1743-3533 (on-line)
doi:10.2495/MPF110261

related impacts as a result of continues use of fossil fuels. Deep saline aquifers and depleted oil and gas reservoirs are potential subsurface deposits for CO_2 [1].

In geological formations, because of hydrostatic pressure and a geothermal gradient, pressure and temperature increase with depth. The critical condition of CO_2 (304 K and 7.38 MPa) is reached at a depth of approximately 700 m, where CO_2 density increases rapidly. However, the specific gravity of supercritical CO_2 at the condition which corresponds to the depths of 1000-3000m is about 0.7. Therefore, buoyancy brings CO_2 upwards. On the other hand, four trapping mechanisms have been identified [1]: structural trapping, capillary trapping, solubility trapping and mineralization trapping. CO_2 injected into the geological formations is expected to be stored by these trapping mechanisms for long time. During a storage period, each trapping mechanism changes relative weight increasing security and stability of storage [2–4].

The mechanism of capillary trapping can be explained as follows from the pore-scale point of view. CO_2 is injected into geological formation by replacing saline water forming a large continuous plume. After the CO_2 injection has stopped, saline water imbibes back into the formation. During this imbibition process, a continuous CO_2 plume is disconnected to form CO_2 bubbles of which volume is in the order of the pore of porous rock. At the pore-scale capillary pressure acting on a CO_2 bubble is expected to be greater than buoyancy and/or viscous force. As a result, CO_2 bubbles are eventually trapped in porous media. Capillary rapping has some advantages. Firstly, it plays an important role from the early stage of CO_2 storage. Secondly, it is capable of storing a large amount of CO_2 per unit volume of the formation [5].

As capillary trapping has been receiving growing attention, residual CO_2 saturation has been investigated experimentally in laboratory core-flooding experiments [5–9]. Because the capillary trapping takes place for a CO_2 bubble at pore-scale, microscopic approaches are taken to investigate its mechanism [10–13]. X-ray computed microtomography has provided a tool for the non-destructive investigation of the three-dimensional microstructure of porous media. Non-destructive direct imaging approaches are attractive because they provide not only a detailed and unique description of the pore-space geometry but also fluid flow within them [14–22]. The X-ray microtomography has been also applied to investigation of CO_2 trapping in porous media [5, 10, 12, 13].

The objective of this article is to study residual gas trapping in packed beds of glass beads at pore scale. A number of gas trapping experiments are conducted with supercritical CO_2 at elevated pressure and temperature and with nitrogen with room condition for various diameters of packing spheres. Characteristics of residual gas trapping of supercritical CO_2 are discussed comparing images obtained with nitrogen at the pore scale. The residual gas saturation has been analysed using the Bond number, which denotes the ratio of the buoyancy to the capillary pressure.

2 Experimental methodology and equipment

2.1 Dimensionless parameters and experimental conditions

A bubble of non-wetting phase in porous media is under the influence of capillary force, buoyancy and viscous pressure gradient. The ratios of gravity to capillary forces and viscous to capillary forces are expressed as dimensionless groups, known respectively as the Bond number, Bo, and capillary number Ca [23]:

$$\mathrm{Bo} = \frac{(\rho_w - \rho_n) g R^2}{\sigma} \tag{1}$$

$$\mathrm{Ca} = \frac{\mu_w u_w}{\sigma} \tag{2}$$

where, ρ is density, g is acceleration due to gravity, R is particle radius, σ is interfacial tension, μ is viscosity, u is displacing fluid velocity, and subscripts w and n denote wetting phase and non-wetting phase, respectively.

Experimental conditions are summarized in Table 1. Experiments of capillary trapping were carried out at two conditions. One is the condition of the pressure of 8.5 MPa and the temperature of 45°C which corresponds to the depth of approximately 850 m. At this condition, CO_2 is in a supercritical state. The other is normal laboratory room condition where we used nitrogen as a non-wetting phase instead of CO_2 to reduce the dissolution in water. Interfacial tension between supercritical CO_2 and water is about half of that between nitrogen and water in laboratory room condition. Because of high density of supercritical CO_2 with respect to nitrogen, buoyancy also is lower for supercritical CO_2. As a result, the Bond number falls in similar range for glass beads radius of 50–300 μm.

Table 1: Experimental conditions [24–26].

		Viscosity	Density	Interfacial tension	Capillary number	Bond number
		μ [μPa·s]	ρ [kg/m^3]	σ [mN/m]	Ca [-]	Bo [-]
Reservoir condition (45°C, 8.5MPa)	$scCO_2$	20	259.6	35.7	1.0×10^{-5} - 4.0×10^{-5}	5.0×10^{-4} - 1.8×10^{-2}
	H_2O	600	993.9			
Lab. room condition (20°C, 0.1MPa)	N_2	17.87	1.123	72.6	1.0×10^{-6}	3.4×10^{-4} - 1.2×10^{-2}
	H_2O	1062	996.7			

2.2 Experimental setup and procedure

Residual gas trapping has been investigated at the microscopic level using a micro-focused X-ray CT scanner (Comscantechno Co. ScanXmate-RB090SS). The magnification ratio of X-ray CT scanners depends on the ratio of the distance between the X-ray detector and the X-ray source, and that between an object and the X-ray source. When a packed bed with a low diameter is placed close to the X-ray source, highly magnified images can be achieved.

Two types of packed bed of glass beads were used for the experiments. One is designed to elevate temperature and pressure which correspond to the reservoir at the depth of about 850 m as shown in Fig. 1a. Glass beads were packed in a titanium tube with inner and outer diameter of 3 mm and 4 mm which could resist pressure up to 20 MPa. The packed bed made of titanium tube was placed in an acrylic resin pipe with outer diameter of 14 mm, where water was circulated to adjust temperature. The other is a simple packed bed used for the experiments at room temperature and pressure. Glass beads were packed in an acrylic resin tube with inner diameter of 10 mm at the height of about 40 mm.

The packed bed was vertically aligned in the X-ray CT scanner. The experimental set-up for high pressure and temperature experiments is shown in Fig. 1b. First, the structure of a backed bed was scanned at dry condition. Next, water was injected vertically upwards. At the same time, pressure was increased to be 8.5 MPa by an automatic back pressure regulator, while temperature was controlled to be 45°C by water circulation. After water injection stopped, a syringe pump forced liquid CO_2 into piping at the flow rate of 1.0 ml/min. to establish the condition of irreducible water saturation. In piping, the CO_2 was heated to a supercritical state of 45°C before arriving at the packed bed. Finally,

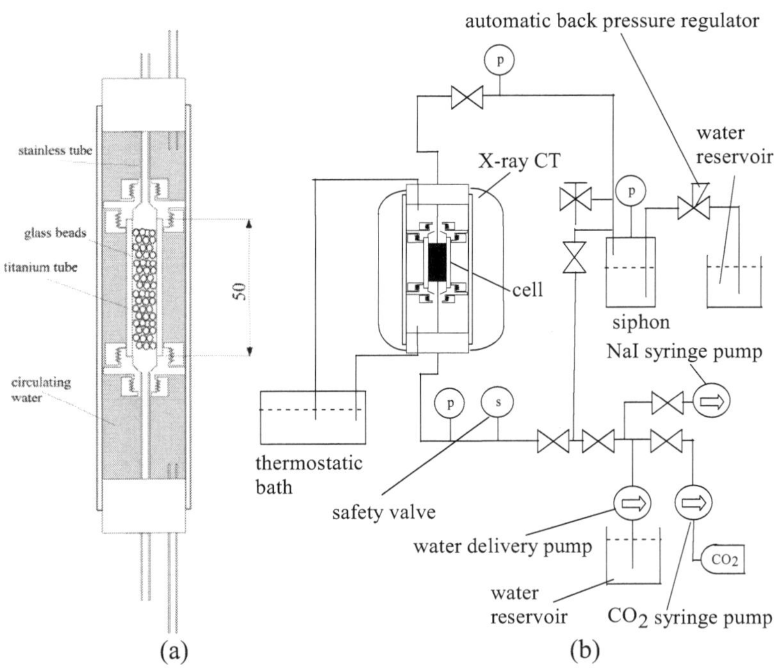

Figure 1: Pressure cell (a) and experimental setup (b) for elevated pressure and temperature.

water was injected into the packed bed at a constant flow rate which corresponds to the capillary number in the range between 1.0×10^{-5} and 4.0×10^{-5}. Water was saturated with CO_2 in advance in order to reduce dissolution of supercritical CO_2 in the packed bed. Between each process, the packed bed was scanned by X-ray CT.

In the case of a simple packed bed, experiments were carried out at laboratory room temperature and pressure. Nitrogen was used for a gas phase to prevent the dissolution of gas in water. Water was injected into the packed bed at the flow rate of 0.322 ml/min. which corresponds to the capillary number of 1.0×10^{-6}.

Water used in the experiments for both packed beds was doped with sodium iodide (NaI) at 7.5 wt% to enhance attenuation of X-ray. Reconstructed three-dimensional images consist of $608 \times 608 \times 610$ pixels at a resolution of 12.66 μm/pixel and 24.91μm/pixel for experiments in reservoir condition and laboratory room condition, respectively, in all directions.

3 Results and discussion

Distribution of trapped bubbles in packed bed of glass beads with a diameter of 200 μm is shown in Fig. 2. The residual gas saturation was 16% and 20% for reservoir and laboratory room conditions, respectively. In the case of nitrogen at laboratory room condition, trapped gas bubbles distribute uniformly throughout the packed bed as shown in Fig. 2b. The size of trapped gas bubbles changes from a small bubble which locates at the centre of single pore to a large bubble which spreads over several pores. The interface between nitrogen and water suggests that nitrogen is perfectly non-wetting to glass beads. In the case of

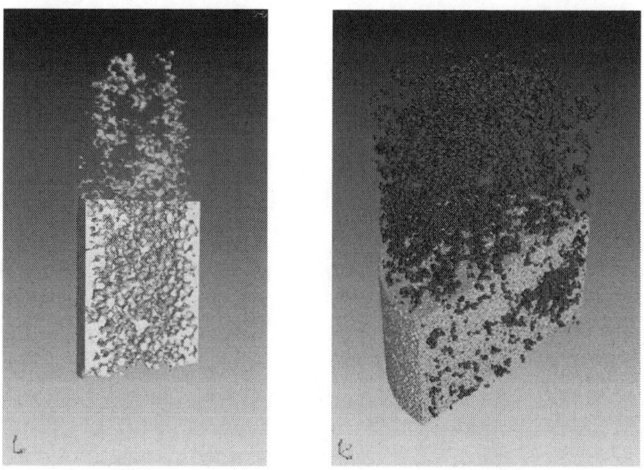

(a) (b)

Figure 2: Distribution of trapped gas bubbles in packed bed of glass beads with the diameter of 200 μm in reservoir condition (a) and laboratory room condition (b).

supercritical CO_2 at reservoir condition, the size of trapped bubbles changes also in the same range with the nitrogen case. The interface between supercritical CO_2 and water suggests that supercritical CO_2 is non-wetting to glass beads at this condition (45°C, 8.5MPa) that is close to a critical point. Pentland *et al.* [9] has been reported that the contact angle of the interface between supercritical CO_2 and de-ionized water increases with a pressure increase up to 40 MPa. Figure 2a visualizes glass beads and CO_2 bubbles in a cylindrical region with the diameter of 3 mm inside of a titanium tube. In the region close to the outer edge, surfaces of glass beads are hard to recognize, because of artefacts caused by a titanium tube.

Based on the CT images, residual gas saturation, which is defined as volume ratio of trapped bubbles to pore, is evaluated for each experiment as shown in Fig. 3. With a decrease in the diameter of glass beads, residual gas saturation increases for both conditions, because capillary pressure tends to be higher compared to buoyancy. Even though the capillary number at laboratory room condition is about one order of magnitude lower than that at reservoir condition, residual saturation can be expressed with simple relationship with the inverse Bond number.

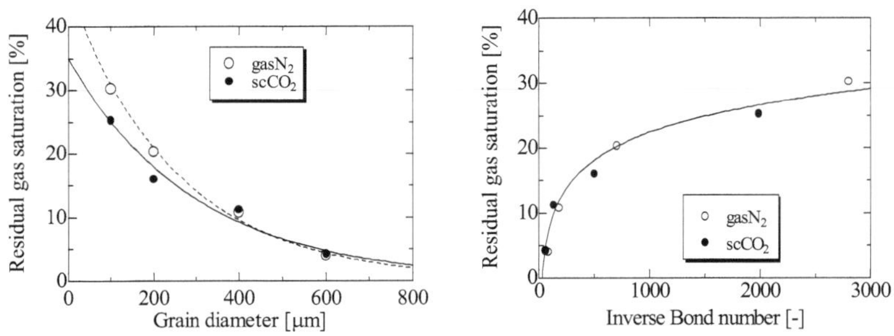

(a) (b)

Figure 3: Residual gas saturation vs. diameter of glass beads (a) and the inverse bond number (b) for reservoir and laboratory room conditions.

The volume of each trapped bubble was evaluated by three-dimensional image analysis software ImageJ [27, 28] with some plug-ins based on CT images such as shown in Fig. 2. The volume of trapped bubble is normalized with the volume of glass beads as shown in Fig. 4. In the packed bed of glass beads with the diameter of 600 μm, it is rare to find large trapped bubbles which spread over several pores for both conditions, because of buoyancy. Therefore, the residual gas saturation is lower for this packed bed. On the other hand, in the packed beds of fine glass beads the volume of some trapped bubbles is more than ten times higher than that of glass breads. As a result, the residual gas saturation is higher for these packed beds.

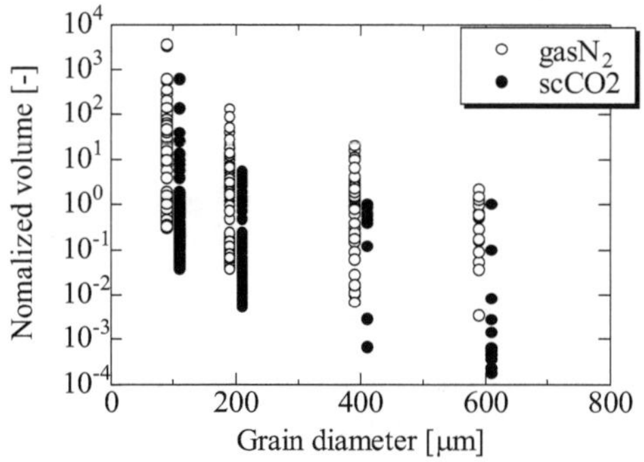

Figure 4: Distribution of the volume of trapped gas bubbles.

Contribution of gas bubbles in each volume range to residual gas saturation is shown in Fig. 5 for packed beds with various diameters of glass beads. The larger trapped bubbles contribute the more to the residual gas saturation for each condition and for each diameter of glass beads. With a decrease in the diameter of glass beads, much larger trapped bubbles contribute to the residual gas saturation for each condition.

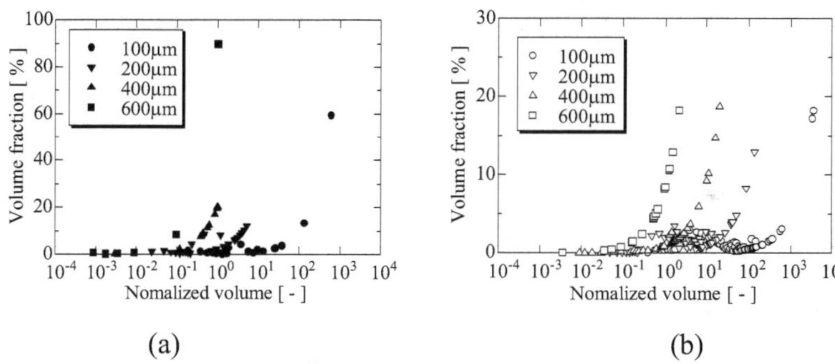

Figure 5: Contribution to residual gas saturation as a function of the volume of each bubble at reservoir condition (a) and laboratory room condition (b).

4 Conclusions

Trapped bubbles in packed beds of glass beads were visualized by means of micro-focused X-ray CT at pore scale for supercritical CO_2 and water systems at reservoir condition and for nitrogen and water systems at laboratory room condition. When the diameter of glass beads is same, distribution of volume of trapped bubbles is similar for each condition. At the pressure of 8.5MPa and the temperature of 45°C, which corresponds to just above a critical point, the morphology of the interface between the supercritical CO_2 and water suggests that CO_2 is non-wetting to glass beads. Residual gas saturation increases with a decrease in a diameter of glass beads, because the capillary pressure is higher with respect to buoyancy. Residual gas saturation can be expressed as a function of the reverse Bond number for each condition. Distribution of trapped bubble volume is identical with each other for each condition at same diameter of glass beads. The large trapped bubbles contribute to the residual gas saturation.

References

[1] Metz, B., Davidson, O., de Coninck, H.C., Loos, M. & Meyer, L.A. (eds.), *IPCC Special Report on Carbon Dioxide Capture and Storage*, Cambridge University Press, Cambridge, pp. 195-276 , 2005.

[2] Doughty, C., Investigation of CO_2 plume behavior for a large-scale pilot test of geologic carbon storage in a saline formation. *Trans. Porous Med.*, **82(1)**, pp.49-76, 2010.

[3] Juans, R., MacMinn, C.W. & Szulczewski, M.L., The footprint of the CO_2 plume during carbon dioxide storage in saline aquifer: storage efficiency for capillary trapping at the basin scale. *Trans.Porous Med.*, **82(1)**, pp. 19-30, 2010.

[4] Bachu, S., CO_2 storage in geological media: role, means, status and barriers to deployment. *Progress in Energy and Combustion Science*, **34**, pp. 254-273, 2008.

[5] Suekane, T., Nobuso, T. Hirai, S. & Kiyota, M., Geological storage of carbon dioxide by residual gas and solubility trapping. *Intern. J. Greenhouse Gas Control*, **2(1)**, pp. 58-64, 2008.

[6] Bennion, D.B. & Bachu, S., Dependence on temperature, pressure, and salinity of the IFT and relative permeability displacement characteristics of CO_2 injected in deep saline aquifers. Presented at the 2006 SPE Annual Technical Conference and Exhibition, San Antonio, Texas, USA, 24-27 September, SPE 102138, 2006.

[7] Iglauer, S., Carbon capture and storage with a focus on capillary trapping as a mechanism to store carbon dioxide in geological porous media (Chapter 4). *Advances in Multiphase Flow and Heat Transfer*, eds. L. Cheng and D. Mewes, 3, pp. 177-197, Bentham e Books [in press].

[8] Al Mansoori, S.K., Itsekiri, E., Iglauer, S., Pentland C.H., Bijeljic, B. & Blunt, M.J., Measurements of non-wetting phase trapping applied to carbon dioxide storage. *Intern. J. Greenhouse Gas Control*, **4**, pp. 283-288, 2010.

[9] Pentland C.H., El-Maghraby, R., Georgiadis, A., Iglauer, & Blunt, M.J., Immiscible displacement and capillary trapping in CO_2 storage. Presented at International Conference on Greenhouse Gas Control Technologies (GHGT-10), Amsterdam, Netherland, 2010.

[10] Zhou, N., Matsumoto, T., Hosokawa, T. & Suekane, T., Pore-scale visualization of gas trapping in porous media by X-ray CT scanning. *Flow Measurement and Instrumentation*, **21**, pp. 262-267, 2010.

[11] Suekane, T., Zhou, N., Hosokawa, T. & Matsumoto, T., Direct observation of gas bubbles trapped in sandy porous media. *Transport in Porous Media*, 82**(1)**, pp. 111-122, 2010.
[12] Wildenschild, D., Armstrong R.T., Herring, A.L., Young, I.M. & Carey, J.W., Exploring capillary trapping efficiency as a function of interfacial tension, viscosity, and flow rate. Presented at International Conference on Greenhouse Gas Control Technologies (GHGT-10), Amsterdam, Netherland, 2010.
[13] Ott, H., de Kloe, K., Marcelis, F. & Makurat, A., Injection of supercritical CO_2 in brain saturated sandstone: pattern formation during salt precipitation. Presented at International Conference on Greenhouse Gas Control Technologies (GHGT-10), Amsterdam, Netherland, 2010.
[14] Prodanović, M., Lindquist, W.B. & Seright, R.S., 3D image-based characterization of fluid displacement in a Berea core. *Advanced in Water Resources*, **30**, pp. 214-226, 2007.
[15] Al-Raoush, R.I. & Willson, C.S., A pore-scale investigation of a multiphase porous media system, *J. Contaminant Hydrology*, **77**, pp. 67-89, 2005.
[16] Bakke, S., & Øren, P.E., 3-D pore-scale modeling of sandstones and flow simulations in the pore networks, *SPEJ*, **2**, pp. 136-149, 1997.
[17] Prodanović, M., Lindquist, W.B. & Seright, R.S., Residual fluid blobs and contact angle measurements from X-ray images of fluid displacement. *Presented at the Computational Methods in Water Resources XVI Conference,* Copenhagen, Denmark, June 19-22, 2006.
[18] Wildenschild, D., Hopmans, J.W., Vaz, C.M.P., Rivers, M.L., Rikard, D. & Christensen, B.S.B., Using X-ray computed tomography in hydrology: systems, resolutions, and limitations. *J. Hydrology*, **267**, pp. 285-297, 2002.
[19] Al-Raoush, R.I. & Willson, C.S., Extraction of physically realistic pore network properties form three-dimensional synchrotron X-ray microtomography images of unconsolidated porous media systems. *J. Hydrology*, **300**, pp. 44-64, 2005.
[20] Øren, P.E. & Bakke, S., Process based reconstruction of sandstones and prediction of transport properties. *Transport in Porous Media*, **46**, pp. 311-343, 2002.
[21] Knackstedt, M.A., Sheppard, A.P. & Sahimi, M., Pore network modelling of two-phase flow in porous rock: the effect of correlated heterogeneity. *Advanced in Water Resources*, **24**, pp. 257-277, 2001.
[22] Soltani, A., Le Ravalec-Dupin, M. Fourar, M. & Rosenberg, E., Three-dimensional characteristics of permeability at the core scale. *Transport in Porous Media*, **84**, pp. 285-305, 2010.
[23] Morrow, N.R. & Songkran, B., Effect of viscous and buoyancy forces on nonwetting phase trapping in porous media. *Surface Phenomena in Enhanced Oil Recovery*, ed. Shah, D.O., Plenum Press, New York, pp. 387-411, 1981.
[24] Hebach, A., Oberhof, A., Dahmen, N., Kögel, A., Ederer, H. & Dinjus, E., Interfacial tension at elevated pressures-measurements and correlations in

the water + carbon dioxide system. *J. Chem. Data*, **47**, pp. 1540-1546, 2002.
[25] Bachu, S. & Bennion, D.B., Dependence of CO_2-brine interfacial tension on aquifer pressure, temperature and water salinity. *Energy Procedia*, 1, pp. 3157-3164, 2009.
[26] JSME, JSME Data Book: Thermophysical Properties of Fluids, Maruzen Co. Ltd., Tokyo, 1983.
[27] Abramoff, M. D., Magelhaes, P. J. & Ram, S. J., Image processing with imageJ. *Biophotonics Int.,* **11**, pp.36-44, 2004.
[28] Rasband, W. S., 1997-2008. ImageJ [Internet]. Bethesda (Maryland, USA): US National Institute of Health. Available from: http://rsbweb.nih.gov/ij/.

Author Index